PROBABILITY THEORY

A HISTORICAL SKETCH

Probability and Mathematical Statistics

A Series of Monographs and Textbooks

Editors   **Z. W. Birnbaum**   **E. Lukacs**

University of Washington   Bowling Green State University
Seattle, Washington   Bowling Green, Ohio

---

1. Thomas Ferguson. Mathematical Statistics: A Decision Theoretic Approach. 1967
2. Howard Tucker. A Graduate Course in Probability. 1967
3. K. R. Parthasarathy. Probability Measures on Metric Spaces. 1967
4. P. Révész. The Laws of Large Numbers. 1968
5. H. P. McKean, Jr. Stochastic Integrals. 1969
6. B. V. Gnedenko, Yu. K. Belyayev, and A. D. Solovyev. Mathematical Methods of Reliability Theory. 1969
7. Demetrios A. Kappos. Probability Algebras and Stochastic Spaces. 1969
8. Ivan N. Pesin. Classical and Modern Integration Theories. 1970
9. S. Vajda. Probabilistic Programming. 1972
10. Sheldon M. Ross. Introduction to Probability Models. 1972
11. Robert B. Ash. Real Analysis and Probability. 1972
12. V. V. Fedorov. Theory of Optimal Experiments. 1972
13. K. V. Mardia. Statistics of Directional Data. 1972
14. H. Dym and H. P. McKean. Fourier Series and Integrals. 1972
15. Tatsuo Kawata. Fourier Analysis in Probability Theory. 1972
16. Fritz Oberhettinger. Fourier Transforms of Distributions and Their Inverses: A Collection of Tables. 1973
17. Paul Erdös and Joel Spencer. Probabilistic Methods in Combinatorics. 1973
18. K. Sarkadi and I. Vincze. Mathematical Methods of Statistical Quality Control. 1973
19. Michael R. Anderberg. Cluster Analysis for Applications. 1973
20. W. Hengartner and R. Theodorescu. Concentration Functions. 1973
21. Kai Lai Chung. A Course in Probability Theory, Second Edition. 1974
22. L. H. Koopmans. The Spectral Analysis of Time Series. 1974
23. L. E. Maistrov. Probability Theory: A Historical Sketch. 1974
24. William F. Stout. Almost Sure Convergence. 1974

In Preparation

Z. Govindarajulu. Sequential Statistical Procedures
E. J. McShane. Stochastic Calculus and Stochastic Models

PROBABILITY THEORY

A HISTORICAL SKETCH

L. E. MAISTROV

Translated and Edited by SAMUEL KOTZ

Department of Mathematics
Temple University
Philadelphia, Pennsylvania

ACADEMIC PRESS   New York and London   1974

A Subsidiary of Harcourt Brace Jovanovich, Publishers

ACADEMIC PRESS, INC.
111 Fifth Avenue, New York, New York 10003

United Kingdom Edition published by
ACADEMIC PRESS, INC. (LONDON) LTD.
24/28 Oval Road, London NW1

Library of Congress Cataloging in Publication Data

Maĭstrov, Leonid Efimovich.
    Probability theory.

        Translation of Teoriia veroiatnosteĭ.
        Bibliography: p.
        1.   Probabilities—History.   I.   Title.
QA273.A4M3513   519.2'09   72-88366
ISBN 0-12-465750-8

PRINTED IN THE UNITED STATES OF AMERICA

Probability Theory—A Historical Sketch. Translated
from the original Russian edition entitled Teoriia
Veroiatnosteĭ Istoricheskiĭ Ocherk, published by
Izdatel'stvo Nauka, Moscow, 1967.

Selections from Oystein Ore, "Cardano: The
Gambling Scholar" (copyright 1953 by Princeton
University Press), pp. 192-241. Reprinted by
premission of Princeton University Press.

Selections on pp. 28-30, 44, 49, 67, and 74 of
this book are from F. N. David: "Games, Gods
and Gambling," 1962; reproduced by permission
of the publishers, Charles Griffin & Company
Ltd., London.

Contents

Foreword   vii
Translator's Preface   ix
Author's Preface   xiii

Introduction   1

I   Prehistory of Probability Theory

1   Basic stimuli in the rise of probability theory   3
2   The role of gambling in the rise of probability theory   7
3   The first problems   15
4   Cardano and Tartaglia   18
5   Elements of probability theory in Galileo's works   28
6   Basic stages in the development of combinatorics   34

II   The First Stage in the Development of Probability Theory

1   De Méré's legend   40
2   Letters between Fermat and Pascal   43
3   Contributions of Huygens to probability theory   48

v

III   The Development of Probability Theory to the Middle of the Nineteenth Century

1	James Bernoulli and his treatise "Ars Conjectandi"	56
2	The development of probability theory in the first half of the eighteenth century	76
3	Simpson's contributions to probability theory	82
4	Thomas Bayes and his theorem	87
5	Euler's and Daniel Bernoulli's contributions	101
6	G. L. Buffon	118
7	Opposition from d'Alembert	123
8	Jean Antoine de Caritat, Marquis de Condorcet	129
9	Laplace and his contributions to probability theory	135
10	Distribution of random errors	148
11	The state of probability theory in Europe prior to the advent of the Russian (St. Petersburg) school	158
12	Probability theory in Russia prior to the advent of the St. Petersburg school	161

IV   Probability Theory in the Second Half of the Nineteenth Century

1	Chebyshev—the originator of the Russian school of probability theory	188
2	Prominent representatives of the St. Petersburg school	208
3	Probability in physics	225
4	Bertrand's paradoxes	234

V   The Axiomatic Foundations of Probability Theory

1	The need for axiomatization	240
2	Prerequisites for axiomatization	248
3	Bernstein's contributions	249
4	The frequency approach of R. von Mises	254
5	The beginning of a new stage in the development of probability theory	259

Bibliography   265

Index   275

Foreword

I am very glad that Samuel Kotz has translated Maistrov's book. There is nothing comparable in English. In 1865 Isaac Todhunter † made a remarkable survey of the early days of probability, but since then we have had nothing so complete or authoritative. Moreover this translation opens up a whole world which has hitherto been closed to those of us who do not read Russian. Chebyshev, Markov, Kolmogorov: these are the great household names of probability, but although individual pieces of work by these men were published in or translated into Western European languages, the story of the St. Petersburg school and its successors has never before been so available.

Probability is not just one mathematical discipline among others. It is part of an extraordinary philosophical success story of modern times. Metaphysics is the study, in general terms, of what there is. Epistemology is the study of how we find out about what there is. Both these branches of philosophy have been conquered by probability. Quantum theory tells us that everything is governed by laws that are irreducibly probabilistic. Many philosophers in recent years have claimed that all our learning from experience must be understood in terms of probability theory. Perhaps no other mathematical concept of recent times has so completely permeated both our practical and our theoretical lives.

Yet probability theory was unknown until the seventeenth century. Why, when people had been gambling for millenia? Maistrov rightly says

---

† "A History of the Mathematical Theory of Probability," Cambridge Univ. Press, London and New York, 1865; Chelsea, New York, 1962.

that the emergence of probability had little to do with gambling. What then got it going? Maistrov urges that economic factors are the clue. Probability theory is an artifact of the new bourgoisie. Much evidence, at least of an anecdotal sort, could be used to support Maistrov's claim—it is a pity that he omits it. Thus Paccioli's problems (p. 17) occur in a book more famous as the origin of double-entry bookkeeping, without which the new accumulations of capital would have been almost impossible. It was an obscure merchant (p. 54) who made people keep what we now call statistics; within a few years of the publication of Graunt's book, the cities of Europe were beginning to gather the statistical data that make probability theory useful. Maistrov seems not to mention that William Petty, an Englishman who made a personal fortune after exploiting the Irish, was the first to urge a central statistical office because of its boon to the new merchant class. Or take again the figures of Hudde (van Hudden) and de Witt so briefly mentioned on pp. 54–55: these are also the paradigms of the new power of the merchant classes in Europe. In short, I think Maistrov has a better case than he gives us in this book, and I hope that readers will take it seriously. My own view is that the origins of probability theory are not primarily economic,† but no one should be confident in the difficult and inconclusive world of prehistory.

Once probability does come into being, it goes with a swing. Maistrov has done a splendid job of giving a sense of the growing momentum. He has omitted one vector. One of the great anomalies of more recent deployment of probability lies in the fact that although Russian mathematicians have so regularly been on the forefront of the formal theory, they have, both in pre- and post-revolutionary times, been less involved with theories on how to analyze statistical data, using the new tools of probability theory. Correspondingly, the great nineteenth century figures who gave us one central use of probability theory such as Quetelet Lexis, Galton, and K. Pearson, get little or no notice in this book. Anyone who would like to supplement this history of pure probability theory with a history of some of its immediate applications may enjoy a survey by Harald Westergaard.‡ If after reading Maistrov you would like to read more history of probability mathematics, the papers in English by the author's compatriot, O. B. Sheynin, are, in my opinion, the best place to start.§

IAN HACKING

† Ian Hacking, "The Emergence of Probability," Cambridge Univ. Press, London and New York, 1974.

‡ "Contributions to the History of Statistics," P. S. King Ltd., London, 1932; Mouton Publ., The Hague, 1969.

§ See pp. 211–215 of the Bibliography; several more of these articles have appeared or are appearing in successive issues of the *Archive for History of Exact Sciences.*

Translator's Preface

In presenting the edited translation of L. E. Maistrov's monograph "The Theory of Probability—A Historical Sketch," the following remarks seem in order.

This monograph is not intended as an encyclopedic work; its purpose is to fill the long existing gap (since Todhunter's classic of 1865) in the literature dealing with the history of probability theory.

The concept of probability, rooted (but dormant) in the ancient history of human civilization, showed a remarkable reawakening in the middle of the seventeenth century and was partially crystallized from the mathematical angle in the first third of the present century. It continues to occupy the minds of the most prominent philosophers, mathematicians, and statisticians. Various theories of probability have been proposed and hotly debated during the past two hundred years, ranging from the equiprobability theory of Pascal and Laplace, through the axiomatic approach of Kolmogorov, the frequency theorem of von Mises, the more recently developed subjective approach, and finally returning to the informal approach (of S. E. Toulmin), which indicates a retreat and abandonment of the hope of reconciling logical, pure-mathematical, and experimental difficulties and contradictions inherent in this concept.

At this juncture of apparent defeatism concerning the feasibility of a unified theory of probability that could fit the multitude of diverse applications and the appearance of strongly voiced—although possibly somewhat sensationally inclined—opinions that probability may, after all be an inadequate and irrelevant tool for inference and decision making, as indicated in the recent book by Professor Terrence L. Fine: "Theories of Probability: An Examination of Foundations," Academic Press, 1973, it is extremely important to trace the historical developments of this elusive although ever-present concept. Maistrov does just this.

He provides a leisurely, but carefully planned and executed, guide and well documented chronological summary of the history of probability theory from its "prehistoric inception" (the author is a specialist in ancient history) up to the axiomatizations of probability initiated in the twenties of the present century. Perhaps some of the author's theses may be debatable, such as the emphasis on economic and scientific factors as the dominant stimulus for the remarkable development of probability theory in the 17–18th centuries; some historical aspects are deliberately ignored, i.e., the concept of probability in the Judaic and Christian theological literature; some of the material based on second-hand sources is presented perhaps less successfully than the original, e.g. the Pascal–Fermat–Méré affair, which is so vividly described by Ore in a paper in the *American Mathematical Monthly;* nevertheless, this is the first book in the field which unifies, emphasizes and reveals many important facts, crucial for understanding the subject matter of probability. The author collects information available hitherto only to a few specialists in the field and scattered in numerous and diverse publications. Among these (to mention but a few) are the discussions of Spinoza and especially Galileo's contributions to probability theory and the detailed description of the ingenious work by R. Adrian, who deduced independently and published in 1808, before Gauss, the famous "law of errors" which serves as the foundation of present-day classical probability and statistics.

Not less important is the appearance for the first time in English of a scholarly "historical sketch" of the Russian School of Probability Theory (which occupies approximately one third of the monograph), in view of the fundamental and pioneering work of the 19th century Russian probabilists in the realm of mathematical probability theory (this feature of Russian mathematics continues to be almost as prominent at present as it was 100 years ago).

The editor has tried to supplement certain omissions and in a few cases to correct minor errors using footnotes, which are especially noticeable in the first part of the monograph. He has also supplemented the author's bibliography with some 45 additional items which reflect, in part, the substantial increase of periodical literature in this subject during the years 1966–1972, as well as certain omissions of topics by the author as indicated above.

This monograph should therefore be of interest and useful to readers of various and diverse categories—student-teachers enrolled in mathematical education programs, instructors giving courses in probability theory (on various levels), historians of mathematics, as well as scientists in various fields of the humanities, behavioral, and natural sciences, who are the main and constant users, whether explicity or tacitly, of probability theory in their efforts to trace further the soundness of our understanding of man, his society, and the universe.

The first part of the manuscript was read by Professor Ian Hacking who contributed a number of very illuminating remarks and comments. The editor is much indebted to Professor Hacking in this connection and regrets that in view of technical difficulties involved in the publication of this translation, he was not able to utilize this material to its fullest extent.

Drs. Florence N. David and Churchill Eisenhart were very helpful in connection with the bibliographical material related to their pioneering research in the field.

I am also indebted to Professor A. Birnbaum for his initiative, advice, and encouragement.

SAMUEL KOTZ

Author's Preface

The history of probability theory is one of the least investigated areas of the history of mathematics. The present book is in the form of a historical sketch and does not pretend to be a systematic exposition of the history of this scientific discipline. For this reason not all periods in the history of probability theory are dealt with equally thoroughly—more attention is devoted to those problems that either were not investigated at all or studied insufficiently in the literature. For example, the space devoted to Bayes' paper is only a little less than that devoted to P. L. Chebyshev's contribution, the reason being that Bayes' paper is discussed for the first time, while Chebyshev's investigation has been thoroughly investigated in the literature.

The exposition is basically chronological and spans the period from the inception of probability theory up to its axiomatization in the 20th century. However, the exposition becomes less complete with the approach of the present period. In particular, when describing the modern period of probability theory, we discuss mainly axiomatization problems. It should be noted that although axiomatization is an important stage in the development of modern science it does not encompass all the new contributions to probability theory obtained in the 20th century.

Section 3 of Chapter III, "Simpson's contributions," was written by O. B. Sheynin.

The author expresses his sincere gratitude to the academician of the Academy of Sciences of the Ukrainian SSR, B. V. Gnedenko as well as to I. L. Kalikhman for their attention to his work, comments and advice, which as far as possible were taken into account.

Introduction

The following stages in the history of probability theory may be distinguished:

1. *The prehistory of probability theory.* In this period, whose beginnings are lost in the dust of antiquity, certain simple problems were posed and solved using elementary calculations. Only later were these categorized as probability theory problems. No specific methods originated during this stage. In the main, materials were accumulated. This period terminated in the sixteenth century with the works of Cardano, Paccioli, Tartaglia, and others.

2. *The origins of probability theory as a science.* During this period the first specific notions (e.g., the concept of mathematical expectation) were developed. The first theorems (such as the addition and multiplication theorems of probability theory) were established. The names Pascal, Fermat, Huygens, etc. are associated with the beginnings of this period. This period continued from the middle of the seventeenth century up to the beginning of the eighteenth. During this time probability theory found its first applications in demography, in the insurance business, and in the estimation of observation errors.

3.   The next period began with the appearance of James Bernoulli's treatise "Ars Conjectandi" ("The Art of Conjecturing") in 1713. In this work a limit theorem pertaining to the simplest case of the law of large numbers was rigorously proved for the first time. Bernoulli's theorem made possible wide applications of probability theory to statistics. The contributions of de Moivre, Laplace, Gauss, Poisson, and others date from this period. The first applications of probability to various areas of the natural sciences were initiated. Various limiting theorems played a central role.

4.   The next period in the development of probability theory was first and foremost associated with the Russian (St. Petersburg) school. Here the names Chebyshev, Markov, and Lyapounov deserve mention.

During this period the extensions of the laws of large numbers and the central limit theorems to various classes of random variables reached their peak. The laws of probability theory were applied to dependent random variables. As a result, probability theory could be applied to many branches of the natural sciences, in particular to physics. Statistical physics was initiated and its development was linked with probability theory.

5.   The modern period in the development of probability theory began with the formulation of axioms. The demand for axiomatization stemmed firstly from practical applications since, in order to apply probability theory successfully to physics, biology, and other branches of science, as well as to engineering and military problems, it was necessary to refine and produce an orderly structure for its basic notions. Due to axiomatization, probability theory became an abstract, deductive mathematical discipline closely connected with set theory and through it with other branches of mathematics. This prompted an unprecedentedly broad variety of investigations in probability theory, ranging from applied problems in economics to subtle problems of cybernetics. Initial contributions to this period were made by Bernstein, von Mises, and Borel. In the thirties of the present century, the axiomatic foundations of probability theory were finalized with the publication of Kolmogorov's postulates which received general recognition.

I

Prehistory of Probability Theory

1 Basic stimuli in the rise of probability theory

Problems that had significant impact on the origin and initial development of probability theory arose in the course of processing numerical data and results of observations in various sciences; this development was also stimulated by the practical requirements of insurance companies and by abstract problems connected with games of chance. These problems were closely associated with combinatorics and with the development of philosophical notions of chance and necessity.

These initial problems of a probabilistic nature, which originated in various fields of human endeavor, gradually crystallized into the notions and methods of probability theory.

In ancient times, the first statistical data were collected in population censuses. The rulers of ancient Egypt, Greece,† and Rome attempted on various occasions to enumerate population, annual quantities of grain,

†For a discussion of the prehistory of probability theory in ancient Greece, see the article by S. Sambursky in *Osiris* **12**, 35–48 (1956) (*Translator's remark*).

taxes, etc. In ancient Rome, beginning with the rule of the Emperor Augustus (27 B.C.–17 A.D.), a population census was carried out in each newly conquered territory.

William the Norman (1027–1087), who conquered England in 1066, carried out an exhaustive economic survey and ordered the compilation of special lists, known as "Doomsday" lists, containing the results of the general census of the land in 1086. The purpose of this census was to establish proper taxation procedures. The population was enumerated according to various categories and occupations. Doomsday lists contained detailed information on royal, Church, and feudal estates with details on their size and number of cattle and stock.

In Russian chronicles related to the tenth century and later periods, there are references to collections of certain statistical data. The Mongols, after the conquest of Russia, carried out a population census for the purpose of collecting tributes and taxes. Cadastres in ancient Russia also served as sources of statistical data. More recently, in the "South-Russian Chronicles" of 1666 we find that a census was conducted in that year of all Russia-Minor by decree of the Moscow Czar [172, pp. 31, 32].

In the Middle Ages detailed agricultural inventories of the estates of major feudal lords were conducted. In Venice in 1268 and 1296 decrees were set forth ordering the governors to compile detailed accounts of their provinces. Diplomatic representatives were obliged to collect intelligence and report, using specially devised forms, to the Senate on the countries of their representation. The Senate organized censuses of population and households and collected data on trade and commercial activities. Venetian reports and other documents provided a very rich source of materials concerning various countries. Mochenigo organized this vast information and presented a detailed report to the Senate in 1421. From that time on, analogous compendia were composed and published in various trade centers of Italy.

During the seventeenth century in Holland, Spain, France, England, and Germany various reference handbooks made their first appearance. In London regular Bills of Mortality were published and parish registers of all weddings, christenings, and burials were kept. These Bills were introduced in times of plague epidemics. The first were dated 1517. The form and contents of these Bills were subject to alterations. In 1532 and 1535 weekly Bills furnished returns of plague deaths classified according to parishes. Annually, starting from 1693, yearly totals were published in December. One of the earliest and very important additions was the inclusion of the cause of death. At first, only two causes were given—disease and accident; however, later the number of causes was substantially increased. As of 1629, sex was also indicated in the records of burials and christenings.

Based on these statistical materials, such notions as the probability of death in a given time period, the probability of survival up to a certain age, etc., originated. Undoubtedly, these developments influenced the formation of basic concepts in probability theory.

Thus, the collection of and some analyses on statistical data were carried out in various historical periods with a certain degree of completeness and regularity. However, systematic and sufficiently extensive statistical investigations began only upon the rise of capitalism, when commerce and monetary transactions, particularly those connected with actuarial operations, were developing and when various new institutions were established.

Statistics was one of the basic stimuli in the initial development of probability theory. An increase in capitalistic relations constantly posed new problems in statistics. As Marx points out: "Although we come across the first beginnings of capitalist production as early as the fourteenth or fifteenth century, sporadically, in certain towns of the Mediterranean, the capitalistic era dates from the sixteenth century" [124, p. 715].

In the fourteenth century, the first marine insurance companies were established in Italy and in Holland. Insurance of goods shipped by sea was followed by insurance of freight across continents, lakes, and rivers. These companies carried out calculations of chances since larger risks made for larger insurance premiums. For shipping by sea, the premiums amounted to about 12–15% of the cost of the goods, while for intracontinental deliveries the rates were about 6–8%. Beginning with the sixteenth century, marine insurance was introduced in many other countries. Other forms of insurance originated in the seventeenth century. The data collected by insurance companies also served as source material utilized in the development of probability theory.

The Renaissance witnessed intensive growth in the natural sciences and an increase in the importance of observations and experiments. The problem of methods of handling the results of observations and, in particular, of estimating random errors occurring in observations became relevant. This also stimulated the development of probability theory.

The interrelation between the chance and the necessary, problems of regularity and causality, and so on, were the subject of investigations in ancient times and were permanently on the philosophers' agenda. Therefore, there can be no doubt that philosophy in the seventeenth century accumulated many ideas which also influenced probability theory during its inception and in the first stages of its development.

The following example may be of interest. In ancient India, even before the Christian era, Jaina religion and philosophy became widespread. The basic constituents of this religion were the logical ideas of *syād* or *syādvāda*.†

---

† "Assertion of possibilities" (see [117, p. 184]) (*Translator's remark*).

This philosophy developed fully in the sixth century B.C.E. and was quite prominent in the medieval period. A treatise dealing with this theory dated 1292 is in existence, as well as even later works.

The basis of *syādvāda* theory is the dialectic of seven predications:

(1)   maybe it is;
(2)   maybe it is not;
(3)   maybe it is, it is not;
(4)   maybe it is indeterminate;
(5)   maybe it is and also indeterminate;
(6)   maybe it is not and also indeterminate;
(7)   maybe it is and it is not and also indeterminate.

These seven categories are, according to *syādvāda*, necessary and completely sufficient to exhaust the possibilities of predications or modes of knowledge.

In the fourth proposition, in addition to the assertions *is* and *is not*, there is also the possibility of the indeterminate. One may observe here the beginnings of the notions that later led to the concept of probability, since the existence of a field for application of probabilistic reasoning is, in fact, asserted here. The well-known modern Indian statistician P. C. Mahalanobis, points out: "I should like to emphasize that the fourth category is a synthesis of three basic modes 'it is,' 'it is not,' and ... 'indeterminate,' ... and supplies the logical foundations of the modern concept of probability" [117, p. 186]. It should, however, be noted that this is only the qualitative aspect of the concept of probability, which does have a quantitative aspect as well.

A related historical anecdote may be of interest in this connection. In a medieval Indian treatise on *syādvāda*, there is a discussion about the practice of giving alms to Brahmins, and a question is raised whether the recipients of the gifts are always deserving persons. It is stated in reply that the practice of giving alms can be recommended because "only ten out of one hundred recipients are undeserving" [117, p. 195]. One can detect a certain probabilistic approach in this reply.

As another example we cite the dictum of Thomas Hobbes (1588–1679), whose thesis was that no matter how long we observe a phenomenon, this is still not sufficient grounds for its absolute and definite knowledge. It is Hobbes' opinion that it is impossible by observation alone to determine all the circumstances which give rise to a particular phenomenon since there are infinitely many such circumstances.

Hobbes writes "for though a man have always seen the day and night to follow one another hitherto; yet can he not thence conclude they shall do so, or that they have done so eternally." From this statement the conclusion follows that "experience concludeth nothing universally. If the signs hit twenty times for one missing, a man may lay a wager of twenty to one of the event; but may not conclude it for a truth" [79, Ch. IV, p. 18].

Only two examples are presented above. However, in the course of human history, many other problems were encountered in the philosophical sciences, and the solutions or discussions of these prompted the development of the basic probabilistic notions.

Probability theory could emerge only after problems connected with probabilistic estimation of certain events started to occur in various areas of human activity and only when the need for solving these problems became more pressing, since they involved the interests of a large number of individuals. These phenomena were characteristic of the period of the collapse of feudal relations, proletarization of the peasants, and the rise of the bourgeoisie, which was also a period of growth in cities and commerce, increase in manufacturing, and so on. As Engels pointed out in his Introduction to "Dialectics of Nature": "Modern natural science ... dates from that mighty epoch which had its rise in the last half of the fifteenth century. Royalty, with the support of the burghers of the towns, broke the power of the feudal nobility" [55, p. 41].

This was the period in which a number of sciences, such as analytical geometry and differential and integral calculus, originated. This was also the age of the birth of probability theory.

Activities in many sciences during that period were confined basically to a collection of facts and their systematization; sporadic discoveries were made, and only initial attempts were undertaken toward the formation of scientific methods. Probability theory also went through this first stage in the course of its development.

2   The role of gambling in the rise of probability theory

Up to the present time there has been a widespread false premise that probability theory owes its birth and early development to gambling. We present several statements asserting this thesis.

> Probability theory originated in the sixteenth century as a result of attempts to develop a theory behind the gambling which was popular in that period. [28, p. 11]

> It is by chance that the invention of probability theory is due to a chance. [177, p. 327]

> Naturally enough, calculators and mathematicians directed their investigations towards the study of gambling; in this manner the foundations of probability theory were laid down. [25]

> The attempt to give a mathematical explanation of certain observed facts [in games of chances] became the immediate cause of the origin

(about 1650) and first development of the Mathematical Theory of Probability. [9, p. 144]

Probability theory was born when Fermat and Pascal started to study games of chance. [156, p. 65]

The first papers in which the basic concept of probability theory originated appeared in the sixteenth and seventeenth centuries and were connected with the attempt to construct a theory of games of chance. [67, p. 6]

This is also the opinion held by the vast majority of investigators. In particular, David and Kendall in *Biometrika* [49, 84], as well as many others, support this thesis.†

This interpretation of the origin of probability theory dismisses the whole prehistory of the theory, when the formulations of basic concepts and initial methods of probability theory actually originated within the framework of other sciences and problems. These ideas acquired independent significance only in the seventeenth century, when probability theory became a science in its own right. At the same time, an undeservedly large role is attributed to gambling as compared with other more important factors which stimulated the creation, as well as early developments of this science.

Similarly, the opinion [75] that the origins of probability theory vanished into antiquity is erroneous.

The following question thus naturally arises. If gambling did not provide the main stimulus in the inception and early development of probability theory, what then actually is its contribution in this connection?

Gambling originated in the early stages of human history. It made its appearance all over the world. Small bones of animals, the astragali,‡ were used as gambling devices. An astragalus of a hooved animal, being almost symmetrical, has only four sides on which it may rest. When throwing an astragalus, the upper side was noted. These sides were enumerated using various methods and no standard enumeration procedure was available. One of the games in ancient Greece may have been that of throwing four astragali together and noting which sides fell uppermost. The best of all throws was the throw of "Venus," when all four sides were different.

In archaeological excavations related to the periods starting from 5000 B.C. (and possibly earlier) it was commonplace to find a preponderance of astragali among the bones of animals. In certain cases they were found up to several dozen times more often than other bones.

The frequency of occurrence of various sides of astragali is rather stable.

---

†See also David's "Games, Gods and Gambling" [189] (*Translator's remark*).

‡The astragalus is a bone in the heel lying above the latus, which in the strict anatomical sense is the heel-bone [189, p. 2] (*Translator's remark*).

Based on numerous experiments with various astragali, we established the following frequencies of occurrences for different sides. The frequency corresponding to the broad, slightly convex side is ≈ 0.39; to the opposite broad side, ≈ 0.37; and to each of the two remaining sides, ≈ 0.12. It is, therefore, evident that "Venus" is not the least probable throw.

In ancient Egypt (at the time of the First Dynasty, *ca.* 3500 B.C.) the game that excavators call "Hounds and Jackals" was in use. A copy of this board game (from the period 1800 B.C.) is located in the Metropolitan Museum of Art in New York. The hounds and jackals were moved according to certain rules by throwing the astragali found with the game.

Different forms and variants of ancient games of chance are known. There were, for example, throwing sticks made of wood or bone. The faces of the sticks were marked with dots from one to four. The game consisted of throwing the stick, which rolled like a pencil, and counting the number of dots appearing on the upper side. These throwing sticks existed among many nations. Several of them (dated at the seventh and eighth centuries A.D.) were found in Asia Minor and are now located in the Hermitage in Leningrad. In the State Historical Museum in Moscow several fragments of such sticks are exhibited; these were found in Chernigov† and are dated at the tenth century. The Hermitage has a sharpened ivory stick that was found in Ol'vija presumably belonging to the period before the third century A.D.

In Pompeii and Kerch‡ plates were found (located in the Hermitage) in the form of thin squares about 1 cm in length and width. One of the sides is marked with one dot and the opposite with six dots. The thin side faces were marked with two, three, four, and five dots, respectively. Apparently these plates were used in a game similar to the modern "heads-or-tails" game. These plates are dated at the beginning of the Christian era.

The most popular games were those using six-sided dice. The faces of a cubical die were marked with from one to six dots. The dice were made of various materials such as clay, ivory, wood, crystal, sandstone, ironstone, etc.

The earliest dice known were excavated in Tepe Gawra (North Iraq) and are dated at the beginning of the third millennium. The Indian dice (excavated in Mohenjo-Daro) belong to the same period.

Sometimes dice with unusual orderings of dots on the faces are found: One dice of the first millennium is said to have 9 opposite 6, 5 opposite 3, and 4 opposite 2; on others, the dots were arranged as 4, 4; 5, 5; and 6, 6. The ancient Greeks (as well as other nations) possessed dice with more than six faces.

---

† An ancient city in Central Ukraine; near the city are mounds dating from the tenth century (*Translator's remark*).

‡ An ancient city in East Crimea (*Translator's remark*).

TABLE I

RESULTS OF INVESTIGATIONS OF SEVERAL DICE

Number	Found	Approximate age	Composition	Length of the face (mm)	Distribution of dots on the face	Present location	Number of tosses	Frequency count of faces					
								1	2	3	4	5	6
1	Egypt; the most ancient die in the USSR	sixteenth century B.C.	Bone	10	1–6; 2–5; 3–4	Hermitage, Leningrad	235	37	17	49	59	28	45
2	Ol'vija	Early A.D.	Bone	10	1–6; 2–5; 3–4	State Historical Museum, Moscow	673	100	110	98	57	137	171
3	Egypt	first–second century A.D.	Glass	15	1–6; 2–5; 3–4	Hermitage	232	49	38	38	29	30	48
4	Egypt	No date available	Stone	13	1–6; 2–5; 3–4	Hermitage	234	43	48	26	22	41	54
5	Ol'vija	Before third century A.D.	Bone	9	1–6; 2–5; 3–4	Hermitage	250	47	38	33	23	40	69

No.	Place	Date	Material		Dice pairs	Museum	Total						
6	Ol'vija	Before third century A.D.	Bone	9	1–2; 3–5; 4–6	Hermitage	240	49	39	31	57	26	38
7	Ol'vija	Before third century A.D.	Bone	11	1–6; 2–5; 3–4	Hermitage	238	48	35	53	33	34	35
8	Pompeii	Before third century A.D.	Bone	11	1–6; 2–5; 3–4	Hermitage	240	38	47	36	31	33	55
9	Novgorod	eleventh–twelfth century A.D.	Bone	10	1–2; 3–4; 5–6	State Historical Museum	408	57	72	48	83	59	89
10	Bulgary	thirteenth–fourteenth century A.D.	Bone	9	1–6; 2–5; 3–4	State Historical Museum	1019	133	178	191	150	186	175
11	Moscow	thirteenth–fourteenth century A.D.	Bone	7	1–6; 2–4; 3–5	Archaeological Institute of the Akudo Science of USSR	1033	214	168	151	181	157	162
12	Moscow	Late fifteenth–early sixteenth century	Bone	3	1–6; 2–5; 3–4	Museum of History and Reconstruction	1000	285	190	78	62	134	251

a For comparison purposes the author carried out 330 tosses on a modern die used in the children's game "Who's First." Here 1 occurred 62 times; 2, 53; 3, 57; 4, 51; 5, 56; and 6, 51. See also [113] where data on four additional dice are presented.

In the various museums are located many dice dated at various periods and originating in different nations. A classification of some of these dice is presented in Table I.

As is seen from this table, the frequency counts of the faces are in many cases close to each other. We also note that the most common dice had faces 9–12 mm long, which is a rather convenient size since it is difficult to render a proper symmetrical form to dice of too small or too large dimensions.

There are 15 ways in which the dots can be placed on the faces of a six-sided die† :

(1)	1–2, 3–4, 5–6;	(6)	1–3, 2–5, 4–6;	(11)	1–5, 2–3, 4–6;
(2)	1–2, 3–5, 4–6;	(7)	1–4, 2–3, 5–6;	(12)	1–5, 2–6, 3–4;
(3)	1–2, 3–6, 4–5;	(8)	1–4, 2–5, 3–5;	(13)	1–6, 2–3, 4–5;
(4)	1–3, 2–4, 5–6;	(9)	1–4, 2–6, 3–5;	(14)	1–6, 2–4, 3–5;
(5)	1–3, 4–5, 2–6;	(10)	1–5, 2–4, 3–5;	(15)	1–6, 2–5, 3–4.

It turns out that the majority of dice have the arrangement 1–6, 2–5, 3–4.‡

The following game of ancient origin is in use on the island of Sumatra. Into one half of a coconut shell two, three, or four gambling sticks are inserted. On the faces of these sticks a number of dots, from one to four, are marked. The sticks are then covered by the second half of the shell and the shell shaken. After it is reopened, the number of dots is counted and noted. The winner is the one who named correctly the sum of the dots or noted correctly their numbers before the shell was reopened. Three of such playing sticks can be found in the Anthropological and Ethnographical Museum of the Academy of Sciences of the USSR in Leningrad. They are made of a thick brown wood. The dots are carved with a knife and the faces are well rounded. Similar sticks are to be found in other museums also (see, e.g., [59]).

In the northern part of Sumatra, among the native Bataks, a primitive roulette was excavated. A copy of this roulette was purchased by the Anthropological and Ethnographical Museum of the Academy of Sciences of the USSR in 1897. This set consists of a hexahedral bone prism with an opening into which a wooden core is inserted, made of a thick southern tree. The sides of the prism are marked with numbers from one to six. This prism is rotated like a top and, when it falls on one of its sides, the side which appears on the top determines the number of points. The top is accompanied by a mat with markings and a saucer on which the top is

---

†1–2 means 1 opposite 2; 3–4 means 3 opposite 4; etc.

‡The so-called two partition of seven arrangement (*Translator's remark*).

rotated. A similar copy of a roulette, also originating in Sumatra, is described in [135].

The rules of the game were as follows: The owner of the roulette accepts bets on the numbers from one to six, which are marked on a mat, as well as on the left-hand and right-hand sides of the mat. Then the top is spun. The winner is the one who bet on the correct number. The left-hand side of the mat corresponds to the first three numbers (1, 2, 3) and the right-hand side to the remaining (4, 5, 6). There is some evidence that the owner used to pay the winner only four times the amount of the stake.

It is quite curious that a completely analogous die (a hexahedral prism with a hole) was found in Kiev, dated at the eleventh to twelfth centuries; the wooden core was not preserved. (This die is kept in the State Historical Museum in Moscow.) We may thus assume that roulette, as well as games with dice, originated all over the world as a natural development of games of chance.

Playing cards were available in various nations from ancient times. Modern cards first appeared in France in the fourteenth century and their use soon became widespread in games of chance. Dice and later playing cards were often used for drawing lots and also for divination purposes. Thus, gaming was widespread throughout the world from ancient times. In the Roman Empire it was found necessary to promulgate laws forbidding gaming, except at certain seasons. In Suetonius' "Life of Augustus" and "Life of Claudius" we are told that they both were devoted to dicing. There exists a vessel (amphora) dated at the sixth century B.C. showing a drawing of Achilles and Ajax casting dice. In medieval European literature, starting from the eleventh and twelfth centuries, we encounter the mention and description of various games of chance. As time passed, the number of such descriptions increased. We find gambling mentioned incidentally in the writings of Dante, François Rabelais, Erasmus, and others. Charles de Coster (1829–1879), who for his famous "La Légende d'Ulenspiegel" widely utilized many archives and museum sources, mentions several times the game of dice. There is substantial evidence that the Christian Church opposed gambling and fought to control the evils associated with it. The attempts were ineffective and gaming was prevalent throughout Europe.

Laws prohibiting gambling were issued by Friedrich II (1232), the Russian Czar Aleksii Mihailovich (1649), Catherine II (1782), and others. Louis IX forbade not only playing, but even the manufacture of dice. Acts of Edward III and Henry VIII added dice and cards to the list of unlawful amusements in order to promote sport and military games, such as practice of archery.

Moreover, in various countries special rules were issued limiting gambling activities. For example, participants in the third Crusade (1190 A.D.), knights and the clergy, according to their briefing instructions, might play but could not lose more than twenty shillings in twenty-four hours.

If we assume, as is commonly done, that probability theory owes its origin to gambling, it would be necessary to explain why gambling, which had been in existence for six thousand years, did not stimulate the development of probability theory until the seventeenth century, while in that particular century the theory originated on the basis of the same games of chance. David [49] suggests, for example, the following explanation: Imperfections in the dice deterred notice of the required regularity. This answer is, however, unsatisfactory.† As is well known, remarkable specimens of art and architecture were created in the ancient world; it is therefore improbable that the manufacture of a perfect cube was impossible in those days. And, indeed, when investigating various dice kept in museums in the USSR, we discovered many perfect dice that date to various periods and nations, including also the ancient world.

Kendall [84] suggests four possible reasons that prevented the development of the calculus of probabilities in more ancient times:

(a) the absence of a combinatorial algebra (or at any rate, of combinatorial ideas);
(b) the superstition of gamblers;
(c) the absence of a notion of chance events;
(d) moral or religious barriers to the development of the idea of randomness and chance.

However, Kendall himself, after putting forward these reasons, expresses doubts whether they could have had a decisive influence on the origin of probability theory. But he nevertheless asserts: "It is in basic attitudes towards the phenomenal world, in religious and moral teachings and barriers, that I am inclined to seek for an explanation of the delay [in the origin of probability theory]" [84, p. 10]. However, all these reasons are rather flimsy. When the needs of practical life require development of a certain new science or some of its parts, new ideas then develop, proper notation is invented, superstitions are overcome, and the barriers that prevented the development of the science are broken. Why then did this not happen with probability theory until the seventeenth century?

A somewhat original answer to this question is given by Hotimskii [80]. He thinks that probability theory originated from gambling as late as the seventeenth century and not before because gambling could acquire popularity only with the increase of monetary transactions and trade, i.e., with the development of bourgeois attitudes, approximately in the sixteenth and seventeenth centuries. This point of view connects the develop-

---

†This is, however, only *one* of two reasons suggested by David. The second is the use of dice in religious ceremonies (see, e.g., [49, pp. 6–7]) (*Translator's remark*).

ment of probability theory with the economic activities of society, but is in contradiction with historical fact. Games of chance were popular in the ancient and medieval period if not more, then certainly not less than in the sixteenth and seventeenth centuries. However, Kolman also maintains a similar opinion. He writes: "Probability theory, having always been closely associated with the origin and development of the capitalistic system, originated as a theory of games of chance. It is true that these games were known in ancient times, however, probability theory actually originated during the period when the economic system with its monetary form of exchange, its markets and competition caused these games to become a mass phenomenon" [89, p. 231].

As we have already pointed out, games of chance were not the main stimulus for the origin and development of probability theory. This does not mean, however, that gambling bore no relation to the appearance of probability theory and to the first stages of its development.

The majority of the early problems in probability theory were associated with gambling games, in form if not in substance. Even now, for methodological and pedagogical purposes, we often revert to gambling for an initial presentation of probability because in such a context it is easy to show how to calculate the probability of this or that outcome. Gambling games played their role in the development of probability because they offered a convenient format and ready terminology, with whose help one could describe many phenomena and solve diverse problems. Of course, practical games of chance motivated problems that stimulated the development of the theory of probability. But we repeat that this was not the decisive stimulus. And what is more, attention was turned to games of chance only after analogous problems were raised in other fields of human endeavor.

3   The first problems

One of the first problems that may be classified as a problem in probability theory is that of calculating the number of various possible outcomes in throwing several dice. The first known calculations of this type for the case of three dice are dated at the tenth or eleventh century. In the Middle Ages (before the fifteenth century) poems were written in which a particular verse corresponded to each one of the possible outcomes with three dice. There were altogether 56 such verses; this is the number of possible outcomes with three dice, not counting repetitions. For example, two deuces and one three are counted as only one possibility, independent of the order of appearance of the numbers.

The earliest known approach to the counting of ways in which three dice can fall (permutations included) appears to have occurred in the Latin poem *De Vetula*. This work was regarded as Ovid's for some time and is included among certain medieval editions of his poems. After a long debate concerning the authorship of this work, the one generally accepted currently is Richard de Fournival (1200–1250), Chancellor of the Cathedral in Amiens. The poem contains a long passage dealing with sports and games. The following excerpt is of special interest (in a free translation):

> If all three numbers are alike there are six possibilities; if two are alike and the other different there are 30 cases, because the pair can be chosen in six ways and the other in five; and if all three are different there are 20 ways, because 30 times 4 is 120 but each possibility arises in six ways. There are 56 possibilities.
>
> But if all three are alike there is only one way for each number; if two are alike and one different there are three ways; and if all are different there are six ways. [84, pp. 5–6]

Here the following calculations are carried out: If all three numbers are different, then, in order to obtain the number of possible outcomes, the 30 possible outcomes, without permutations in the case when two numbers are equal, should be multiplied by 4 since two of the numbers are already included in the triple. But then the 120 possible outcomes with permutations are obtained. In each triple there are six possible permutations. Hence, without permutations there will be $120/6 = 20$ triples of different numbers. We thus arrive at $56 = 6 + 30 + 20$, which is the previously encountered number of all possible outcomes without permutations with three dice.

It follows, but is not stated in the preceding text, that the total number of ways is

$$6 \cdot 1 + 30 \cdot 3 + 20 \cdot 6 = 216.$$

It is of interest that the simplest method of arriving at this number, namely, by multiplying all the possible outcomes in throwing two dice (36) by 6 escaped the attention of this poet. In spite of the fact that the number of outcomes was correctly calculated in this particular poem, it was yet common to encounter errors in calculating all the possible outcomes with several dice for a long period of time.

During the period 1307–1321 "The Divine Comedy" by Dante Alighieri was written. The sixth canto of the "Purgatorio" starts with the verses:

> The loser, when the game at dice breaks up,
>   lingers despondent, and repeats the throws
>   to learn, in grief, what made his fortune droop;
> with the other all the folk depart; one goes
>   in front, one plucks him from behind, one near
>   his side protests he knew him once: to those,

then these, while halting not, he lends an ear;
any he gives his hand to, cease to crowd:
thus from the jostling he at last wins clear. [48, p. 295]

In 1477 in Venice Benvenuto d'Imola published "The Divine Comedy"
with his commentaries. Dživelegov refers to Benvenuto d'Imola as a prominent interpreter of the "Comedy." d'Imola used to explain it publicly in
Latin and for a number of years taught a course on Dante's work at Bo-
logna [53, p. 383].

In his commentary on the passage quoted above d'Imola writes:

Concerning these throws it is to be observed that the dice are square
and every face turns up, so that a number which can appear in more
ways (sc. as the sum of points on three dice) must occur more frequently,
as in the following example: with three dice, three is the smallest
number which can be thrown, and that only when three aces turn up;
four can only happen in one way, namely, as two and two aces.

This commentary is often quoted in various texts on the history of mathematics (see, e.g., [84, p. 5]).

Benvenuto committed an error by assuming that the sum 4, similarly to
the sum 3, can be obtained with three dice only in one way. The sum 3
can indeed be obtained only in one way, but 4 comes up in three different
ways: 2,1,1; 1,2,1; and 1,1,2. The same is true of 17 and 18. Benvenuto did
not distinguish between the cases with repetitions and therefore would not
have been able to calculate the total number of possible outcomes in a case
of several dice.

In spite of the relative unimportance of these commentaries as far as
probability theory is concerned, they traditionally receive greater attention
than they deserve (see, e.g., [182, p. 421]).

A more specific problem is that of the fair division of stakes between two
players when the match is stopped before one of the players wins a definite
number of rounds agreed in advance. This problem was considered by
Luca dal Borgo, or Paccioli† (or Paciolo) (ca. 1445–ca. 1514), but was known
even before. Paccioli's basic volume "Summa de Arithmetica, Geometria,
Proportioni et Proportionalità" was completed in 1487 and printed in
Venice in 1494 in Italian. The book was an encyclopedia of mathematical
knowledge of its period. In the section devoted to "unusual" problems
Paccioli presented the following:

1.   A team plays ball so that a total of 60 points is required to win and
the stakes are 22 ducats. Due to circumstances, they cannot finish the game

† Paciuoli is also found in the literature (*Translator's remark*).

and one side has 50 points, and the other 30. What share of the prize money belongs to each side?

2. Three compete with a cross bow. The one who obtains six first places wins; they stake 10 ducats. When the first has four best hits, the second three, and the third two, they decide not to continue but to divide the prize fairly. What should the share of each be?

For the first problem, Paccioli carries out the following calculation: $\frac{5}{11} + \frac{3}{11} = \frac{8}{11}$; since $\frac{8}{11}$ corresponds to 22 ducats, $\frac{5}{11}$ will correspond to $13\frac{3}{4}$ ducats, and $\frac{3}{11}$ to $8\frac{1}{4}$ ducats. The stake, according to Paccioli, should be divided proportionally to the scored points. In other words, if two players A and B play among themselves and up to the moment of termination of the game they win p and q rounds correspondingly, then the stakes should be divided in the ratio $p:q$.

4 Cardano and Tartaglia

The treatise by Gerolamo Cardano (1501–1576) entitled "Liber de Ludo Aleae" ("The Book on Games of Chance")† was an important contribution toward the development of probabilistic notions. The manuscript was discovered posthumously in 1576 among Cardano's papers and was first published in the first volume of his collected works which appeared in 1663. Cardano mentions (in Chapter 20 of the book) that this treatise was written in 1526.

In Chapter XI, entitled "On the cast of two dice," Cardano states: "In the case of two dice, there are six throws with like faces, and 15 combinations with unlike faces, which, when doubled, give 30, so that there are 36 throws in all" [141, p. 195].

In this manner Cardano obtains the total number of possible cases in throwing two dice. He clearly distinguishes between different permutations. "Doubled" throws, according to Cardano, are those which, for example, give 2 on the first die and 3 on the second or, conversely, 3 on the first die and 2 on the second.

Furthermore, Cardano enumerates the number of possible occurrences of a given number of points (an ace, say) on at least one die when two dice are cast.

The total number of such casts is 11. Indeed, writing down, as an example, all the cases where an ace occurs at least once, we have: 1,1; 1,2; 1,3; in total 11 cases. Next, Cardano writes: "This number is somewhat more than

---

†The English translation of this book was published as a supplement to [141]. It should be noted that in Ore's otherwise well-edited book, some misprints in certain tables have slipped in.

half of equality,† and in two casts of two dice the number of ways of getting at least one ace twice is more than $\frac{1}{6}$, but less than $\frac{1}{4}$ of equality."‡

When casting two dice, there are 36 possible outcomes: 30 with unlike faces and six with like faces. Without repetitions there are 15 outcomes and, by adding three outcomes with like faces to this number, we obtain 18, i.e., one-half of the possible outcomes. Of these 18, there are nine with unlike faces and a different sum of dots on the two dice. Cardano expresses this in the following manner: "As for the throws with unlike faces, they occur in pairs in the eighteen casts of equality, so that equality for such a throw consists of nine casts" [141, p. 195].

Subsequently, Cardano calls the "set of circuits" (or "the circuit") the 36 casts possible with two dice. Concerning his conclusion about six cases of the like faces and 30 cases of unlike faces he writes: "So the whole set of circuits is not inaccurate, except insofar as there can be repetitions...in one of them. Accordingly, this knowledge is based on conjecture which yields only an approximation...; yet it happens in the case of many circuits that the matter falls out very close to conjecture" [141, p. 196].

Here Cardano asserts that when the number of observations is small, the frequency can deviate substantially from the "portion," i.e., from the probability; however, when the number of trials is large this deviation is insignificant. In this manner Cardano appears to have approached an understanding of statistical regularity and the law of large numbers.

In Chapter XII "On the cast of three dice" Cardano writes:

> Throws with three alike are the same, except in one respect, as the throws with two alike in the preceding chapter; thus there are six of them. The number of different throws of three dice with doublets and one different point is 30, and each of these occurs in three ways, which makes 90. Again the number of different throws with three different faces is 20, each of which occurs in six ways, which makes 120. Thus the circuit of all of them will be 216 and equality will be 108. [141, p. 196]

We may thus conclude that the calculations of the number of various outcomes in the case of three dice are carried out correctly.

It is of interest to note that Cardano obtains the number 216 not as the product of $36 \cdot 6$, but by other methods.

Next it is explained in Cardano's text how all of the numbers mentioned

† That is, more than nine (*Translator's remark*).

‡ That is, of the half of the total number of possible outcomes. This statement is correct, although the limits given are only rough estimates (see [141, p. 155]). (The actual probability is $(\frac{11}{36})^2 = 1/10 \cdot 7$ which is between $\frac{1}{2} \cdot \frac{1}{4}$ and $\frac{1}{2} \cdot \frac{1}{6}$.) The remark in the Russian original, concerning the incorrectness of Cardano's assertions is probably based on the misinterpretation of Cardano's text (*Translator's remark*).

above are obtained. For example: "In the case of a doublet and one different, there are six possible values for the like faces and five for the odd point; so, since there are six faces, there will be 30 variates of throws. Also each of them can be varied in three ways, making 90" [141, pp. 196–197]. Other numbers are explained analogously.

In this chapter several other problems connected with the calculation of various outcomes are also solved. The number of outcomes for two different points (i.e., 1 and 2) when three dice are cast is determined: "But two distinct faces, as (1, 2), we shall distinguish thus: that, if an ace is adjoined, it will be done in three ways, and if a deuce, then in the same number of ways, therefore already six. Moreover, it happens in four other ways: but the latter are varied, each one, in six different ways; therefore, there will be 24" [141, p. 197].

Further in this chapter the numbers of certain outcomes in casting two and three dice are compared: "But three different faces, as (1, 2, 3), bear to the number of equality† exactly in the same proportion of doublets from two dice" [141, p. 197], i.e., $\frac{6}{108} = \frac{2}{36}$.

In Chapter XIII, "On composite numbers up to six and beyond and for two and three dice," Cardano comes very close to a proper method of calculating probabilities of various events. He writes:

> The point 10 consists of (5, 5) and of (6, 4), but the latter can occur in two ways, so that the whole number of ways of obtaining 10 will be $\frac{1}{12}$ of the circuit . . . . Again, in the case of 9, there are (5, 4) and (6, 3), so that it will be $\frac{1}{9}$ of the circuit . . . . The 8 points consist of (4, 4), (3, 5) and (6, 2). All five possibilities are thus about $\frac{1}{7}$ of the circuit . . . . The point 7 consists of (6, 1), (5, 2) and (4, 3). Therefore, number of ways of getting 7 is 6 in all, $\frac{1}{6}$ of the circuit. The point 6 is like 8, 5 like 9, 4 like 10, 3 like 11 and 2 like 12. [141, p. 192]

Actually, he determines probabilities of obtaining a given number of points in the case of two dice: $\frac{1}{9}, \frac{1}{7}, \frac{1}{6}$ are the probabilities of obtaining 9, 8, 7 points correspondingly. Then he remarks that $P(8) = P(6) \approx \frac{1}{7}$; $P(9) = P(5) = \frac{1}{9}$, and so on. However, his probability is not a ratio as yet, but a part of the "circuit."

Next a table of "*sors* with two dice"‡ is presented

2	12	1	3	11	2	4	10	3
5	9	4	6	8	5	7	8	18

---

† That is, half of the total number (*Translator's remark*).

‡ Coincidence of chances with two dice. *Sors* may perhaps be best translated as outcome or point and refers to the probability of obtaining a particular total on the dice (*Translator's remark*).

The table is constructed as follows: 2 and 12 can occur only in one way; 3 and 11 can occur in two ways; 4 and 10 in three ways, and so on. In the last three numbers there seems to be an error. Instead of 7, 8, and 18, it should be 7, 6, and 18. This would mean that 7 can occur in six ways, and 18 is the "equality," i.e., half of the circuit which often appears in Cardano's text.

Next Cardano presents a table of "*Sors* with three dice" (Coincidence of chances with three dice):

| 3 | 18 | 1 | 5 | 16 | 6 | 7 | 14 | 15 | 9 | 12 | 25 |
| 4 | 17 | 3 | 6 | 15 | 10 | 8 | 13 | 21 | 10 | 11 | 27 |

The first two columns show sums of points obtained when casting three dice. The third column shows the number of ways in which a given sum can be obtained. Cardano clearly distinguishes the number of cases in which 3 and 4 can be obtained, 18 and 17 correspondingly, and so on.

The table is followed by a detailed explanation of how these numbers were obtained. For example: "The point 6 in *sors* (a single cast with three dice) can be obtained in 10 ways, namely, 3 deuces, 2 aces with a 4; and (3, 2, 1)" [141, p. 199].

In Chapter XIV, "On combining points," Cardano calculates the number of cases when casting two dice, in which an ace occurs at least once, or an ace or a deuce occur, and so on.

"The ace occurs in 11 casts, and the deuce likewise, and the trey and so on for each of them, but the ace or deuce do not occur in 22 casts, but only in 20. For the ace occurs in 11 and the deuce in 9 more. Thus, if the trey is added, there will not be 29, nor 31, but 27" [141, p. 202].

To simplify the computations the following table is presented:

| 20 | 27 | 32 | 35 | 36 |
| 11 | 9 | 7 | 5 | 3 |

Here 11 is the number of cases using two dice in which 1 is obtained at least once; 20 is the number of ways in which 1 or 2 is obtained at least once; 27 is the number of ways in which 1, 2 or 3 occurs at least once, and so on.

The numbers 3, 5, 7, . . . are the consecutive odd numbers, which should be added to 11 to obtain the required numbers: $11 + 9 = 20$; $20 + 7 = 27$; $27 + 5 = 32$; $32 + 3 = 35$; $35 + 1 = 36$. The case of ace being evident, is not indicated in the table. This pattern can be obtained from the following observations: Record all the outcomes obtained in the case of two dice.

1,1	1,2	1,3	1,4	1,5	1,6
2,1	2,2	2,3	2,4	2,5	2,6
3,1	3,2	3,3	3,4	3,5	3,6
4,1	4,2	4,3	4,4	4,5	4,6
5,1	5,2	5,3	5,4	5,5	5,6
6,1	6,2	6,3	6,4	6,5	6,6

The number of at least one occurrence of an ace is separated by the upper line $5 + 6 = 11$. 20 is obtained by adding the preceding smaller odd number (i.e. 9) which indicates the number of occurrences of 2 at least once, and so on.

Following the table, Cardano remarks:

> If, therefore, someone should say, I want an ace, a deuce, or a trey, you know that there are 27 favorable throws, and since the circuit is 36, the rest of the throws in which these points will not turn up will be 9; the odds will therefore be 3 to 1. Therefore, in four throws, if fortune be equal, an ace, deuce, or trey will turn up three times and only one throw will be without any of them; if, therefore, the player who wants an ace, deuce, or trey were to wager three ducats and the other player one, then the former would win three times and would gain three ducats; and the other once and would win three ducats; therefore in the circuit of four throws they would always be equal. So this is the rationale of contending on equal terms; if, therefore, one of them were to wager more, he would strive under an unfair condition and with loss; but if less, then with gain. [141, p. 200]

The case of 1, 2, 3, or 4 points is considered next, and he concludes that "the same reasoning is to be observed in the case of three dice" [141, p. 201].

Here Cardano in fact approaches the definition of a fair game, i.e., a game in which the mathematical expectation of the gain is equal to the amount of the stakes. In this particular instance Cardano does not mention that the assertions presented above are valid in the case when the game is continued for a sufficiently long time, but he repeatedly emphasizes this point in his book.

In deriving the rule of calculating the amount of the stakes, Cardano defines the probability essentially in terms of a ratio of equally likely events:

> So there is one general rule, namely, that we should consider the whole circuit, and the number of those casts which represents in

how many ways the favorable result can occur, and compare that number to the remainder of the circuit, and according to that proportion should the mutual wagers be laid so that one may contend on equal terms. [141, p. 202]

In this passage Cardano asserts that if the possible number of trials is n and the number of favorable trials is m, then the stakes should be set in the ratio $m/(n - m)$. Since all his examples are taken from dice games, the equiprobability of events is tacitly assumed. Cardano uses the term "approximately" the same proportion of stakes. The fact of the matter is that, in some problems discussed in the book, the ratio $m/(n - m)$ cannot be expressed in round numbers of monetary units. For example: "In order that an ace may turn up, since there are 11 favorable throws, the odds will be 25 to 11, a little more than 2 to 1 [141, p. 201].

Next the problem of the occurrence of a given number of points (2) at least once in a single series of three casts is considered. This is followed by the problem of the occurrence of a given number of points at least once in each one of two series of three casts and in each one of the three series of three casts.

Thus, if it is necessary for someone that he should throw an ace twice, then you know that the throws favorable for it are 91 in number, and the remainder is 125; so we multiply each of these numbers by itself and get 8281 and 15,625, and the odds are about 2 to 1. Thus, if he should wager double, he will contend under an unfair condition, although in the opinion of some the condition of the one offering double stakes would be better. In three successive casts, therefore, if an ace is necessary, the odds will be 753,571 to 1,953,125, or very nearly 5 to 2, but somewhat greater. [141, p. 202]

Here $753,571 = 91^3$; and $1,953,125 = 125^3$. In this problem Cardano utilizes the multiplication rule for probabilities. For example, when computing the amount of stakes in two series, he squares the quantity $m/(n - m)$ obtaining $m^2/(n - m)^2$, which means that the corresponding probabilities are obtained using the multiplication rule:

$$p_1 = \frac{m}{n}\frac{m}{n} = \frac{m^2}{n^2}; \qquad q_1 = \frac{n - m}{n}\frac{n - m}{n} = \frac{(n - m)^2}{n^2},$$

and the stakes which are proportional to these probabilities are

$$\frac{p_1}{q_1} = \frac{m^2/n^2}{(n - m)^2/n^2} = \frac{m^2}{(n - m)^2}.$$

In the case of three sets of casts he proceeds in a similar manner.

Two further chapters (XXX and XXXI) are connected with the history of gambling. Cardano writes that the inventor ("as they say") of gambling was Palamedes during the Trojan wars. The siege of this city lasted ten years and various games were invented to bolster the soldiers' morale since they suffered from boredom.

Cardano presents many accounts and statements by the ancients concerning the game of dice and relates various methods of cheating in different games.

Next a game with astragali (knuckle-bones) is discussed and it is assumed that the various sides of an astragalus have the values 1, 3, 4, and 6, respectively. Four such astragali are cast and the number of favorable cases for various outcomes is calculated.

Cardano states† that a throw which results in at least two aces is named a dog "because whatever the points on the two other astragali may be, the total sum cannot exceed the average number" [141, pp. 240–241] which is 14 for the case of four astragali. This is indeed correct under the assumption that all the sides of an astragalus are equally probable (which they are not).

In 1570 a book by Cardano was published containing three treatises. The first one is entitled "Opus novum de proportionibus numerorum" ("New Work on Proportionalities"). In this treatise there are several problems connected with combinatorics. For example, he explicitly notes all fifteen combinations of two elements out of six; referring to Stifel, he writes down the binomial coefficients and asserts, without a proof, that the number of all possible combinations of n elements is

$$2^n - 1, \quad \text{i.e.,} \quad \binom{n}{1} + \binom{n}{2} + \cdots + \binom{n}{n} = 2^n - 1.$$

In another treatise entitled "Practica arithmeticae generalis" ("Practical General Arithmetic") published in 1539, Cardano validly argues with Paccioli concerning the latter's solution of the problem of "fair division of stakes." He observes that Paccioli divides the stakes in proportion to the number of games already won, but does not take into account the number of games yet to be won by each of the players. He pointed out (correctly) this error in Paccioli's work but was unable to obtain the correct solution of the problem. Thus he assumes that if S is the total number of rounds needed for completion of the game (according to the rules) and p and q are the number of games won by each of the players, then the stakes should be divided in the ratio

$$[1 + 2 + 3 + \cdots + (S - q)] : [1 + 2 + 3 + \cdots + (S - p)].$$

---

†This is entirely his own speculation (see [141, p. 171]) (*Translator's remark*).

By adding the chances, Cardano actually uses the addition rule for probabilities. He was aware that in order to use this rule the events must be disjoint.

Actually, Cardano also uses the multiplication rule for probabilities for the case of independent events. To be sure, in all these cases he was not computing the probabilities, but the stakes in fair games which are proportional to the probabilities. However, the rule for computing the amounts of stakes as formulated by Cardano is quite similar to the so-called classical definition of probabilities.

Cardano's work was a substantial step forward in the development of probabilistic notions and concepts. He carried out correct calculations of the number of possible outcomes with and without repetitions in the case of two and three dice. He approached the notion of statistical regularity and stated a number of observations concerning problems directly related to the law of large numbers. Finally, he came close to a definition of probability in terms of the ratio of equiprobable events and, using the idea of mathematical expectation, introduced what amounts to the notion of a fair game.

In Niccolo Tartaglia's (ca. 1499–1557) work "Trattato generale di numeri e misure" ("General Treatise on Number and Measure") published in Venice in 1556, various problems of combinatorics and probability theory are closely interrelated. Tartaglia considers the following problem: "Ten people are to be seated and served as many dishes as there are different ways in which they can be seated, so that their seating arrangement the second time is different from the first." Solving this problem he obtains $n = 1 \cdot 2 \cdots 10$, i.e., $n = 10!$ Then Tartaglia asserts: "I shall adhere to this mode of operation whether there are as many as 1000 people or any other number, since the rule approaches infinity" [175, Pt. II, Chap. 16, § 10]. Tartaglia, however, does not prove this rule in its general form.

Further the following heading is presented in Tartaglia's treatise: "A general rule of the author discovered on the first day of Lent, 1523 in Verona for obtaining the number of variations in positions of any number of dice cast." Tartaglia then relates a story that in 1523, when he visited Verona, a group of young as well as mature people were extracting answers from a book called "Book of Happiness," using throws of three dice. The author observed that there were 56 variations in throwing three dice and decided to find corresponding rules to derive this number not only for three dice, but for any number. He spent the whole night pondering this problem and, on the next day, which was the first day of Lent, found that these rules are formed from "special kinds of progressions." Here are some of his calculations: One die can land in six ways: $1 + 1 + 1 + 1 + 1 + 1 = 6$. Two in 21 ways (Tartaglia counts here the number of falls without repetition). 21 is obtained as a sum of six numbers, where each number is the sum of the corresponding number of the terms in the preceeding series. The number of various falls for three dice is obtained in a similar manner. Thus,

$$
\begin{array}{lll}
(1) & 1 + 1 + 1 + 1 + 1 + 1 & = \quad 6; \\
(2) & 1 + 2 + 3 + 4 + 5 + 6 & = \quad 21; \\
(3) & 1 + 3 + 6 + 10 + 15 + 21 & = \quad 56; \\
(4) & 1 + 4 + 10 + 20 + 35 + 56 & = \quad 126; \\
(5) & 1 + 5 + 15 + 35 + 70 + 126 & = \quad 252; \\
(6) & 1 + 6 + 21 + 56 + 126 + 252 & = \quad 462; \\
(7) & 1 + 7 + 28 + 84 + 210 + 462 & = \quad 792; \\
(8) & 1 + 8 + 36 + 120 + 330 + 792 & = 1287.
\end{array}
$$

Tartaglia concludes this chapter with the remark that, in order to explain in detail in written form how the six terms of all the progressions given above originated, it would be necessary to compile an entire book. However, following the rule described above, one can find out how many variations there are in casting 10,000 dice.

It is easily seen that the sums in question are actually of the following form:

$$
\begin{array}{ll}
(1) & C_0^0 + C_1^1 + C_2^2 + C_3^3 + C_4^4 + C_5^5 = 6 = C_6^5; \\
(2) & C_1^0 + C_2^1 + C_3^2 + C_4^3 + C_5^4 + C_6^5 = 21 = C_7^5; \\
(3) & C_2^0 + C_3^1 + C_4^2 + C_5^3 + C_6^4 + C_7^5 = 56 = C_8^5; \\
(k) & C_{k-1}^0 + C_k^1 + C_{k+1}^2 + C_{k+2}^3 + C_{k+3}^4 + C_{k+4}^5 = C_{k+5}^5 = C_{6+k-1}^k.
\end{array}
$$

In the case of k dice the final number is equal to the sum of the first six terms of the difference series of the $(k - 1)$th order or to the sixth term of the difference series of the kth order, which is equal to

$$
\binom{k + 5}{5} = \binom{6 + k - 1}{k}.
$$

In Section 20, entitled "Error di Fa Luca dal Borgo,"† Tartaglia deals with the division of stakes problem. He reproduces the text and solution as given by Paccioli which states that the stakes should be divided in the proportion of "rounds" already won.

In this connection he remarks: "His rule seems neither agreeable nor good, since, if one player has, by chance, ten points and the other no points, then following this rule, the player who has the ten points would take all the stakes which obviously does not make sense."

Further he writes: "Therefore I say that the resolution of such a question is judicial rather than mathematical, so that in whatever way the division is made there will be cause for litigation."

---

† Paccioli's error (*Translator's remark*).

He then proceeds with solutions (which, in his opinion, are the least controversial) of the following two problems:

1.   In a match going to 60 games, A has won ten and B zero games. How should the stakes be divided if each player puts up 22 ducats? The solution is as follows:

$$\tfrac{10}{60} = \tfrac{1}{6} \text{ part of 22 ducats} \qquad \text{or} \qquad \tfrac{22}{6} = 3\tfrac{2}{3} \text{ ducats}.$$

Player A therefore should receive $22 + 3\tfrac{2}{3} = 25\tfrac{2}{3}$ ducats, and player B $18\tfrac{1}{3}$ ducats.

2.   Under the same conditions, player A won 50 games and player B 30. How should the stakes be divided? Solution:

$$50 - 30 = 20; \qquad \tfrac{20}{60} = \tfrac{1}{3}; \qquad \tfrac{22}{3} = 7\tfrac{1}{3}.$$

Player A thus receives $23 + 7\tfrac{1}{3} = 29\tfrac{1}{3}$ ducats, and player B $14\tfrac{2}{3}$ ducats.

Tartaglia also solves the following problem: "In a match going to six games A has won five and B three games. How should the stakes be divided?

Solution: The difference between A's score (five) and B's score (three) is two, which is one-third of the number of games needed to win (six). Therefore, A should take two-thirds of the stake and B one-third, i.e., the total stake should be divided in the ratio 2:1."

Tartaglia concludes this section with the following remark: "Paccioli's solution is a subject making little sense and will be substantial cause for litigation."

As opposed to Paccioli, who recommends that we divide the stakes in proportion to the number of rounds won by each player, Tartaglia suggests that the deviation from half of the stake should be proportional to the difference of the rounds won by the players. Moreover, one player receives half of the stake plus the additional amount calculated according to the above rule, while the second player gets half of the stake minus the same amount. Both of these solutions are incorrect, since a fair division of the stakes should be proportional to the probability of winning the whole stake (assuming the game to be continued).

Two years after Tartaglia's treatise, in 1558, a short work of G. F. Peverone "Due Brevi e Facili Trattati, il Primo d'Arithmetica, l'Altro di Geometria" ("Two Short and Easy Treatises, the First on Arithmetic, the Second on Geometry") appeared. In the first part of this work Peverone considers a similar problem of division of stakes: "In a match going to ten games, A has won seven and B nine games. How should the stakes be divided?" He argues that the division should be in the ratio 1:6. Here he is quite close to the correct solution which is 1:7.

5   Elements of probability theory in Galileo's works

The most complete solution of the problem on the number of all possible outcomes with three dice was given by Galileo Galilei† (1564–1642) in his treatise "Considerazione sopra il Giuco dei Dadi"‡ ("On Outcomes in the Game of Dice"). The date of this paper remains unknown. It was first published in Florence in 1718. The method proposed here by Galileo can quite easily be generalized for the case of a larger number of dice.

He considers the following problem: Three dice are thrown simultaneously and the sum of the scores is noted. In this case

> although 9 and 12 can be made up in as many ways as 10 and 11, and therefore they should be considered as being of equal utility to these, yet it is known that long observation has made dice players consider 10 and 11 to be more advantageous than 9 and 12. And it is clear that 9 and 10 can be made up by an equal diversity of numbers (and this is also true of 12 and 11): since 9 is made up of 1,2,6; 1,3,5; 1,4,4; 2,2,5; 2,3,4; 3,3,3; which are six triple numbers; and 10 of 1,3,6; 1,4,5; 2,2,6; 2,3,5; 2,4,4; 3,3,4; and in no other ways, and these also are six combinations. [189, p. 192]

Why is 10 then preferable to 9? Someone presumably posed this question to Galileo, since he writes further§:

> Now I, to oblige him who has ordered me to produce whatever occurs to me about such a problem, will expound my ideas, in the hope not only of solving this problem but of opening the way to a precise understanding of the reasons for which all the details of the game have been with great care and judgment arranged and adjusted. [189, p. 193]

He obtains the total number of outcomes using the simplest method: for two dice this number is $6 \cdot 6 = 36$ and for three, $36 \cdot 6 = 216$.

After detailed discussion of all possible cases Galileo arrives at the following three basic propositions:

> Therefore, we have so far declared these three fundamental points: first, that the triples, that is the sum of three-dice throws, which are made up of three equal numbers, can only be produced in one way; second,

---

† See [84, p. 8] (*Translator's remark*).

‡ Galileo's original title was "Sopra le Scoperta dei Dadi" (see [189, p. 64]) (*Translator's remark*).

§ David [189, p. 65] conjectures that this was his mentor, the Grand Duke of Tuscany (*Translator's remark*).

that the triples which are made up of two equal numbers and the third different, are produced in three ways; third, that those triples which are made up of three different numbers are produced in six ways. From these fundamental points we can easily deduce in how many ways, or rather in how many different throws, all the numbers of the three dice may be formed. [189, p. 194]

He concludes his paper with the table shown here as Table II.

TABLE II

	10		9		8		7		6		5		4		3
1															
3															
6															
10	631	6	621	6	611	3	511	3	411	3	311	3	211	3	111
15	622	3	531	6	521	6	421	6	321	6	221	3			
21	541	6	522	3	431	6	331	3	222	1					
25	532	6	441	3	422	3	322	3							
27	442	3	432	6	332	3									
—	433	3	333	1											
108															
108		27		25		21		15		10		6		3	1
—															
216															

The top row containing the numbers 10, 9, and so on shows the sum of points on three dice. The first column under number 10 indicates the ways in which 10 can be obtained with three dice (6,3,1; 6,2,2; and so on). The next column indicates the number of ways in which 10 is obtained: if the three points on each of the dice are different, then there are six ways; if two points are the same, then three ways are available, and if all three points are the same (as in the case of nine, six, and three), there is only one way to produce the corresponding heading. Number 27 on the bottom of the column indicates the total number of dice throws available to obtain the sum of 10 points. Other columns are constructed in a similar fashion. In the left-hand column the total number of all outcomes is tallied—the number being 108. If outcomes with the sum of points greater than 10 are considered, an additional 108 outcomes are obtained. Therefore, the total number of all possible throws is 216 and Galileo's table is essentially one-half of the complete table.

In this connection, Galileo points out:

And there being a similar number of throws for the higher sums of points, that is, for the points 11, 12, 13, 14, 15, 16, 17, 18, one arrives at the sum of all the possible throws which can be made with the faces of the three dice, which is 216. [189, p. 195]

This problem, which had a history of many centuries, was solved by Galileo in a most detailed manner:

The following comment would seem to be in order: The probability of obtaining the sum 10 with three dice is equal to $\frac{27}{216} = 0.125$ and the sum 9 is $\frac{25}{216} = 0.116$. The difference is $0.125 - 0.116 = 0.009$. Such a small difference could hardly be detected in a practical game because it would take a very large number of throws using the same dice under the same conditions, not to mention that all the dice would have to be perfectly balanced. Thus Galileo's reference to the fact that this difference was detected by the players is only a tribute to a traditional allegation which exists up to the present time. The type of problem Galileo solved was theoretical in nature. That such theoretical results could be utilized to a certain degree in actual games of chance is merely incidental.

In [62, pp. 231–284], among other matters, some of Galileo's correspondence is presented. In particular, letters which he exchanged with a priest called Nozzolini and also a letter of Benedetto Castelli. In these letters the following problem is discussed: A horse really worth 100 *scudi* was valued by one man at 1000, and by another at 10. Which of the two was the better estimate and which of the two judged more extravagantly? Galileo suggests that the two estimates are equally extravagant because the ratio of a thousand to a hundred is the same as the ratio of a hundred to ten. Nozzolini considers the higher estimate to be more extravagant than the other, since the difference $1000 - 100$ is greater than $100 - 10$. It appears from Castelli's letter that Galileo at first was of the same opinion as Nozzolini, but afterward changed his mind.

With the invention of the telescope, astronomical observations using various equipment became more widespread and the problem of estimating errors in observations became more pressing. Galileo was one of the first investigators to pose this problem in his writings. Although he did not arrive at a quantitative or analytic solution of the problem, many of his propositions and remarks have greatly influenced the development of the problem of error estimation, as well as working out the basic notions of probability theory. These problems were discussed in detail in his "Dialogue of the Great World Systems" which were first published in 1632.†

In the discussion of the third day in this book the problem is examined whether the new star of 1572 is located below the moon, above the moon, or on the sphere of the firmament.

In brief the history of this problem is as follows: On 11 November 1572, Tycho Brahe (1546–1601) observed over the skies of Denmark a bright

†English translations are available by Salusbury (17th century work reprinted by Univ. of Chicago Press, 1953) and by Drake [63] (*Translator's remark*).

star in the constellation Cassiopeia. This new star was as bright as Venus and was even visible by day. This remarkable phenomenon created a great stir all over the world. Some believed that the new star belonged to the sublunar world of the variable earthly elements; others suggested that this was a comet, condensed from fiery vapors. There were also those who conjectured that the new star belonged to the world of fixed stars. The latter were against Aristotle's doctrine of the invariability and eternity of the celestial vault. Tycho Brahe's conclusion, however, was that the new star did belong to the world of fixed stars and that, contrary to widespread opinion, changes in the celestial vault were possible. He published his conclusions in a book "De Nova Stella" (1573).

Shortly after its appearance, the new star began to decline in brightness and by the end of March 1574 it had disappeared.† However, heated discussions concerning this star continued for a long time.

In 1628 a book by Chiaramonti appeared entitled "De Tribus novis stellis quae Annis 1572, 1600, 1604" (On the Three New Stars Which Appeared in the Years 1572, 1600, 1604"). In this book the author defends the opinion that the distance to the star Nova 1572 is less than the distance to the moon. Galileo in his "Dialogue" challenged this conclusion.

At first Galileo remarks that if the new star were located "among the other fixed stars, its meridian altitudes taken in diverse elevations of the pole ought necessarily to differ from each other with the same variations that are found among those elevations themselves."

"But, if the star's distance from the earth was but very little, its meridian altitudes ought, approaching to the north, to increase considerably more than the polar altitudes" [63, p. 301].‡ By means of this increment in altitudes (which Galileo calls the differences of parallaxes or simply parallax) it is easy to compute the star's distance from the center of the earth.

Chiaramonti using 13 observations of the new star carried out by various astronomers, combines, according to his own judgment, 12 pairs of observations. Based on parallaxes in these pairs, he computes the distance to the star which turns out to be from $\frac{1}{48}$ of the radius up to 32 times the radius of the earth, which is less than the distance to the moon. From these results he concludes that the actual distance to the star is less than the distance to the moon.

The majority of pairs of observations [and there are $\binom{13}{2} = 78$ of them] result in distances greater than the distance to the moon, but Chiaramonti believes that the observations resulting in these large distances are in error.

----

† See [146, pp. 207, 208] for more details.

‡ "Dialogue Concerning the Two Chief World Systems—Ptolemaic and Copernican" by Galileo Galilei, 2nd rev. ed. (Transl.: Stillman Drake), 1953, 1962. Originally published by the University of California Press; reprinted by permission of the Regents of the University of California.

In the course of discussion with Chiaramonti, Galileo arrives at a number of important basic propositions. He notes that all the 12 pairs of observations utilized by Chiaramonti yield different distances. This is not due to "any defects in the rules of calculations, but [depends] upon errors made in determining those angles and distances" [63, p. 303].† Next, Galileo writes that the observations are always accompanied by errors. "In each combination of observations there is some error, which I believe to be absolutely unavoidable. ... We know too well that in the taking of only one altitude of the pole, with the same instrument, in the same place, by the same observer who has repeated the observation a thousand times, there will still be a variance of one, or sometimes of many minutes" [63, p. 305].†

Galileo, thus, arrives at the conclusion that random errors are inescapable in instrumental observations. This idea is repeated by him on many occasions in the "Dialogue." Next, Galileo poses the question of how to correct the observed values in order to obtain reliable results. In other words, how can the random errors be taken into account?

It is Galileo's opinion that since smaller errors are more frequent than the larger ones, one should correct the smaller errors rather than the larger. Galileo repeatedly emphasizes that the probability of small deviations is greater than large ones. This problem was discussed in later years by many mathematicians.

Further Galileo suggests that all the observations which yield impossible results should be discarded, i.e., one should disregard all those observations which result in values that are far beyond the majority of other observations.

Following these remarks, Galileo discusses the problem of the signs of the errors appearing in the observations:

> First I ask you whether astronomers, in observing with their instruments and seeking, for example, the degree of elevation of the star above the horizon, may deviate from the truth by excess as well as by defect; that is, erroneously deduce sometimes that it is higher than is correct, and sometimes lower? Or must the errors be always of one kind, so that when they err they are always mistaken by an excess, or always by a defect and never by an excess? [63, p. 307]†

Galileo has a definite reply to these questions: "I do not doubt that they are equally prone to err in one direction or the other." [63, p. 309],† i.e., these errors have the same probabilities. Here Galileo had in mind the fact that the distribution law of the errors is symmetric, although this was not expressed in a sufficiently clear manner.

Next, Galileo discusses the popular erroneous view that one can, based

†"Dialogue Concerning the Two Chief World Systems—Ptolemaic and Copernican" by Galileo Galilei, 2nd rev. ed. (Transl.: Stillman Drake), 1953, 1962. Originally published by the University of California Press; reprinted by permission of the Regents of the University of California.

on the size of the error obtained after performing the observations and corresponding computations, judge and estimate the errors of the instrument with which the observations were obtained and, conversely, using the errors in the measuring instruments, estimate the size of the ultimate errors. Galileo rejects this opinion: "know that it may be (and it happens more often than not) that an observation which gives you the star at the distance of Saturn, for example, with the addition or subtraction of a single minute of elevation to that taken by the instrument, will send the star to an infinite distance. ... And while I say one minute, a correction of one-half that, or one-sixth, or less may suffice" [63, p. 308].†

The following conclusion is then presented: "The size of the instrumental errors, so to speak, must not be reckoned from the outcome of the calculation, but according to the number of degrees and minutes actually counted on the instrument" [63, p. 309].†

Galileo next affirms that the greatest number of measurements ought to be concentrated around the true value. Concerning the question of the distance to Nova 1572 he writes: "And among the possible places, the actual place must be believed to be that in which there concur the greatest number of distances, calculated on the most exact observations" [63, p. 309].†

Concluding the detailed discussion of the problem of the distance to Nova 1572, Galileo remarks:

> It is obvious that the corrections to be applied to observations which give the star as at an infinite distance will, in drawing it down, bring it first and with least amendment into the firmament rather than below the moon. Hence everything supports the opinion of those who place it among fixed stars [63, p. 309].†

The validity of this conclusion follows from the fact that the majority of the observations place the star infinitely high above and in order to place the new star among the fixed stars, significantly smaller corrections in the measurements are needed as compared with the corrections required to place the star on any other level; and smaller errors, which require smaller corrections, are more probable than the large ones.

Galileo's final conclusion is thus completely justified: "You can see ... with how much greater probability it is implied that the distance of the star placed it in the most remote heavens" (among fixed stars) [63, p. 309].†

Thus Galileo arrived at the conclusion that errors in measurements are inevitable, that the errors are symmetrically distributed, that the probability of error increases with the decrease of the error size, and that the majority of observations cluster around the true value. Moreover, errors obtained

†"Dialogue Concerning the Two Chief World Systems—Ptolemaic and Copernican" by Galileo Galilei, 2nd rev. ed. (Transl.: Stillman Drake), 1953, 1962. Originally published by the University of California Press; reprinted by permission of the Regents of the University of California.

in observation are not to be compared with the final errors, which arise as a result of computations based on the observed values.

In these conclusions Galileo revealed a number of characteristic features of the normal probability distribution law which later became one of the basic laws of probability theory (see [116], which also deals with Galileo's contributions to probability theory).

Galileo's ideas deeply and pointedly reflected the essence of the problem as can be seen, for example, from the fact that over a century later, in 1765, Lambert in his treatise "On the Problem of Application of Mathematics" in the chapter devoted to "Reliability of observations and experiments" arrives at the same conclusions. Observing the distinction between systematic and random errors, Lambert asserts that random errors are inevitable, that the same deviations are equiprobable on both sides from the center, and that smaller errors are more frequent than the large ones. As regards the probability curve (which he refers to as "possibility"), he states that the curve is symmetric and that its largest ordinate is located on the axis of the symmetry; the curve has two inflection points and approaches asymptotically the abscissa axis (see also [12]).

6 Basic stages in the development of combinatorics

Until differential and integral calculus came into use in the theory of probability, the basic tool of this theory was combinatorics. Almost all problems were solved using combinatorial methods. Therefore the development of combinatorics influenced to some extent the development of probability theory, especially in its initial stages.

As ancient scholars as the Pythagoreans (*ca.* 540 B.C.) investigated triangular numbers which are closely connected with the notion of combinatorics. These numbers are $1; \ 3 = 1 + 2; \ 6 = 1 + 2 + 3; \ 10 = 1 + 2 + 3 + 4;$ and in general $n(n + 1)/2 = 1 + 2 + 3 + \cdots + n$. Hence, triangular numbers represent the number of combinations of two elements. Indeed $\binom{n+1}{2} = n(n + 1)/2$. In the first centuries A.D. more complex numbers were considered—the tetrahedrals, which correspond to the number of combinations of three elements.

The so-called Pascal arithmetic triangle was in use by the Hindus as early as 200 B.C. who were also familiar with the rule for constructing its elements; in particular the identity

$$1 + \binom{n}{1} + \binom{n}{2} + \cdots + \binom{n}{n} = 2^n$$

was known.

Bhâskara II (in 1114 and later in 1178) in his treatise "Lilavati" ("The Noble Science," *ca.* 1150) describes methods for computing permutations and combinations. He was also aware of the general formula for the number of combinations $C_n{}^p = \binom{n}{p}$.

The Hindu mathematician Nârâyana (fourteenth century A.D.) computing the size of a herd of cows originating with a single cow during a period of 20 years essentially arrived at the method of computing the number of combinations with repetitions of k elements out of n.

In India combinatorics was applied to and possibly even stemmed from the computation of different combinations of long and short syllables in an n-compound foot.

Tables of binomial coefficients up to the eighth power are found in the work of the Chinese mathematician Chu Shih-Chien "The Precious Mirror of the Four Elements," written in 1303. It is plausible that the general formula for $C_n{}^m$ was also available in that period. There are references indicating that these types of tables were known in China as early as the twelfth century. The general theorem on binomial expansion for the case of positive integral exponents was apparently first introduced in the works of Al-Kashi (fourteenth century), but was probably known also to Omar Khayyam (died *ca.* 1214 A.D.)

A systematic investigation of combinatorial problems is found in a Hebrew arithmetical work by Levi ben Gerson, written in 1321. It contains a recursion formula for the number of arrangements of n things taken p at a time $(A_n{}^p)$ and, in particular, the number of permutations on n objects. Ben Gerson formulates rules which are equivalent to the following formulas

$$\binom{n}{p} = \frac{A_n{}^p}{p!}; \qquad \binom{n}{n-p} = \binom{n}{p}.$$

However, his manuscripts were apparently unknown to his contemporaries and all these results were later rediscovered.

Michael Stifel (1486–1567) in his book "Arithmetica Integra" (1544) constructed the table of coefficients in the expansions

$$(a + b)^2, (a + b)^3, \ldots, (a + b)^{17}.$$

These coefficients are

		1		2		1		for $(a + b)^2$,
	1		3		3		1	for $(a + b)^3$,
1		4		6		4		1	for $(a + b)^4$,

etc. up to the 17th power.

In the same treatise Stifel investigates and compares the series of numbers in arithmetic with the corresponding numbers in the geometric progressions.

In 1634 P. Herigone in the second volume of his "Cursus mathematicus novus" entitled "Arithmetica Practica" ("Practical Arithmetic") determines quite correctly the number of combinations of m elements out of n.

A correct solution of this problem is also given by Tacquet in his "Theory and Practice of Arithmetic" (1656). In this volume a short chapter is included devoted to combinations and permutations.

During this period many other works were published containing studies of various problems in combinatorics. A substantial contribution to the development of combinatorics is due to Leibniz (1646–1716). A dissertation was published by Leibniz in 1666,† entitled "Dissertatio de Arte Combinatoria" [99, pp. 27–102]; part of it had been previously published in the same year under the title "Disputatio arithmetica de complexionibus." (In the literature his work is often referred to as "Ars Combinatoria"). In this work Leibniz extensively develops combinatorial methods primarily for the purpose of logical deductions, which is closely connected with investigations aimed at the construction of the "universal characteristic."

Moreover, he considers various combinations and permutations, linear as well as circular, and many other problems. He actually utilizes such formulas as:

$$\binom{n}{k} = \binom{n-1}{k} + \binom{n-1}{k-1}; \qquad C_n{}^m = \binom{n}{m}.$$

A table is presented in his book which is analogous to Pascal's arithmetical triangle.‡ Leibniz notes the following properties which are corollaries from the table:

$$\binom{n}{k} = 0 \;\; \text{if } n < k; \qquad \binom{n}{n} = 1; \qquad \binom{n}{n-1} = n;$$

$$\binom{n}{k} = \binom{n}{n-k};$$

as well as several other identities. Next he utilizes (without proof) the relation:

$$\binom{n}{1} + \binom{n}{2} + \cdots + \binom{n}{n} = 2^n - 1.$$

In one of the problems he obtains the number of permutations on a given integer and presents a table of the number of permutations up to 24. In particular, it is noted that the number of permutations of 24 objects 6,204,484,017,332,394,339,360,000; i.e., 24! is equal to this number.

---

†Written at the age of twenty (*Translator's remark*).

‡Todhunter ([176, p. 32]) points out that the mathematical treatment of the subject of combinations (as given by Leibniz) "is far inferior to that given by Pascal" (*Translator's remark*).

Leibniz presents the following properties of permutations:

1. The number of permutations is always even.

2. From the number of permutations of $n - 1$ elements and from the given number n, one can compute the number of permutations of n elements.

3. If the number of permutations on n elements is divided successively by integers from 1 up to the number of the elements n, one obtains the harmonic progression.

For example, if we divide 120 successively by 1, 2, 3, 4, 5, we obtain the terms of the harmonic progression 120, 60, 40, 30, and 24.

4. The relation $2P_n - (n - 1)P_{n-1} = P_n + P_{n-1}$. Indeed:

$$2P_n - (n - 1)P_{n-1} = 2n! - (n - 1)(n - 1)!$$
$$= 2(n - 1)!n - (n - 1)(n - 1)!$$
$$= (n - 1)!2n - (n - 1)$$
$$= (n - 1)!(n + 1) = (n - 1)!n + (n - 1)!$$
$$= n! + (n - 1) = P_n + P_{n-1}.$$

5. The relation: $P_n \cdot P_n/(n - 1)! = P_{n+1} - P_n$. Indeed:

$$P_n \cdot P_n/(n - 1) = (n!n!)/(n - 1)! = n \cdot n!;$$
$$P_{n+1} - P_n = (n + 1)! - n!$$
$$= n!(n + 1) - n!$$
$$= n!(n + 1 - 1) = n \cdot n.$$

Leibniz also obtains the number of circular permutations:

$$Q_n = P_n/n.$$

In his later works Leibniz repeatedly returns to combinatorial problems. He deals with calculating the number of different outcomes in various games of dice. In particular, he finds that the number of outcomes with m dice, in which the particular side occurs k times equals $\binom{m}{k} 5^{m-k}$. His results for the number of outcomes (without repetitions) with 1, 2, 3, etc., dice can be expressed in the following manner:

$$\text{1 die} \quad \binom{6}{1} = 6,$$

$$\text{2 dice} \quad \binom{1}{0}\binom{6}{1} + \binom{1}{1}\binom{6}{2} = \binom{7}{2} = 21,$$

$$3 \text{ dice} \quad \binom{2}{0}\binom{6}{1} + \binom{2}{1}\binom{6}{2} + \binom{2}{3}\binom{6}{3} = \binom{8}{3} = 56 \,,$$

and so on up to six dice.

Leibniz extends these results obtained for six-sided dice for other types of polyhedra.

In Leibniz's treatise "Ars Combinatoria" all† the previously known results related to combinatorics were considered and collected. Moreover, further developments in the field were also presented. He introduced new types of combinatorial problems in which one must consider combinations and arrangements with infinite repetitions.

We observe from the above that combinatorics, which up to a certain period of time served as the basic tool for solving probabilistic problems, was the subject of investigation of numerous mathematicians. Even at the present time combinatorial methods constitute a significant tool in various applications of probability theory.‡

We conclude this chapter with a very perceptive quotation from a paper by Ostrogradskiĭ [77]:

> Probability theory should be classified as a modern science since it did not really originate until the second half of the seventeenth century. It is true that certain topics related to this science were known in ancient times, calculations based on average duration of life were carried out, marine insurance was in use and the number of chance occurrences in hazard games had been calculated, if only in the simplest cases, the division of stakes in certain fair games were determined —all these results were not based on any rules whatsoever. Nevertheless, probability theory is considered a modern science and its origin is attributed to the middle of the seventeenth century because the problems concerning probability were not subject to mathematical analysis and no rigorous general rules for their solution were available at that period. [77, p. 120]

---

†This assertion is not strictly true (for example, Pascal's results were certainly known but not all included) (*Translator's remark*).

‡More details on the history of combinatorics and Leibniz's contributions to this field are presented in [25] and [26] (the latter discusses an unpublished manuscript of Leibniz's on combinatorics written in 1676) (*Translator's remark*).

II

The First Stage in the Development
of Probability Theory

Up to the middle of the seventeenth century no general method of solving probabilistic problems was available, and certainly no complete mathematical theory—only specific problems were solved. However, quite a substantial amount of material had accumulated by then in various branches of human activity related to probabilistic topics.

In the middle of the seventeenth century several prominent scientists became involved in the development of probability theory. First and foremost were Pascal, Fermat, and Huygens. They applied the addition and multiplication rules of probability, were familiar with notions such as dependence and independence of events, and introduced into practice one of the basic and characteristic notions of probability theory: the concept of mathematical expectation.

At this stage new methods of solving problems in probability theory were being developed, the realm of problems relevant to this new science was being determined, and the basic notions of probability were being constructed; all these activities transformed probability theory into a bona fide science.

As Engels points out in his "Dialetics of Nature": "the first stage of of the new natural sciences in the inorganic world was concluded with Newton. This was a period of mastering the available materials" [55, p. 153].

1   De Méré's legend

In 1654 correspondence started between Pascal and Fermat in connection with a number of problems, among them the problem of division of stakes. Many authors consider Chevalier de Méré's role decisive in the initiation of this correspondence, as well as in the origin of probability theory. We cite the following examples† :

> Profiteering of the worldly gambler, Chevalier de Méré during the reign of Louis XIV produced probability calculus or at least directed Pascal's and Fermat's ideas in this direction. [10, p. 249]

> The beginnings of probability calculus arose from a problem on division of the winnings between two players in case they terminate the game before its completion. [183, p. 5]

This problem was suggested to Pascal by Chevalier de Méré. With the passage of time additional details embellished the story :

> Chevalier de Méré, a fervent gambler, on one occasion set before Pascal a problem, which had greatly concerned him and which apparently was of some applied significance. de Méré's problem was as follows: Two players agree to play a number of games. The winner is the one who first wins S games. The game is interrupted when one of the players has won a $(a < S)$ games and the other b $(b < S)$ games. The question is how the stakes should be divided. [69, p. 342]

Here Chevalier de Méré becomes a fervent gambler.
The legend about Chevalier de Méré is related in a most detailed manner in the third volume of the "Children's Encyclopedia" in an article written by Khinchin and Yaglom entitled "The science of chance" [87].
This story also appeared in the magazine *Knowledge is Strength* (1960, No. 2) under the title "The story of the Knight de Méré."
We present the story with some omissions :

> A French Knight, Chevalier de Méré, was an ardent dice gambler. He tried to become rich by this game and was constantly thinking of various complicated rules which he hoped would help him reach his goal.

---

†See also the Introduction to Adler's popular elementary text on probability theory, *"Probability and Statistics for Everyman"* [184], where Chevalier de Méré is introduced as an enthusiastic gambler who initiated a new branch of mathematics (*Translator's remark*).

For example, de Méré thought of the following rule: He proposed to throw one die four times in a row and wagered that at least one six would appear; if no six turned up then the opponent won. de Méré assumed that he would win more often than lose, but, nevertheless, he approached his friend B. Pascal (1623–1662) and asked him to calculate the probability of his winning this game.

We present Pascal's calculations:

In one toss the probability of getting six is equal to $\frac{1}{6}$ and the probability of not getting six is $\frac{5}{6}$. The probability that in four throws six will not appear is equal to $\left(\frac{5}{6}\right)^4 = \frac{625}{1296}$. Thus, the probability of losing for Chevalier de Méré will be $\frac{625}{1296} < \frac{1}{2}$. Hence the probability of winning is greater than $\frac{1}{2}$. In other words, there is more than half a chance that the Knight would win each game and, when the game is repeated many times, he will almost certainly be the winner. Indeed, the more he played, the more he won. Chevalier de Méré was very happy and thought that he had found a sure method of striking it rich. However, the other players soon discovered that the game was not fair and they stopped playing with de Méré. It was time to think of some other rules and de Méré devised a new game. He proposed to throw two dice 24 times and bet that two fives would turn up at least once. But here the Knight made a mistake. The probability that two fives would turn up in throwing two dice equals $\frac{1}{36}$. The probability that they would not appear is $\frac{35}{36}$. The probability that no two fives would appear together in 24 throws of two dice equals $\left(\frac{35}{36}\right)^{24} > \frac{1}{2}$. Therefore the chance of losing is greater than one-half. It means that the more the Knight played, the more he was bound to lose. And this is what happened. The more he played, the more he ruined himself and ultimately he ended up impoverished. The most interesting side of this historical anecdote is that, due to these peculiar "practical inquiries," the theory of calculating random phenomena was initiated. In the seventeenth and eighteenth centuries scientists considered this type of example as "amusing happenstances" of the application of mathematical knowledge to phenomena which do not have wide appeal.

In all these assertions we encounter the familiar incorrect assumption that probability theory originated from games of chance in the seventeenth century. To these is also added the "fatal" role of de Méré. But, as we have seen, the problem of stakes division had a long history before Chevalier de Méré, who was not the first to deal with this problem. de Méré (1607–1684) was a philosopher and a man of letters, who rapidly became a prominent figure at the court of Louis XIV. He knew and corresponded with almost all the leading mathematicians of his time, including Pascal (see, e.g.,

[142, p. 409]). He actively participated in the solution of various problems which were known in those days to all mathematicians. This enabled him† to write in one of his letters to Pascal:

> You know that I have discovered such rare things in mathematics that the most learned among the ancients have never discussed them and they have surprised the best mathematicians in Europe. You have written on my inventions, as well as Monsieur Huygens, Monsieur de Fermat, and many others who have admired them. You may conclude from this that I do not propose to anyone to scorn this science and truly, it may be of service provided one does not attach oneself too closely to it, for ordinarily, that which one seeks with so much curiosity appears useless to me and the time spent at it could be better employed. [142]

De Méré refers here to probability theory, which, while it appeared "useless" to him, his contemporaries, in particular Huygens, understood and properly evaluated its great importance.

In a letter to Fermat dated 29 July 1654, Pascal writes: "I have seen several people obtain the solution for dice, like M. le Chevalier de Méré, who first posed these problems to me and also to M. de Roberval" [142].

In another letter‡ Pascal relates this story in somewhat greater detail:

> He [de Méré] told me that the figures were wrong for the following reason: If one wants to throw a six with one die, one has an advantage in four throws, as the odds are 671 to 625. If one throws two sixes with two dice, there is a disadvantage in having only 24 throws. However, 24 to 36 (the number of cases for two dice) is as four to six (the number of cases on one die). [142, p. 411]

The probability of no sixes in four throws equals $(\frac{5}{6})^4 = \frac{625}{1296}$, and the probability of at least one six is $1 - \frac{625}{1296} = \frac{671}{1296}$.

In the case of two dice, however, the probability of two sixes in at least one throw will, in fact, be smaller than $\frac{1}{2}$ in 24 throws.

Indeed, $(\frac{35}{36})^{24} = 0.509$ is the probability of not having "double six" in 24 throws, and the probability of at least one throw of "double six" is therefore $1 - 0.509 = 0.491$, which is less than $\frac{1}{2}$.

De Méré's reasoning at the end of the passage cited above was unintelligible to Pascal, as well as to many of the future biographers of Pascal. de Méré's objection in the letter to Pascal is based, however, on the following gambling rule used also by Cardano:

> If in one case there is one chance out of N_0 in a single trial, and in

---

†Although with little justification (see, e.g., [142]) (*Translator's remark*).
‡Actually toward the end of the same letter (*Translator's remark*).

another one chance out of N_1, the ratio of the corresponding critical numbers is as $N_0 : N_1$. That is, we have $n_0 : N_0 = n_1 : N_1$. [142, p. 414]

Using this proportion, de Méré obtains that if $n_0 = 4$, $N_0 = 6$, and $N_1 = 36$, then $n_1 = 24$.

De Méré assumed this proportion to be exact. About 60 years later this problem was investigated by de Moivre in his "Doctrine of Chances" (1716); he proposed the following formula:

$$n = (\ln 2) N = 0.69N.$$

Using this formula, we obtain in our case:

$$n = 0.69 \cdot 36 = 24.84 \approx 25.$$

In general, de Moivre's rule gives very good values when N is fairly large, but it is of course only an approximation. de Moivre applied this formula to the so-called Royal Oaks lottery in London; here there was one chance in 32 to win and thus, by this rule, n is found to be $0.69 \cdot 32 = 22.08$. The actual value of n is obtained by observing that

$$\left(\tfrac{31}{32}\right)^{22,134} > \tfrac{1}{2}, \qquad \text{but} \qquad \left(\tfrac{31}{32}\right)^{22,135} < \tfrac{1}{2},$$

i.e., $n = 22.135$.

These observations all positively indicate that de Méré did not turn to Pascal with a problem from an actual gambling experience, but with a purely theoretical question. And evidently it is not those questions posed by de Méré that laid the foundation of probability theory.

2 Letters between Fermat and Pascal

The correspondence between Blaise Pascal (1623–1662) and Pierre de Fermat (1601–1665) was a substantial step forward in the development of probability theory. This correspondence is dated 1654 and was published in 1679 in Toulouse. Unfortunately, not all of the correspondence survived. In these letters both Fermat and Pascal arrive—each one in his own way— at a correct solution of the problem of division of stakes. Pascal's method is evident from his letter to Fermat dated 29 July, 1654†:

> Here, more or less, is what I do to show the fair value of each game, when two opponents play, for example, in three games, and each person has staked 32 pistoles.

†There are at least two English translations of these letters. The first (by V. Sandford) is included in Smith's "A Source Book for Mathematics" [214] and the second by M. Merrington) in the Appendix to David's "Games, Gods and Gambling" [189] (*Translator's remark*).

Let us say that the first man had won twice and the other once; now they play another game, in which the conditions are that, if the first wins, he takes all the stakes, that is 64 pistoles, if the other wins it, then they have each won two games, and therefore, if they wish to stop playing, they must each take back their own stake, that is, 32 pistoles each.

Then consider, Sir, if the first man wins, he gets 64 pistoles, if he loses he gets 32. Thus if they do not wish to risk this last game, but wish to separate without playing it, the first man must say: "I am certain to get 32 pistoles, even if I lose I still get them; but as for the other 32, perhaps I will get them, perhaps you will get them, the chances are equal. Let us then divide these 32 pistoles in half and give one half to me as well as my 32 which are mine for sure." [189, p. 231]

The first man should receive 48 pistoles and the second 16. In the same letter another case was also considered: The first player had won two games and the other none. Then, as Pascal writes, the first player says:

If I win it I will have all the stakes, that is 64; if I lose it, 48 will legitimately be mine; then give me the 48 which I have in any case, even if I lose, and let us share the other 16 in half, since there is as good a chance for you to win them as for me. [189, p. 232]

It follows therefore that the first will have $48 + 8 = 56$ pistoles.
The next case to be considered is as follows:

Let us suppose, finally, that the first had won one game and the other none. You see, Sir, that if they begin a new game, the conditions of it are such that, if the first man wins it he will have *two* games to *none*, and thus by the preceding case, 56 belongs to him; if he loses it, they each have one game, then 32 pistoles belong to him. So he must say: "If you do not wish to play, give me 32 pistoles which are mine in any case, let us take half each of the remainder taken from 56. From 56 set aside 32, leaving 24, then divide 24 in half, you take 12 and give me 12 which with 32 makes 44." [189, p. 232]

Pascal suggests that we divide the stakes proportionally to the probabilities of winning, provided the game is continued. Pascal's solution is ingenious, but hardly suitable for more complicated cases.

Fermat's method can be established from a letter by Pascal to Fermat dated 24 August 1654. The original letter in which Fermat describes his solution has not survived. Fermat's problem was the following:

If two players, playing several games, find themselves in that position when the first man needs *two* games and the second needs *three* for gaining the stake, how do we find the fair division of stakes? [189, p. 239]

Fermat's argument is as follows: The game can continue for at most four games. What are the possible outcomes of these games? Denote a win for player A by + and for player B by −. There are the following 16 possibilities:

1	2	3	4	5	6	7	8	9	10	11	12	13	14	15	16
+	+	+	+	−	+	+	−	+	−	−	−	−	−	+	−
+	+	+	−	+	+	−	+	−	+	−	−	−	+	−	−
+	+	−	+	+	−	+	+	−	−	+	−	+	−	−	−
+	−	+	+	+	−	−	−	+	+	+	+	−	−	−	−

Out of 16 possible outcomes 11 (from 1 to 11) are favorable for A to win the stake, while B has only five favorable outcomes (from 12 to 16).

Thus, the $\frac{11}{16}$ of the stake is due to A and the remaining $\frac{5}{16}$ to B. As we see, Fermat suggests that the stake be divided in proportion to the probability of winning the whole match (and hence the whole stake).

In spite of the correctness and originality of the solution of these problems, no general method was devised at that time.

In 1653 Pascal informed his friends of his manuscript "The Arithmetic Triangle." This work was published only posthumously in 1655 under the title: "Traité du triangle arithmétique" [147]. In this treatise an exposition of the properties and relations between the terms of progressions and between the binomial coefficients with the appropriate proofs is given.

In the section entitled "Utilization of the arithmetic triangle to determine the number of games required between two players who play a large number of games" Pascal applies the arithmetic triangle (later to be called the Pascal triangle) to solutions of various games connected with the calculation of chances:

```
1   1   1   1    1    1    1    1   1   1  ...
1   2   3   4    5    6    7    8   9  ...
1   3   6  10   15   21   28   36  ...
1   4  10  20   35   56   84  ...
1   5  15  35   70  126  ...
1   6  21  56  126  ...
1   7  28  84  ...
1   8  36  ...
1   9  ...
1  ...
```

In this table the coefficients of the expansion $(a + b)^n$ for $n = 1, 2, \ldots$ are presented. The first row consists of ones, the second contains the sums of numbers of the first row (up to the last number), in the third row the sum of the numbers of the second row are given, and so on. The numbers on the diagonal represent the binomial coefficients.

As we have already seen these numbers were actually used by Tartaglia, Stifel, and others.

The numbers in Pascal's table (the binomial coefficients) represent the number of combinations of n elements taken k at a time.

If we use the symbol $(r)_k$ to denote the number in the rth column and kth row, then the rule for constructing the entries in Pascal's table is as follows: $(r)_k = (r)_{k-1} + (r-1)_k$ and the number $(r)_k$ is equal to $\binom{r+k-2}{k-1}$. As a corollary Pascal writes that in any arithmetic triangle the numbers which are the same distance from the extreme terms of the diagonal are equal, i.e., $(r)_k = (k)_r$. Pascal also notes that the terms in the horizontal row and the corresponding vertical column with the same index are equal: "In any arithmetic triangle the horizontal series and the vertical series of the same number consist of entries such that each of the entries for the one series is equal to the corresponding entry of the other series" [147, p. 247]. It follows from this fact that

$$(r)_1 + (r)_2 + \cdots + (r)_k = (k)_1 + (k)_2 + \cdots + (k)_r.$$

In the treatise "Application of the arithmetic triangle to the theory of combinations" Pascal uses the term "combination" in its modern sense [147].

In this work he actually obtains the following relations:

$$(r)_k = \binom{k + r - 2}{r - 1} = \binom{k + r - 2}{k - 1};$$

$$\binom{k + r - 2}{k - 1} = \binom{k + r - 3}{k - 1} + \binom{k + r - 3}{k - 2}.$$

Here he also applies the arithmetic triangle to the binomial expansion with integral values of the powers.

By means of the triangle, Pascal solves the "problem of points" (the division of stakes) in a general form. The method of his solution is: First the number of points that each one of the players lacks to win is added. Next, the diagonal of the table is chosen such that the total number of terms in the diagonal is equal to the sum obtained. Then the first player's part of the stake is equal to the sum of terms in this diagonal starting from 1, and the number of summands from the diagonal is equal to the number of points that the second player lacks to win. For the second player, his part of the stake is equal to a similar sum with the exception that the number of summands is equal here to the number of points which the first player lacks to win.

For example, if player A lacks three points to win, and player B four points, then $3 + 4 = 7$. We write down the diagonal consisting of 7 entries;

these are: 1, 6, 15, 20, 15, 6, 1. Hence the part of the stake belonging to A will be $1 + 6 + 15 + 20 = 42$, and the part of B is $1 + 6 + 15 = 22$. The stakes should thus be divided in the ratio $\frac{42}{22} = \frac{21}{11}$. This solution implies that the stakes are divided in proportion to the probabilities of winning the whole stake for each one of the players. Indeed, in the case of this particular problem, if the game is continued, at most six rounds may be played. The number of favorable outcomes for player A will then be

$$1 + C_6{}^1 + C_6{}^2 + C_6{}^3 = 42,$$

where 1 represents the possible outcome when A wins all the remaining six rounds. $C_6{}^1$ represents the six possible outcomes in which A wins five rounds out of six; $C_6{}^2$ represents the 15 possible outcomes in which A wins four rounds; and $C_6{}^3$ the 20 possible outcomes in which A wins three rounds.

The total number of outcomes is $2^6 = 64$. Therefore, the probability that A gets the whole stake is equal to $\frac{42}{64}$. Analogously we obtain the probability that B gets the whole stake to equal $\frac{22}{64}$. Hence, the stakes should be divided in proportion $\frac{42}{64} : \frac{22}{64} = \frac{42}{22} = \frac{21}{11}$.

Using Pascal's rule as cited above we may write the following general formula for division of stakes: If player A lacks m points to win and B n points to win, then the stakes should be divided in the ratio

$$\frac{P(A)}{P(B)} = \frac{\binom{m+n-1}{0} + \binom{m+n-1}{1} + \binom{m+n-1}{2} + \cdots + \binom{m+n-1}{n-1}}{\binom{m+n-1}{0} + \binom{m+n-1}{1} + \binom{m+n-1}{2} + \cdots + \binom{m+n-1}{m-1}}.$$

We thus see that Pascal obtained two solutions for the problem of division of stakes: one for a particular case and the other for the general case [147, p. 261].

It follows from Pascal's letter to Fermat dated 27 October 1654 that he considers his method of solving this problem different from Fermat's method. Pascal writes:

Monsieur,

Your last letter satisfied me perfectly. I admire your method for the problem of the points, all the more because I understand it well. It is entirely yours, it has nothing in common with mine, and it reaches the same end easily. Now our harmony has begun again. [214, p. 564]

However, upon more detailed observation it becomes evident that both methods are essentially identical. The only difference is that Pascal utilized his arithmetic triangle in order to calculate the number of various outcomes during the continuation of the game. This is of course a more general form and allows the problem of division of stakes to be solved for various numbers of "lacking" points for each one of the players. Fermat calculated the number of the various outcomes directly by enumerating all the possible outcomes, which is more cumbersome and less efficient.

Although a number of mathematicians of that period devoted significant attention to problems connected with games of chance, actual gambling was, as a rule, condemned. These games did serve as a convenient and readily understandable scheme for handy illustration of various probabilistic propositions.

3   Contributions of Huygens to probability theory

In 1655, Christiaan Huygens (1629–1695) visited France† and stayed in Paris where he met many prominent scientists. He was very impressed with the new problems investigated by Pascal and Fermat related to him by Mylon and Roberval; he also heard about the problem of fair division of stakes. He was not, however, told of the solutions to these problems, nor of the methods followed by Fermat and Pascal. On his return to Holland, at the end of 1655, Huygens set to work himself on these problems. His investigations resulted in the treatise "De Ratiociniis in Ludo Aleae" ("About Dice Games") which was published as an appendix in Latin to a volume entitled "Exercitationes Mathematicae" ("Mathematical Studies") by Francis van Schooten, which appeared in 1657.‡

Although Huygens' work was published after the correspondence between Pascal and Fermat took place, these letters had no influence on Huygens' investigations, since they were published as late as 1679. In place of a foreword, a letter from Huygens to van Schooten, dated 27 April 1657, is presented. In the letter he writes: "I would like to believe that in considering these matters closely, the reader will observe that we are dealing not only with games but rather with the foundations of a new theory, both deep and interesting." [81, p. 58]§

In the same letter, Huygens also explains the background for writing this book:

> It should be said, also, that for some time some of the best mathematicians of France have occupied themselves with this kind of calculus so that no one should attribute to me the honor of the first invention. This does not belong to me. But these savants, although they put each other to the test by proposing to each other many questions difficult to

†To receive a doctorate in Law from the Protestant University in Angers (*Translator's remark*).

‡Van Schooten translated Huygens' treatise into Latin. In 1660 this work was published in the original Dutch version.

§In volume XIV of Huygens' collected works, both the Dutch original and French translation are presented. This volume [81] serves as our reference source.

solve, have hidden their methods. I have had therefore to examine and to go deeply for myself into this matter by beginning with the elements, and it is impossible for me for this reason to affirm that I have even started from the same principle. But finally I have found that my answers in many cases do not differ from theirs." [81, p. 58]

Huygens' treatise was warmly received by contemporary mathematicians, went through several editions, and up to the beginning of the eighteenth century served as the classical introduction to probability theory.

The book consists of a short preface and 14 propositions.

Proposition I "To have equal chances of getting a and b is worth $(a + b)/2$" [81, p. 62].

Proposition II "To have equal chances of getting a, b, or c is worth $(a + b + c)/3$" [81, p. 64].

In **Proposition III** Huygens writes: "To have p chances of obtaining a and q of obtaining b, chances being equal, is worth $(pa + qb)/(p + q)$" [81, p. 66]. This is actually a definition of the mathematical expectation for discrete random variables. In the sequel Huygens extensively utilizes this notion.

In **Proposition IV** the problem of division of stakes is investigated:

> Suppose I play against an opponent as to who will win the first three rounds and that I have already won two and he one. I want to know what proportion of the stakes is due to me if we decide not to play the remaining rounds ... It should be noted firstly that it is sufficient to take into account the number of games each one of the players lacks. It is true that if our game consists in who will win the first 20 rounds and if I have won 19 and my opponent 18, I would have the same advantage as in the case stated above, where in three rounds I have won two and he one, because in both cases I lack one point and my opponent two. Next, to calculate the proportion due to each of us, one should consider what would happen if the game was continued. It is true, that if I win the first round, then I will finish the game and would thus win the stake in full, which I would call a. But if my opponent wins the first round, our chances would be equal from now on, considering that each one lacks one point; therefore each one would have a claim on $\frac{1}{2}a$. Evidently, I have the same chance of winning as losing the first round. Thus I have equal chances of getting a and $\frac{1}{2}a$, which, according to Proposition I is equivalent to the sum of halves, i.e. $\frac{3}{4}a$, so that my opponent remains with $\frac{1}{4}a$. [81, p. 68; 189, p. 116]

We have presented Proposition IV almost in its entirety because his method of solution of other problems on division of stakes is analogous. We therefore do not present the solutions, but merely formulate the problems and supply the answers.

Proposition V 1. Suppose I lack one point and my opponent three. What proportion of the stakes is due to me and to my opponent? (*Answer*: $\frac{7}{8}a$ and $\frac{1}{8}a$).
2. Suppose I lack one point and my opponent four? (*Answer*: $\frac{15}{16}a$ and $\frac{1}{16}a$).

Proposition VI 3. Suppose I lack two points and my opponent three? (*Answer*: $\frac{11}{16}a$ and $\frac{5}{16}a$).

Proposition VII 4. Suppose I lack two points and my opponent four? (*Answer*: $\frac{13}{16}a$ and $\frac{3}{16}a$).

Proposition VIII 5. Suppose now that three people play together and that the first and second lack one point each and the third two points? (*Answer*: $\frac{4}{9}a$, $\frac{4}{9}a$, and $\frac{1}{9}a$).

Having presented a detailed solution of these problems, Huygens gives the following suggestion in **Proposition IX**:

> In order to calculate the proportion of stakes due to each of a given number of players who are each given numbers of points short, it is necessary, to begin with, to consider what is owing to each in turn in the case where each might have won the succeeding game. These parts should be added together and the sum obtained should be divided by the number of players which will give the required part. ... First the simplest cases should be investigated. ... Using this method one can calculate all the cases given in the table and in infinitely many other cases. [81, p. 74]

Next he presents the above-mentioned table:

Number of points short	1,1,2	1,2,2	1,1,3	1,2,3	1,1,4	1,1,5	1,2,4	1,2,5
Parts of stakes due to the players	$\frac{4,4,1}{9}$	$\frac{17,5,5}{27}$	$\frac{13,13,1}{27}$	$\frac{19,6,1}{27}$	$\frac{40,40,1}{81}$	$\frac{121,121,1}{243}$	$\frac{178,58,7}{243}$	$\frac{542,179,8}{729}$

We do not present Huygens' table in its entirety; it is continued up to the case of 2, 3, 5 points short with the corresponding parts of stakes $\frac{1433}{2187}$,

$\frac{635}{2187}$, and $\frac{119}{2187}$. The upper row of the table gives the number of points short for each partner to win the stake. For example, 1, 1, 2 denotes that the first two partners are one point short each and the third two. The notation $\frac{4,4,1}{9}$ in the lower row indicates that each of the partners should receive $\frac{4}{9}$, $\frac{4}{9}$, and $\frac{1}{9}$ respectively of the stakes and similarly for other entries in this row.

Huygens' solution of the problem of division of stakes is correct. His basic assumption is that the stakes should be divided in proportion to the probabilities of winning the whole stake if the play is continued.

Huygens' book was essentially the only text on probability theory prior to the appearance of James Bernoulli's "Ars Conjectandi" [19]. As mentioned before, it gained widespread popularity and substantially influenced many scientists dealing with problems in probability theory.

Huygens was actually the first to introduce the notion of mathematical expectation and to apply it. Mathematical expectation is a generalization of the arithmetic mean. The latter was widely utilized in commercial and industrial problems in determining mean prices, mean profits, etc.

With the development of commerce and industry various financial transactions acquired great importance. Holland was the leader in the development of commercial and bank accounting procedures. Marx points out that "Holland, the first country to develop the colonial system to the full, had attained the climax of its commercial greatness as early as the year 1648" [124, Vol. 2]. In any type of accounting the arithmetic mean is repeatedly utilized. From these applications this method was introduced into science. Marx also observes that "commercial profiteering in its calculus of probabilities proceeds from the mean prices, which are taken as the center of variations, as well as from upper-mean and lower-mean prices or variations in the prices up and down from this center" [125, p. 47].

Huygens' terminology in probability theory was directly influenced by commercial jargon. He considered mathematical expectation as the "value of the chance" to win in a fair game and concluded that the fair value is the mean value. He calculates "the fair price for which I would be willing to yield my place in a game." He does not refer to mathematical expectation as "expectation," but designates it "the value (price) of the chance." The term "expectation" appears for the first time in van Schooten's translation.†

Huygens did not utilize combinatorial methods when solving probabilistic problems; it is for this reason that his solutions are rather cumbersome. His methods are therefore hardly applicable to solving problems in a general form. And no such solutions are given in Huygens' treatise.

---

†Ian Hacking points out that Huygens and van Schooten argued a great deal about how to translate Dutch terminology, and he expresses the opinion that "expectatio" did appear in Huygens' drafts for translation (*Translator's remark*).

At the end of his treatise Huygens suggests five problems for the reader without solutions. (Of these problems, two were suggested by Fermat and one by Pascal.) The solutions to these problems were published by him eight years later in 1665. These solutions were given without explanation, containing only mathematical calculations.

His problems were as follows:

1. A and B play with two dice on the condition that A gains if he throws six, and B gains if he throws seven. A first has one throw, then B has two throws, then A two throws, and so on until one or the other wins. Show that A's chance is to B's as 10,355 to 12,276.

2. Three gambers A, B, and C take 12 balls of which four are white and eight black. They play with the rules that the drawer is blindfolded, A is to draw first, then B, and then C; the winner is the one who first draws a white ball. What is the ratio of their chances?

3. A wagers B that, given 40 cards of which ten are of one color, ten of another, ten of another, and ten of yet another, he will draw four so as to have one of each color. Here A's chance is to B's as 1000 to 8139.†

4. Twelve balls are taken, eight of which are black and four white. A plays with B and undertakes in drawing seven balls blindfolded to obtain three white balls. Compare the chances of A and B.

5. A and B take each twelve counters and play with three dice on the condition that if eleven is thrown, A gives a counter to B, and if fourteen is thrown B gives a counter to A; and he wins the game who first obtains all the counters. Show that A's chance is to B's as 244,140,625 is to 282,429,536,481.

Huygens states at the end of his book that he does not present his solutions to the problems because it is too difficult "to properly set forth the reasoning which led to the solutions." Moreover, he considers these problems a good exercise for the reader. The date of completion is then given as The Hague, 14 April, 1657.

Many mathematicians in the seventeenth century were engaged in the solutions of these problems.‡

In 1687, approximately ten years after Spinoza's death, a thin volume appeared in The Hague containing a twenty-page essay on the rainbow "Steelkonstige Reeckening van den Regenboog" (Part I) and an eight-page pamphlet on "Mathematical Probability"—"Reeckening van Kanssen" (Part II). This work was republished in 1884 by the Dutch mathematician D. Bierens de Haan with an introductory article entitled "Two Works of Benedict Spinoza" ("Twee Zeldzame Werken van Benedictus Spinoza").

---

†David [189, p. 119] points out that the correct solution is 1000 to 9139 (*Translator's remark*).

‡A recent (1967) doctoral dissertation by K. Kohli (Zürich) is devoted to the history of these problems (*Translator's remark*).

In this paper he attributes to Spinoza the authorship of the above mentioned volume. Some years later, Gebhardt investigated in detail the evidence attesting to Spinoza's authorship. Concluding his investigation he states that there appears to be little doubt that Spinoza actually wrote the essays.†

In 1953 an English translation by Jacques Dutka of "Reeckening van Kanssen" was published in *Scripta Mathematica* [54]. The pamphlet consists of a statement of five problems that had appeared without solution in Huygens' tract on probability discussed above and a detailed solution of the first problem.

Before presenting a solution to Huygens' first problem, Spinoza solves a different simpler problem: "*B* and *A* play against each other with two dice on the condition that *B* shall win if he throws seven points and *A* if he throws six points, that each shall have two throws, one after the other, and that *B* shall throw first.

Their chances are:

$$\frac{\begin{matrix} B & \quad A \\ 14256 & \quad 8375 \end{matrix}}{22631}.\text{ "}$$

Spinoza solves this problem as follows: Let A's chance be worth x; if the stake is taken as a, then B's chance is worth $a - x$.

"Each time that B's turn comes up, A's chance must again be worth x, but every time it is A's turn to throw, his chance should be greater."

B must throw first. There are six out of 36 ways of throwing two dice to get seven points, i.e., with probability $\frac{1}{6}$ in one throw; the probability of not getting seven is $1 - \frac{1}{6} = \frac{5}{6}$. The probability of not getting seven in two throws is equal to $(\frac{5}{6})^2 = \frac{25}{36}$, and the probability of getting seven at least once is equal to $1 - \frac{25}{36} = \frac{11}{36}$.

Thus in the first two throws B has 11 chances to win and 25 to lose. But these are the 25 chances for A to win. Their worth for A is greater. Let their value be y. Then

$$y/x = 36/25; \qquad y = 36x/25. \tag{II.1}$$

There are five ways for A to win in one throw, i.e., the probability of winning is $\frac{5}{36}$ and of losing is $1 - \frac{5}{36} = \frac{31}{36}$. The probability of losing in two throws is $(\frac{31}{36})^2 = \frac{961}{1296}$; the probability of winning in two throws is $1 - \frac{961}{1296} = \frac{335}{1296}$. The value of the chance for A before the throw consists of the "sure win" $\frac{335}{1296} a$ plus the "problematical win" $\frac{91}{1296} x$; hence

$$y = (335a - 961x)/1296.$$

---

†Most recent investigations cast doubt on the authenticity of this fragment of Spinoza's (*Translator's remark*).

Substituting this value in (II.1) we obtain

$$\frac{36x}{25} = \frac{335a + 961x}{1296} \;\; ; \quad x = \frac{8375a}{22,631},$$

which is the value of A's chance; B's chance would be worth $a - x = \frac{14,256}{22,631}a$ "so that the chance of A stands to that of B as 8375 to 14,256."

After solving this problem, Spinoza proceeds with a solution of Huygens' first problem. His argument is as follows:

The probability that A wins stake a in the first throw equals $\frac{5}{36}$, i.e., the value of the chance is $\frac{5}{36}a$. With probability $\frac{31}{36}$, A passes on to the situation of the preceding problem, i.e., his chances to win will be worth $\frac{31}{36}\cdot\frac{8375}{22,631}a$. From the addition theorem we obtain the value of A's chance as:

$$\frac{5}{36}a + \frac{31}{36}\cdot\frac{8375}{22,631}a = \frac{10,355}{22,631}a.$$

The value of B's chance will be

$$a - \frac{10,355}{22,631}a = \frac{12,276}{22,631}a.$$

Hence the ratio of A's chance to that of B is as 10,355 to 12,276.

Here Spinoza's pamphlet ends. It is not known whether he attempted to solve any of the other problems in Huygens' tract.

We observe from this paper that Spinoza was quite a good mathematician who was able to use freely the methods required for solving Huygens' problems in probability. In the process of his solution Spinoza takes into account variations in the value of the chance depending on the course of the game.

In a letter dated 16 May 1662, the president of the London Royal Society asked Huygens his opinion of the tract by John Graunt published in London in 1662.† Graunt's tiny book was devoted to various problems of vital statistics. In his reply dated 9 June 1662, Huygens gave a highly favorable appraisal of Graunt's pamphlet.

In 1669, based on Graunt's work, Huygens constructed a mortality curve and properly defined the notions of mean and probable duration of life. He was the first to apply the methods of probability theory meaningfully to demographic statistics.

In 1671 a Dutch mathematician van Hudden (1628–1704), who served as burgomaster of Amsterdam, approached Huygens with a number of problems connected with research on annuities. Van Hudden collaborated

†Entitled "Natural and Political Observations Made upon the Bills of Mortality."

in a project conducted by Johan de Witt† on this subject. In his reply of 3 October 1671, Huygens gave his approval of the work.

We thus observe that as early as the seventeenth century quite a variety of problems on probability theory had been solved. A number of important notions and theorems were known to mathematicians. Among those were the widely utilized addition and multiplication rules of probabilities. The notion of probability itself started to attain more meaningful and tangible substance. The concept of mathematical expectation, one of the important notions of probability theory, was introduced into scientific methodology in the guise of the term "true worth of a chance." Probability theory at that time was closely interrelated with another branch of mathematics, that of combinatrics. Probability theory made its first appearance in applications to statistics, physics, and astronomy.

However, as a mathematical discipline, probability theory was merely at the initial stage of its development. Although only individual questions and propositions were tackled and solved, they were by then unified in the general framework of probabilistic problems. An ever increasing interest in this new science was characteristic of that period.

It should be noted that in the literature on the history of mathematics the role of Pascal and Fermat as the founders of probability theory has often been overestimated. This has resulted in another, in our opinion, more important misconception: that Huygens' contributions were only secondary; consequently, his role in the development of probability theory was deemphasized. However, it must be acknowledged that Huygens wrote the first book in probability theory, in which the notion of mathematical expectation, among others, was introduced. His book greatly influenced many scientists. James Bernoulli, who laid the foundation for a new era, highly valued Huygens' work and paid tribute to his influence.

In the conclusion of this section devoted to Huygens' contributions we note that the collection of his works consisting of 22 volumes was in the process of publication from 1888 up to 1950. His treatise "About Dice Games" was included in the fourteenth volume where the original text in Dutch is supplemented by a French translation. In this volume the nine propositions written by Huygens in the course of a number of years, the last as late as in 1688, are presented as well. This betokens the constant interest on Huygens' part in the problems of probability theory.

---

† Jan (Johan) de Witt (1625–1672), the Grand Pensioner of Holland as of 1653, wrote one of the first papers on applications of probability theory to annuities. [The English translation appeared in 1853 (*Translator's remark*).]

III

The Development of Probability Theory
to the Middle of the Nineteenth Century

1 James Bernoulli and his treatise "Ars Conjectandi"

James (Jacques) Bernoulli (1654–1705) was an offspring of a family of famous Swiss mathematicians. He held a chair in mathematics at the University of Basel from 1687 on and as of 1699 was a member of the Paris Academy of Sciences.

In 1713 his book "Ars Conjectandi" ("The Art of Conjecturing") [19] was published. This book played a significant role in the history of probability theory. We can state that thanks to Bernoulli's contributions, probability theory was raised to the status of a science and began a new era in its development.

In "Ars Conjectandi" the first limit theorem of probability theory is rigorously proved. We shall refer to this theorem as J. Bernoulli's theorem. In later years problems connected with limit theorems became the focus of attention in probability theory.

Bernoulli's book [19] was published in Latin eight years posthumously by his nephew, Nicholas Bernoulli.†

The first part of Bernoulli's treatise consists of a reprint of Huygens' "De Ratiociniis in Ludo Aleae" accompanied by a commentary on all but one of Huygens' propositions. This first part is entitled "A Treatise on Possible Calculations in a Game of Chance of Christianus Huygens with J. Bernoulli's Comments."

We shall now describe the most important additions made by James Bernoulli. In reference to Huygens' Proposition 1:

> The author of this treatise presents . . . here and in two succeeding propositions the basic principle of the art of conjecture. Since it is very important that this principle be well understood, I shall attempt to demonstrate it using calculations more common and comprehendable to all, by starting entirely from the axiom or proposition that everyone should expect or suppose to expect as much as he is going invariably to receive.
>
> The word "expectation" is not meant here in its usual sense in which "to expect" or "to hope" refers to the most favorable outcome, although the least favorable may occur; we should understand this word here as the hope of getting the best diminished by the fear of getting the worst. Thus the value of our expectation always signifies something in the middle between the best we can hope for and the worst we fear. We shall interpret this word thusly here and in further discussions.

After considering Proposition 3, Bernoulli remarks:

> From this discussion . . . it is evident that here there exists a similarity to the rule called in arithmetic the rule of association, which consists of finding the value of a mixture composed from definite quantities of different things with different values. Or rather that the calculations are absolutely the same. Namely, similar to the fact that the sum of the products of the quantities of the substances in the mixture by its corresponding values divided by the sum of the quantities yields the required value which is always between the extreme values; in the same manner the sum of the products of outcomes by the corresponding gains divided by the number of all outcomes gives the value of the expectation, which is consequently in the middle between the largest and the smallest of the gains. [19; 176, pp. 25–26]

---

†Nicholas Bernoulli also dealt with problems of probability theory. In 1709 he defended a dissertation "Specimina Artis conjectandi ad quoestiones Juris applicatae" ("On utilizing the art of conjecture in juristic matters"). This dissertation was highly praised by Leibniz. In this essay N. Bernoulli applies mathematical calculations to problems of debt payments, reliability of evidence of witnesses, and so on.

The above passage gives a suitable explanation of the notion of the arithmetic mean and its relation to the weighted arithmetic mean.

Proposition 4 concerns a problem on division of stakes (or problem of points). In this connection he comments: "When computing shares, one should pay attention only to the forthcoming games and not to the games already completed."

Next, in a very original manner, Bernoulli explains the application of the addition rule for probabilities, in particular its inapplicability in the case of nondisjoint events.

> If two persons, doomed to die, are ordered to throw dice under the condition that the one who gets the smaller number of points will be executed, while the one who obtains a larger number of points will be spared, and both will be spared if the number of points is the same, we then find that the expectation of one of them is $\frac{7}{12}$ or $\frac{7}{12}$ of life ..., but it does not follow from here that the expectation of the other is $\frac{5}{12}$ of life, since, clearly, here the fate of both is the same and the other also has the expectation of $\frac{7}{12}$, which gives for the two of them $\frac{7}{6}$ of the life, i.e. more than the whole life. The reason is that there is no outcome such that at least one of them is not spared, while there are several outcomes when both of them are spared. [19; 176, pp. 26–27]

As an addition to Huygens' propositions on the problem of points (division of stakes), Bernoulli constructs tables that furnish the chances of two players and the ratio of stakes between the two players under various conditions.

In the commentary on Huygens' ninth proposition Bernoulli discusses in detail the various throws that can be made with two or more dice and the number of cases favorable to each outcome and constructs a special table equivalent to the following theorem: The number of ways in which m points can be obtained throwing n dice is equal to the coefficient of x^m in the expansion

$$(x + x^2 + x^3 + x^4 + x^5 + x^6)^n.$$

Bernoulli's commentary on the eleventh proposition is as follows:

> The author has established ..., that one can with a gain undertake to throw a six in four throws with a single die. He now demonstrates, however, that one cannot, without a loss, undertake to throw two sixes in 24 throws with two dice. This may sound absurd to many people, since there exists the same relation between 24 throws and 36 outcomes with two dice as between four throws and six outcomes with one die [19; 176, p. 26].

In his commentary on Proposition 12 Bernoulli obtains the result referred to nowadays as Bernoulli's formula. He establishes the probability that event A (where $P(A) = p$) will occur m times during n trials. The corresponding formula given in all textbooks on probability theory is

$$P_{m,n} = \binom{m}{n} p^m q^{n-m} \qquad (q = 1 - p).$$

As we observe from the examples given above, Huygens' problems served for Bernoulli in many cases as a pretext for expressing his own opinions and as a basis for obtaining new formulas.

The second part of the "Ars Conjectandi" is entitled "The Doctrine of Permutations and Combinations" and consists of the following nine chapters:

1. Permutations
2. Combinations in general (combinations without repetitions)
3. Combinations (without repetitions) of a particular class of things; figurate numbers and their properties
4. The number of combinations (without repetitions) of one particular class: a number which indicates how many times a given object occurs separately or in combination with others
5. The number of combinations with repetitions
6. The number of combinations with restricted repetitions
7. Variations without repetitions
8. Variations with repetitions
9. The number of variations with restricted repetitions.

Probability theory at that time required the theory of combinations for the solution of many of its problems. Prior to the penetration of methods of infinitesimal analysis into probability theory, which happened somewhat later, the theory of combinations was the basic tool. By tracing the development of both combinatorics and probability theory, we observe to what extent these two branches of mathematics interacted.

In his treatise Bernoulli indicates that he is familiar with the investigations on combinatorics carried out by a number of well-known mathematicians such as Leibniz, Schooten, Wallis, and others.

He supplements the works of his predecessors with some new results, the most important of which were, in his opinion, the investigations on figurate numbers. Bernoulli remarks that there is no complete exposition of the theory of combinations available so far and he therefore presents all the necessary detailed information on this theory from the very beginning.

The theory of combinations was widely used in constructing anagrams as well as in compiling Proteus verses.† In Europe during the sixteenth and

†Proteus verses are verses constructed from words of a given verse. Proteus, in Greek mythology, was a prophetic old man of the sea who tended the seals of Poseidon. He could change himself into any shape he pleased, but if he was nevertheless seized and held, he would foretell the future (*Translator's remark*).

seventeenth centuries anagrams of proper names often served as pseud-
onyms. They were also used in order to conceal a new method or invention.
Anagrams occurred particularly often in religious literature. The first pages
of the second part of "Ars Conjectandi" are devoted to these problems. We
note in passing that these topics may be of some interest for modern mathematical linguistics also.

The first chapter of the second part is devoted to the theory of permutations. Permutations, according to Bernoulli, are those variations in which
the number of elements remains the same, but the order is changed in
various ways. He distinguishes between the cases when all the elements are
different and when there are several identical elements. Bernoulli obtains
the number of permutations on n different elements in the following manner:
Since any element may appear at the first position, there are therefore n
possibilities for the first position. For the second position, there are $n - 1$
possibilities. Thus the first two positions may be filled in $n(n - 1)$ ways.
Continuing this type of argument, Bernoulli obtains the number of permutations on n elements to be equal to $1 \cdot 2 \cdots (n - 1) \cdot n$. He presents a table of
permutations from 1 to 12:

1	2	3	4	5	6	7	8	9	10	11	12
1	2	6	24	120	720	5040	40320	362880	3628800	39916800	479001600

The lower row represents the values of the factorials of the corresponding
number in the top row.

There is no symbolic notation for the number of permutations, combinations, or arrangements in Bernoulli's treatise, although, as we have seen, the
terms "permutation" and "combination" are used in accordance with
modern terminology. Bernoulli also distinguishes between permutations
or combinations with or without repetitions. In place of the term "arrangements" he uses the expression "combinations together with their permutations." The following final result for the number of permutations with
repetitions was obtained by Bernoulli:

$$P_n(\alpha_1, \alpha_2, \ldots, \alpha_k) = n!/\alpha_1! \, \alpha_2! \ldots \alpha_k!.$$

As examples he considers the number of permutations of letters in various
words. The letters of the word *Roma* yield $1 \cdot 2 \cdot 3 \cdot 4 = 24$ permutations,
the word *Leopoldus* produces

$$1 \cdot 2 \cdot 3 \cdot 4 \cdot 5 \cdot 6 \cdot 7 \cdot 8 \cdot 9/2 \cdot 2 = 362,880/4 = 90,720$$

permutations, and the word *Studiosus* $362,880/2 \cdot 6 = 30,240$ permutations.

Next he proceeds to the theory of combinations. A combination, according
to Bernoulli, is a collection of elements in which certain elements are
extracted from a given set of elements and combined without paying atten-

tion to the order.† The index of the class of combinations is the number of things taken at a time. Among the various classes investigated by Bernoulli there also appears the zero class, i.e. the class without elements. Bernoulli distinguishes between combinations with repetitions and without. For various elements, Bernoulli constructs the table of combinations (which is probably borrowed from Schooten's "Exercitationum Mathematicarum Libri Quinque" [176, pp. 30–31]). Bernoulli's table for five elements a, b, c, d, and e is as follows:

$$a;$$
$$b, ab;$$
$$c, ac, bc, abc;$$
$$d, ad, bd, cd, abd, acd, bcd, abcd;$$
$$e, ae, be, ce, de, abe, ace, bce, ade,$$
$$bde, cde, abce, abde, acde, bcde, abcde.$$

Bernoulli asserts that this table can be extended. From it, using the method of mathematical induction, Bernoulli proves the theorem that the number of all possible combinations of all "classes" is equal to the product of as many twos as there are elements, minus 1. Using present-day notation, this result states

$$\binom{n}{1} + \binom{n}{2} + \cdots + \binom{n}{n} = 2^n - 1.$$

This theorem appeared without proof even earlier in the works of Cardano, Stifel, and Leibniz.

Next Bernoulli presents a table for the number of combinations:

	I	II	III	IV	V	VI	VII	VIII	IX	X	XI	XII
1	1	0	0	0	0	0	0	0	0	0	0	0
2	1	1	0	0	0	0	0	0	0	0	0	0
3	1	2	1	0	0	0	0	0	0	0	0	0
4	1	3	3	1	0	0	0	0	0	0	0	0
5	1	4	6	4	1	0	0	0	0	0	0	0
6	1	5	10	10	5	1	0	0	0	0	0	0
7	1	6	15	20	15	6	1	0	0	0	0	0
8	1	7	21	35	35	21	7	1	0	0	0	0
9	1	8	28	56	70	56	28	8	1	0	0	0
10	1	9	36	84	126	126	84	36	9	1	0	0
11	1	10	45	120	210	252	210	120	45	10	1	0
12	1	11	55	165	330	462	462	330	165	55	11	1

† A more precise definition of this concept may be given as follows: A set of k things chosen from a set of n things is called a combination of n things taken k at a time (*Translator's remark*).

This table differs from an analogous table of Leibniz's only in the arrangement of the entries, and is essentially a Pascal triangle. Every number in this table is obtained by adding the numbers in previous rows of the preceding column. Bernoulli lists various properties of this table. We shall present some of these here:

1.   The second column starts with one zero, the third with two zeros, the fourth with three and, in general, the Cth with $C - 1$ zeros.

4.   Every term of the table is equal to the sum of all preceding terms of the preceding column.

8.   If, starting from the beginning, we take a certain number of rows and add the terms columnwise, then we obtain the terms of the next row without the first term. For example:

1	0	0	0	0
1	1	0	0	0
1	2	1	0	0
1	3	3	1	0
1	4	6	4	1
5	10	10	5	1

12.   The sum of a certain number of terms (including zeros) in every column is related to the sum consisting of the same number of identical summands each one of which is equal to the last term, as 1 (is related) to the number of the column. Analogously, the sum of a certain number of terms of any given column starting from 1 is related to the sum of the same number of summands each one of which is equal to the term following the last summand as 1 (is related) to the number of the column. For example,

0 3	1 5	0 10	1 56
1 3	2 5	0 10	4 56
2 3	3 5	1 10	10 56
3 3	4 5	4 10	20 56
		10 10	35 56

$6:12 = 1:2$	$10:20 = 1:2$	$15:60 = 1:4$	$70:280 = 1:4$

Bernoulli proves that the sum of n terms in the kth column or, equivalently, the term located in the $(n + 1)$th row and $(k + 1)$th column is equal to

$$n(n - 1) \cdots (n - k + 1)/1 \cdot 2 \cdots k,$$

i.e., to the number of combinations of n things taken k at a time.

In the fifth chapter Bernoulli constructs tables of combinations. For example, the following combinations of classes from one to three can be constructed from the elements a, b, c, and d:

> *a, aa, aaa*;
> *b, ab, bb, aab, abb, bbb*;
> *c, ac, bc, cc, aac, abc, bbc, acc, bcc, ccc*;
> *d, ad, bd, cd, dd, aad, abd, bbd, acd, bcd, ccd*;
> *adb, bdd, cdd, ddd.*

This table can be continued even further. If we write down the number of elements in each row according to classes, we then obtain the following results. For class I: 1, 1, 1, 1, ... ; for class II: 1, 2, 3, 4, ... ; for class III: 1, 3, 6, 10, ...

If we continue this list and summarize the results in a table, we obtain a table of the number of combinations with repetitions, which coincides with Pascal's arithmetical triangle.

Bernoulli observes the following properties of this table:

1.   The columns and rows consist of the same numbers.

2.   The sum of the first n terms in the kth column is equal to the term located in the $(k + 1)$th column and the nth row.

3.   The sum of the first n terms in the kth column (or row) is related to the sum of the same number of summands each one of which is equal to the term following the last summand as 1 [is related] to the number of the column (or row). For example,

$$
\begin{array}{cc}
1 & 35 \\
4 & 35 \\
10 & 35 \\
20 & 35 \\
\hline
\end{array}
$$

$$35 : 140 = 1 : 4$$

Next Bernoulli proves that the sum of the first n terms of the kth column, or, equivalently, the number of combinations with repetitions of n things taken k a time is equal to

$$\frac{n(n + 1) \cdots (n + k - 1)}{1 \cdot 2 \cdots k} = \binom{n + k - 1}{k}.$$

Let $a_{n,k}$ be the nth term of the kth column. The sum of the first n terms of the first column is equal to $n/1!$. By property 2 this number is equal to the nth term of the second column, i.e., $a_{n,2} = n/1$, moreover

$$a_{n+1,2} = (n + 1)/1; \qquad a_{n+2,2} = (n + 2)/1;$$

and so on. In view of the same property 2 we obtain

$$\sum_{i=1}^{n} a_{i,2} = a_{n,3}$$

and in view of property 3 we have

$$\sum_{i=1}^{n} a_{i,2} : na_{n+1,2} = 1:2,$$

hence

$$a_{n,3} = na_{n+1,2}/2! = n(n + 1)/2.$$

Analogously one finds that $a_{n,4} = n(n + 1)(n + 2)/3!$.

Let the formula be valid for an arbitrary n and $k + 1$:

$$a_{n,k+1} = \frac{n(n + 1)(n + 2) \cdots (n + k - 1)}{1 \cdot 2 \cdots k}.$$

We now verify its validity for an arbitrary n and $k + 2$. In view of property 2 we have

$$a_{n,k+2} = \sum_{i=1}^{n} a_{i,k+1};$$

and in view of property 3

$$\sum_{i=1}^{n} a_{i,k+1} : na_{n+1,k+1} = 1:(k + 1).$$

Hence

$$a_{n,k+2} = \frac{na_{n+1,k+1}}{k + 1} = \frac{n(n + 1)(n + 2) \cdots (n + k)}{1 \cdot 2 \cdots k(k + 1)}.$$

Bernoulli was the first to consider the problem of combinations with restricted repetitions. However, he was successful in solving this problem in the general case. He constructed a table for particular cases and shows the method for determining the overall number of classes.

In Chapter 7 Bernoulli discusses the problem of the number of arrangements without repetitions. It is here that he refers to arrangements as "combinations together with their permutations."

Bernoulli finds that the number of arrangements of the kth class from n elements (i.e., of n things taken k at a time) is equal to

$$\frac{n(n - 1) \cdots (n - k + 1)}{1 \cdot 2 \cdot 3 \cdots k} k(k - 1) \cdots 3 \cdot 2 \cdot 1 = n(n - 1) \cdots (n - k + 1).$$

Next he finds that the number of arrangements of the kth class with repetitions from m different elements is equal to m^k. These computations are then followed by the sum of all arrangements of classes from 1 to the kth class obtained from m different elements. Bernoulli expresses this number as a sum of geometric progression with the common ratio m, i.e.,

$$m + m^2 + m^3 + \cdots + m^n = m(m^n - 1)/m - 1.$$

The ninth chapter is devoted to the problem of the number of arrangements with restricted repetitions. This is discussed not in the general form, but by means of a particular example for which a table is constructed. Bernoulli, apparently, was the first to deal with the problem of arrangements and combinations with restricted repetitions. He considers sequences of the so-called figurate numbers, presents them in the form of a table and establishes a number of properties of these numbers. Based on these properties Bernoulli finds formulas for the sum of equipowered integers up to the tenth power inclusive:

$$S(n) \quad = \tfrac{1}{2}n^2 + \tfrac{1}{2}n:$$
$$S(n^2) = \tfrac{1}{3}n^3 + \tfrac{1}{2}n^2 + \tfrac{1}{6}n;$$
$$S(n^3) = \tfrac{1}{4}n^4 + \tfrac{1}{2}n^3 + \tfrac{1}{4}n^2;$$
$$S(n^4) = \tfrac{1}{5}n^5 + \tfrac{1}{2}n^4 + \tfrac{1}{3}n^3 - \tfrac{1}{30}n;$$
$$S(n^{10}) \doteq \tfrac{1}{11}n^{11} + \tfrac{1}{2}n^{10} + \tfrac{5}{6}n^9 - n^7 + n^5 - \tfrac{1}{2}n^3 + \tfrac{5}{66}n.$$

Formulas for $S(n)$, $S(n^2)$, and $S(n^3)$ were known as far back as ancient Greece. The expression for $S(n^4)$ was derived in the middle ages. Fermat was aware of the formula

$$S(n^i) = \frac{1}{i+1}n^{i+1} + a_{i,1}n + a_{i,2}n^2 + \cdots + a_{i,i}n^i,$$

which was proved later by Pascal, but the rule for determination of the coefficients $a_{i,k}$ was not available to Pascal and to previous investigators.

Based on an analogy and trial and error, Bernoulli writes without proof the following general formula:

$$S(n^i) = \frac{1}{i+1}n^{i+1} + \frac{1}{2}n^i + \frac{1}{2}C_i^1 An^{i-1}$$

$$+ \frac{1}{4}C_i^3 Bn^{i-3} + \frac{1}{6}C_i^5 Cn^{i-5} + \frac{1}{8}C_i^7 Dn^{i-7} + \cdots,$$

where C_n^m are the binomial coefficients. He observes that starting with the third term the power of n constantly decreases by 2 and each formula terminates either with the term involving n^2 or the term involving n.

The numbers $A, B, C, D, \ldots$, which Euler called Bernoulli numbers, are equal to the coefficients of n in $S(n^2)$, $S(n^4)$, $S(n^6)$ respectively. It thus follows that $A = \tfrac{1}{6}$; $B = -\tfrac{1}{30}$, $C = \tfrac{1}{42}$.

Bernoulli's numbers appear in many formulas and problems of mathematical analysis and the theory of numbers. Problems connected with these numbers were the subject of investigation of many famous mathematicians including Euler, Laplace, Ostrogradskiï, and others.

Part II of "Ars Conjectandi" was a valuable treatise in its period on combinatorics. This work served as a textbook in this area of mathematics during the eighteenth century.

Part III of Bernoulli's treatise is called "Application of the Theory of Combinations to Different Games of Chance and Dicing." This part consists of twenty-four problems with detailed solutions. We present statements of some of these problems:

1.   A person puts two balls into an urn (a white ball and a black one) and offers a bonus to the three players under the condition that the first person who draws the white ball will receive this bonus, but if none of them draws the white ball, the bonus will be withdrawn. Player A was the first to draw a ball, next in line was player B and player C was the last. What are the chances of each one of the players?

5.   A wagers with B that he will draw from 40 playing cards (consisting of ten cards from the four different suits) four cards of different suits. What are the chances of each of the players?

12.   In throwing a six-faced die six times a person desires to get all the six faces in the following order: one point at the first throw, two points at the second throw, and so on. How large is his expectation?

In a number of problems one is required to calculate the expectation of winning in a certain game of chance. As we have seen, such problems were quite prevalent in that period.

In the first part of this book, Bernoulli rebukes Huygens on several occasions for solving only numerical problems, rather than problems in their general form using algebraic symbols which would make it possible to discover the general rules.

In the third part of "Ars Conjectandi" a number of elementary, but sufficiently complex, problems are solved in their general form. For example, Problem 22: A game of chance is structured so that the total number of cases is a, of which there are b cases of a certain particular type, and the number of remaining cases is $a - b = c$. Titus, by paying Gaius several coins, obtains the right to throw a die several times. If he gets one of the outcomes contained in case b, he will receive m coins from Gaius; however, if the outcome is one of the c cases, no payment is made. But if he throws n times in a row one of the c cases, Titus will get his n coins back from Gaius. What are the expectations of winning for Titus and Gaius?

The three parts of Bernoulli's book examined above are of definite interest to the history of mathematics. Some problems of probability theory which by that time had become standard were examined from a new angle. The role of combinatorics in probability theory was fully appreciated. For the first time, the theory of combinations was treated in a systematic manner and many new properties and formulas were obtained. Some very interesting results were also derived in other areas. The first three parts per se were thus a significant contribution in the development of mathematics in general and the theory of probability in particular.

However, the basic part of the book that actually constitutes the start of a new era in the history of probability theory is the fourth part, entitled "Pars Quarta, tradens usum et applicationem proecedentis Doctrinae in Civilibus, Moralibus et Oeconomicis" ("Application of the Previous Study to Civil, Moral and Economic Problems") [20].

This part contains the proof of Bernoulli's theorem, i.e., the law of large numbers in its simplest form. This part was unfortunately left incomplete by the author; the treatise ends with the proof of Bernoulli's theorem. It follows, however, from the title that Bernoulli also considered examining applications of probability theory to civil, moral, and economic problems. This was also mentioned by Nicholas Bernoulli in his preface to James Bernoulli's book.

Many general and philosophical problems connected with probability theory are touched upon in "Ars Conjectandi" and this is especially true of the fourth part. Bernoulli is clearly an advocate of metaphysical determinism. Moreover, the so-called Laplacian determinism, which became popular later, is found in Bernoulli's work consistently and precisely.† He writes in the first chapter of this part:

> If something which is destined to occur is not certain to occur, then it is unclear how the praise of omniscience and omnipotence of the Almighty can remain unshattered ... Given the position of the die, its speed and distance from the board, when the die leaves the hand of the tosser, it will undoubtedly not fall differently than it actually falls.
>
> Similarly, for a given composition of the air and given masses, positions, directions, and speed of the winds, vapor, and clouds and also the laws of mechanics which govern all these interactions, tomorrow's weather will be no different from the way it should actually be. So these phenomena follow with no less regularity than the eclipses of heavenly bodies. It is, however, the usual practice to consider an eclipse as a regular event, while the fall of a die or tomorrow's weather as chance events. The reason for this is exclusively that succeeding actions in nature are not sufficiently well known. And even if they were known, our mathematical and physical knowledge is not sufficiently developed and so, starting from initial causes, we cannot calculate these phenomena, while from the absolute principles of astronomy eclipses can be precalculated and predicted. ... The chance depends mainly on our knowledge. [20, 189]

† Even the terminology is very similar to Laplace's (*Translator's remark*).

Chapter II of "Pars Quarta" starts with the following definition:

> The art of conjecturing is defined here as the art of measuring probabilities of things, as precisely as possible, in order to be able in our judgments and actions always to choose or follow the path which is found to be the best, most satisfactory, easy, and reasonable. The whole wisdom of the philosopher and the prudence of the statement consists solely in this. [20]

Before proceeding with his main task, Bernoulli writes that it is "useful to assume certain general rules and axioms." He presents nine such rules. To understand the nature of the rules, we present some of them here:

1. There is no place for conjecture where complete certainty can be reached.

6. Something which in certain cases is useful and never harmful should always be preferred to something which is never useful or harmful.

7. One should not appraise human action on the basis of its results, etc.

After stating these rules, Bernoulli observes that each person can compile for himself many similar rules.

Only after these rather detailed introductory remarks, the author proceeds in Chapters III and IV of the treatise to the statement of his basic task. He writes: "The strength of a proof which corresponds to a certain argument depends on the number of cases for which it may or may not exist, be proven or even disproved." [20; 187, p. 1452]

Next he turns to one of the central points in his book. Here he actually presents quite a satisfactory explanation of the statistical interpretation of probability.

> We have now reached the point where it seems that, to make a correct conjecture about any event whatever, it is necessary only to calculate exactly the number of possible cases, and then to determine how much more likely it is that one case will occur than another. But here at once our main difficulty arises, for this procedure is applicable to only a very few phenomena, indeed almost exclusively to those connected with games of chance. The original inventors of these games designed them so that all the players would have equal prospects of winning, fixing the number of cases that would result in gain or loss and letting them be known beforehand, and also arranging matters so that each case would be equally likely. But this is by no means the situation as regards the great majority of the other phenomena that are governed by the laws of nature or the will of man. . . . But what mortal, I ask, could ascertain the number of diseases, counting all possible cases, that afflict the human body in every one of its many parts and

at every age, and say how much more likely one disease is to be fatal than another—plague than dropsy, for instance, or dropsy than fever. [The results in these cases] depend on factors that are completely obscure, and which constantly deceive our senses by the endless complexity of their interrelationships, so that it would be quite pointless to attempt to proceed along this road.

There is, however, another way that will lead us to what we are looking for and enable us at least to ascertain a posteriori what we cannot determine a priori, that is, to ascertain it from the results observed in numerous similar instances. It must be assumed in this connection that, under similar conditions, the occurrence (or non-occurrence) of an event in the future will follow the same pattern as was observed for like events in the past. . . . This empirical process of determining the number of cases by observation is neither new nor unusual; . . . and in our daily lives we can all see the same principle at work. [20; 187, pp. 1452–1453]

Following these remarks, Bernoulli develops his ideas in a more detailed manner:

This type of prediction requires "a large number of observations" . . . but though we all recognize this to be the case from the very nature of the matter, the scientific proof of this principle is not at all simple, and it is therefore incumbent on me to present it here. To be sure I would feel that I were doing too little if I were to limit myself to proving this one point with which everyone is familiar. Instead there is something more that must be taken into consideration—something that has perhaps not yet occurred to anyone. What is still to be investigated is whether by increasing the number of observations we thereby also keep increasing the probability that the recorded proportion of favorable to unfavorable instances will approach the true ratio, so that this probability will finally exceed any desired degree of certainty, or whether the problem has, as it were, an asymptote. This would imply that there exists a particular degree of certainty that the true ratio has been found which can never be exceeded by any increase in the number of observations.

Lest this matter be imperfectly understood, it should be noted that the ratio reflecting the actual relationship between the numbers of the cases—the ratio we are seeking to determine through observation—can never be obtained with absolute accuracy; . . . The ratio we arrive at is only approximate: it must be defined by two limits, but these limits can be made to approach each other as closely as we wish.

This is therefore the problem that I now want to publish here, having considered it closely for a period of twenty years. [20; 187, p. 1455]

Only after this rather detailed explanation does Bernoulli proceed in Chapter V to the proof of the theorem. At first several lemmas are proved.

Lemma 1   Consider the two series

$$0, 1, 2, \ldots, r - 1, r, r + 1, \ldots, r + s;$$

$$0, 1, 2, \ldots, nr - n, \ldots, nr, \ldots, nr + n, \ldots, nr + ns.$$

It is asserted that as n becomes larger, the number of terms between nr and $nr + n$; nr and $nr - n$; $nr + n$ and $nr + ns$; and between nr and 0 also increases. Moreover, no matter how large n is, the number of terms exceeding $nr + n$ will not be greater than $s - 1$ times the number of terms between nr and $nr + n$ or between nr and $nr - n$, and, moreover, the number of terms below $nr - n$ will not exceed $r - 1$ times the number of terms between the same numbers (nr and $nr + n$ or nr and $nr - n$).

Lemma 2   The number of terms in the expansion of $(r + s)^n$, where n is an integer, is $n + 1$.

Lemma 3   In a binomial $r + s$ raised into a power of at least $t = r + s$ or $nt = nr + ns$, a certain term M will be the largest if the terms immediately preceding it and immediately following it are in the ratio $s : r$, or equivalently if in this term the powers of the letters r and s are in the ratio of the quantities r and s themselves; the closer term from each side being larger than the more distant from the same side; but the ratio of the term M to the closer term is less than the ratio of the closer term to a more distant one, provided the number of the intermediate terms is the same.

PROOF

$$(r + s)^{nt} = r^{nt} + \frac{nt}{1} r^{nt-1} s + \frac{nt(nt - 1)}{1 \cdot 2} r^{nt-2} s^2$$

$$+ \frac{nt(nt - 1)(nt - 2)}{1 \cdot 2 \cdot 3} r^{nt-3} s^3 + \cdots.$$

It is noted that the coefficients of the terms equidistant from the ends are the same. The total number of the terms is $nt + 1 = nr + ns + 1$. The largest term is

$$M = \frac{nt(nt - 1) \cdots (nt - ns + 1)}{1 \cdot 2 \cdots ns} r^{nr} s^{ns}$$

$$= \frac{nt(nt - 1) \cdots (nr + 1)}{1 \cdot 2 \cdot 3 \cdots ns} r^{nr} s^{ns}.$$

M can also be recorded in a somewhat different form using the formula $C_{nt}^{ns} = C_{nt}^{nt-ns} = C_{nt}^{nr}$:

$$M = \frac{nt(nt-1)\cdots(nt-nr+1)}{1\cdot 2\cdots nr}r^{nr}s^{ns} = \frac{nt(nt-1)\cdots(ns+1)}{1\cdot 2\cdots nr}r^{nr}s^{ns}.$$

The neighboring term from the left is

$$\frac{nt(nt-1)\cdots(nr+2)}{1\cdot 2\cdots(ns-1)}r^{nr+1}s^{ns-1};$$

and the one from the right

$$\frac{nt(nt-1)\cdots(ns+2)}{1\cdot 2\cdots(nr-1)}r^{nr-1}s^{ns+1}.$$

The term succeeding it from the left is

$$\frac{nt(nt-1)\cdots(nr+3)}{1\cdot 2\cdots(ns-2)}r^{nr+2}s^{ns-2};$$

and the one from the right

$$\frac{nt(nt-1)\cdots(ns+3)}{1\cdot 2\cdots(nr-2)}r^{nr-2}s^{ns+2};$$

and so on. By means of appropriate divisions and comparisons all the assertions of the lemma can easily be proved.

Lemma 4 Let r, s, n, and t all be positive integers with $t = r + s$. Consider the expansion of $(r + s)^{nt}$. One can choose n so large, that the ratio of the largest term M to the two terms L and λ which are n terms immediately preceding it and n terms immediately following it, may be made as large as desired.

PROOF

$$M = \frac{nt(nt-1)\cdots(nr+1)}{1\cdot 2\cdots ns}r^{nr}s^{ns}$$

$$= \frac{nt(nt-1)\cdots(ns+1)}{1\cdot 2\cdots nr}r^{nr}s^{ns};$$

$$L = \frac{nt(nt-1)\cdots(nr+n+1)}{1\cdot 2\cdots(ns-n)}r^{nr+n}s^{ns-n};$$

$$\lambda = \frac{nt(nt-1)\cdots(ns+n+1)}{1\cdot 2\cdots(nr-n)}r^{nr-n}s^{ns+n}.$$

We must show that

$$\lim_{n \to \infty} (M/L) = \infty \qquad \text{and} \qquad \lim_{n \to \infty} (M/\lambda) = \infty.$$

For the case of the first limit, Bernoulli proceeds with the solution of this problem in the following manner: First he performs the algebraic transformations

$$\frac{M}{L} = \frac{nt(nt-1)(nt-2)\cdots(nr+1)\cdot 1 \cdot 2 \cdots (ns-n)\, r^{nr} s^{ns}}{nt(nt-1)\cdots(nr+n+1)\cdot 1 \cdot 2 \cdots ns \cdot r^{nr+n} s^{ns-n}}$$

$$= \frac{(nr+n)(nr+n-1)\cdots(nr+1)\, s^n}{(ns-n+1)(ns-n+2)\cdots ns \cdot r^n}$$

$$= \frac{(nrs+ns)(nrs+ns-s)\cdots(nrs+s)}{(nrs-nr+r)(nrs-nr+2r)\cdots nrs}.$$

Next he writes: "But these ratios [also the ratio M/λ] will be infinitely large when n is assumed to be infinite, since in this case the numbers 1, 2, 3, and so on disappear as compared with n, and the numbers $nr \pm n \mp 1$, $nr \pm n \mp 2$, $nr \pm n \mp 3$, and so on, $ns \mp n \pm 1$, $ns \mp n \pm 2$, $ns \mp n \pm 3$, and so on will have the same value as $nr \pm n$ and $ns \mp n$." Now discarding these numbers and carrying out appropriate cancellations by n, we obtain

$$\frac{M}{L} = \frac{(rs+s)(rs+s)\cdots rs}{(rs-r)(rs-r)\cdots rs}.$$

The number of factors in the numerator and denominator equals n. "As a result this ratio becomes an infinite power of $(rs+s)/(rs-r)$ and therefore infinitely large" [20]. For those who have some doubts concerning this conclusion, Bernoulli presents another argument and a proof of the fact that $(rs+s)/(rs-r)$ raised to the nth power becomes infinitely large as n is assumed to be infinite. "It is thus shown that, when the binomial is raised to an infinitely high power, the ratio of the largest term to another term L exceeds any given ratio."

The proof of the assertion $\lim_{n \to \infty}(M/\lambda) = \infty$ is carried out analogously.

Lemma 5   The ratio of the sum of all terms from L up to λ to the sum of all the remaining terms may be made arbitrarily large as n increases.

PROOF   Let M be the largest term in the expansion. Let its neighboring terms on the left be $F, G, H, \ldots$; let the neighboring terms of L on the left be $P, Q, R, \ldots$. From Lemma 3 we have

$$M/F < L/P; \qquad F/G < P/Q;$$

$$G/H < Q/R, \cdots \qquad \text{or} \qquad M/L < F/P < G/Q < H/R < \cdots.$$

Since, by Lemma 4, as n becomes infinitely large, the ratio M/L is infinite, the ratios $F/P, G/Q, H/R, \ldots$, becomes infinite all the more and therefore the ratio $(F + G + H + \cdots)/(P + Q + R + \cdots)$ is also infinite, i.e. the sum of the terms between the largest M and the limiting L is infinitely larger than the sum of the same number of terms beyond the limiting L and adjacent to it. And since the total number of terms beyond L exceeds, in view of Lemma 1, no more than $s - 1$ times (i.e. a finite number of times), the number of terms between this limit and the largest term M, and the terms decrease in magnitude with the increase of their distance from the limit L, it follows from the first part of Lemma 3, that the sum of all terms between M and L (even not including M) will be infinitely larger than the sum of all the terms beyond the limit L.

An analogous assertion may be proved concerning the terms between M and λ. These two assertions provide the proof of the lemma.

After presenting an additional method to show the validity of the lemmas, Bernoulli proceeds to the main topic of his treatise. He formulates what he refers to as the "main proposition" (or the "golden theorem").

Finally, the proposition follows for which all the previous results were given and whose proof follows solely from the application of the previous lemmas. . . . Let the number of "happenings" (favorable cases) of an event be related to the number of failings as r to s, exactly or approximately, or be related to the whole number of trials as r to $r + s$ or r to t; this ratio lies between $(r + 1)/t$ and $(r - 1)/t$. One must show that the number of trials can be taken such that for any number c, we have the odds of c to 1 and that the number of happenings will be within these limits, i.e. the ratio of the number of times the event happens to the whole number of trials lies between $(r + 1)/t$ and $(r - 1)/t$.

Evidently this statement is equivalent to Bernoulli's theorem presented in modern books on probability theory.

PROOF Let the number of observations required be nt. The probability that all the observations are favorable is

$$P_{nt,nt} = \left(\frac{r}{t}\right)^{nt}$$

and that all except one are favorable is

$$P_{nt-1,nt} = \frac{nt}{1} s \frac{r^{nt-1}}{t^{nt}},$$

and that all except two are favorable is

$$P_{nt-2,nt} = \frac{nt(nt-1)}{1\cdot 2} s^2 \frac{r^{nt-2}}{t^{nt}},$$

and so on. But these are exactly the terms in the expansion of the binomial $(r + s)$ raised to the power nt (divided by t^{nt}), which were investigated in previous lemmas. All further conclusions are based on these lemmas. The number of trials with ns failings and nr happenings yield the term M. The number of trials corresponding to $nr + n$ or $nr - n$ "happenings" are expressed by the terms L and λ which are n terms distant from M. Hence, the number of trials for which the number of happenings is no less than $nr + n$ and no more than $nr - n$ is expressed by the sum of the terms between L and λ. The total number of trials for which the number of happenings is either larger than $nr + n$ or smaller than $nr - n$ is expressed by the sum of the terms placed to the left of L and to the right of λ. Since the power into which the binomial is raised

can be chosen so large that the sum of terms contained between both limits L and λ will exceed more than c times the sum of all the remaining terms placed outside of these limits, as follows from Lemmas 4 and 5, one can therefore choose the number of observations so large that the number of trials for which the ratio of the number of happenings to the whole number of trials is contained between the limits $(nr + n)/nt$ and $(nr - n)/nt$ or $(r + 1)/t$ and $(r - 1)/t$, will exceed more than c times the number of other trials, i.e. the odds would be greater than $c:1$ that the ratio of the number of happenings to the whole number is contained within the limits $(r + 1)/t$ and $(r - 1)/t$. [189, pp. 136–137]

Q.E.D.

For comparison, we present the modern statement of Bernoulli's theorem: If the probability of occurrences of an event A in a sequence of independent trials is constant and equals p, then for any positive ε, one can assert with probability as close to 1 as desired, that for a sufficiently large number of trials n the difference $m/n - p$[†] is less than ε in absolute value:

$$Pr\{|(m/n) - p| < \varepsilon\} > 1 - \eta,$$

where η is an arbitrarily small number.

To clarify the substance of Bernoulli's theorem we add several remarks: It may always happen that for n no matter how large, the difference $|(m/n) - p|$ will be larger than ε in a given sequence of n trials. However, according to Bernoulli's theorem, we can assert that if n is sufficiently large and if a sufficiently large number of sequences of n trials were performed, then in the vast majority of these series the inequality $|(m/n) - p| < \varepsilon$ will be satisfied.

Bernoulli's theorem in no way asserts that as n—the number of trials—

[†]m denotes the number of occurrences of event A in n trials (*Translator's remark*).

increases indefinitely, the frequency m/n tends to p, i.e., that $\lim_{n \to \infty} m/n = p$; it merely asserts that the probability of large deviations of frequency m/n from probability p is small provided n is sufficiently large.

Bernoulli's theorem constitutes an immense contribution to the theory of probability, and is of first-rate significance in various practical applications of probability theory.

This theorem was repeatedly verified using specially designed experiments and first and foremost using coin-tossing experiments.

J. Bernoulli concludes his treatise with the statement later adopted by many scientists, including Laplace, as the basic assertion of deterministic philosophy. It is Bernoulli's opinion that from his theorem the following "remarkable" conclusions can be deduced:

> If all events from now through eternity were continually observed (whereby probability would ultimately become certainty), it would be found that everything in the world occurs for definite reasons and in definite conformity with law, and that hence we are constrained, even for things that may seem quite accidental, to assume a certain necessity and as it were, fatefulness. [187, p. 1455]

At this point "Ars Conjectandi" abruptly concludes. It may be inquired why the last chapter of this work in which Bernoulli promised to apply probability theory to various civil and economical problems remained unfinished. One may speculate that Bernoulli did not envision serious and significant applications of probability theory to the above-mentioned problems.

Bernoulli's works were always evaluated very highly.[†] In 1913, on the two-hundredth anniversary of its initial publication, the fourth part "Pars Quarta" of the "Ars Conjectandi" was translated into Russian.[‡] In the Introduction to the translation, Markov writes that in this treatise "is published and proved for the first time the famous ... theorem which laid the foundation of the law of large numbers. ... J. Bernoulli gave a precise statement and a completely rigorous proof of his theorem" [20, Introduction pp. 1–2].

Kolmogorov observes that Bernoulli proved his limiting theorem with "complete analytical rigor." [90, p. 56]

[†] Note, however, the opinion of Pearson:
  All Bernoulli achieved was to show that by increasing the number of observations the results would undoubtedly fall within certain limits, but he failed entirely to determine what the adequate number of observations was for such limits. That was entirely de Moivre's discovery. ... The "Pars Quarta" of the "Ars Conjectandi" has not the importance which has often been attributed to it [207, pp. 208, 210]. Sheynin [211, 213] polemizes with this conclusion of Pearson (*Translator's remark*).

[‡] A special meeting of the Academy of Sciences in St. Petersburg was devoted to this occasion (*Translator's remark*).

2   The development of probability theory
in the first half of the eighteenth century

The beginning of the eighteenth century was marked not only by the appearance of J. Bernoulli's treatise on probability theory, but also by the works of Montmort, de Moivre, and others which were published during that period.

Pierre-Rémond de Montmort (1678–1719) was a French mathematician as well as a student of philosophy and religion. He was in correspondence with a number of prominent mathematicians (N. Bernoulli, J. Bernoulli, Leibniz, etc.) and was a well-established and authoritative member of the scientific community. In particular, Leibniz selected him as his representative at the commission set up by the Royal Society to rule on the controversy between Newton and Leibniz concerning priority in the discovery of differential and integral calculus.† His basic work on probability theory was the "Essai d'Analyse sur les Jeux de Hazard." It went through two editions, the first of which was printed in Paris in 1708. It is commonly reckoned that the second edition of the "Essai d'Analyse" [134] appeared in 1713, although Todhunter claims that "The date 1714 is on the title page of my copy which appears to have been a present to 's Gravesande from the author" [176]. The second edition is much more extensive (414 pages) than the first (189 pages) and contains a series of letters between Montmort and Nicholas Bernoulli as well as one letter from John Bernoulli, James' brother.

Montmort's treatise (second edition) is divided into four parts. The first part contains the theory of combinatorics; the second discusses certain games of chance with cards; the third deals with games of chance with dice, the fourth part contains the solution of various problems including the five problems proposed by Huygens; these are followed by the correspondence mentioned above. In the preface Montmort briefly presents the outline of J. Bernoulli's treatise, of which he was aware, from the summaries by Fontenelle‡ and Saurin.§

Montmort was under the impression that Bernoulli's treatise would not be published posthumously. He writes:

> I have come to the conclusion that one can proceed a long way into this unexplored area and discover a great number of truths equally surprising and novel. This gave me the idea of getting to the bottom of this problem and the desire to make up to the public in some fashion for the loss sustained in being deprived of the excellent work of M. Bernoulli. [176]

† Although Montmort favored Newton in this dispute (*Translator's remark*).

‡ Who gave a summary of the contents of "Ars Conjectandi" in his eulogy of James Bernoulli in 1706 (*Translator's remark*).

§ Who wrote briefly about this work in his eulogy also in 1706 [176].

In the preface Montmort points out that mathematics has penetrated into the natural sciences, firstly into physics, where it achieved a great many successes. "It would have been a great commendation for this science if it could also serve for the determination of judgments and human behavior in practical life" [134]. He then points out that this was attempted by James Bernoulli, but his untimely death prevented his completing this undertaking. Montmort knew but little about Bernoulli's book:

> M. Bernoulli divided his work into four parts. The first three are devoted to the solution of various problems in games of chance. There should be many new results on infinite series, combinations and permutations together with solutions of problems proposed by the mathematician Huygens some time ago. In the fourth part he uses the methods presented in the first three parts in the solution of various civil, moral, and political problems. We are not aware what the games were whose stake divisions were determined by this author, neither do we know which problems of morals and politics he intended to clarify, but no matter how remarkable the project was, there is every reason to assume that the learned author would have performed his task admirably.... I am certain that he would have carried out everything that is promised in the title of this book. [134]

It should be noted that Montmort does not include applications of probability theory to moral, political, economic, and other similar problems in his work. He writes:

> If I were going to follow M. Bernoulli's project I should have added a fourth part where I applied the methods contained in the first three parts to political, economic, and moral problems. What has prevented me is that I do not know where to find the theories based on factual information which would allow me to pursue my researches. [134]

Montmort (as well as Bernoulli) does not find well-grounded applications of probabilistic considerations to the moral sciences. It should be observed that Montmort adheres to the philosophy of metaphysical determinism in his discussion of general methodological problems.

Next in his preface Montmort presents a long discussion concerning the applicability of probability theory to human behavior, but all these discussions are very general and indefinite. He refers to the works of Halley, Petty, Huygens, and to the correspondence between Pascal and Fermat. At the conclusion of his preface he writes: "In this treatise I had in mind foremost the enjoyment of the mathematicians and not the advantages of the players; it is our opinion that those who waste time on games fully deserve to lose their money as well" [134].

The first part of the book is entitled "Traité des Combinaisons" [134, pp. 1–72]. In this section Montmort investigates the arithmetic triangle, math-

ematical expectations, and other topics. The binomial theorem is proved
for the case of $(a + b)^4$ by the following argument: Suppose we have four
counters each having two faces, one black and one white. If the four counters
are thrown "promiscuously," there is one way in which all the faces presented
will be black, four ways in which three faces will be black and one white,
six ways in which two faces will be black and two white, and so on. Therefore,
$(a + b)^4$ must contain a^4 and b^4 which corresponds to taking the four black
faces and four white faces; next there should be coefficient 4, which corresponds to the number of occurrences of three black faces and one white
face, and also three white faces and one black; next it must be followed by
coefficient 6 which corresponds to the number of occurrences of two white
and two black faces. Hence, Montmort concludes that the coefficients in the
binomial theorem must be 1, 4, 6, 4, 1.

   He then proceeds with a discussion of various problems. Among these
the following were considered: "There are p dice (each having the same
number of faces); find the number of ways in which, when they are thrown at
random, we can have a aces, b twos, c threes, and so on."

   The second part of Montmort's treatise [134, pp. 73–172] is related to
games of chance involving cards, among these the games *Pharaon* and
Bassette.

   Games of chance involving dice are considered in the third part [134,
pp. 173–215] of Montmort's work. Here the problem of fair division of
stakes (the problem of points) is also discussed: "To carry out in the general
case the division of stakes among several players, playing several games,
under fair conditions of the game" [134]. In this connection Montmort
presents a detailed discussion of the correspondence between Pascal and
Fermat, reproducing in particular the full text of Pascal's letter dated
24 August, 1654 [134, pp. 233–244]. First of all in this part Montmort
considers the following problem: "Three players (Pierre, Paul, Jacques)
agreed to play three games with the understanding that if Pierre, who needs
only one game [to win], wins it before any one of the remaining players wins
two games, he will then be declared the winner; and he becomes the loser
if one of the remaining players, who lack two games, wins the two games
before he wins one" [134]. What are the probabilities of winning for each one
of the players? Montmort solves the problem by means of the following
table:

	Pierre			Paul	Jacques
aaa	*abc*	*bab*	*cac*	*bba*	*cca*
aab	*aca*	*bac*	*cba*	*bbb*	*ccc*
aac	*acb*	*bca*		*bbc*	*ccb*
aba	*acc*	*caa*		*bcb*	*cbc*
abb	*baa*	*cab*		*cbb*	*bcc*

Out of 27 different arrangements of three dice, 17 lead to Pierre's win, five are favorable to Paul's win, and five to that of Jacques. Next he presents the general rule that one should "consider the number of throws necessary for completion of the game, take as many dice as the number of these throws and assign to these dice as many faces as there are players; then one would need only determine out of all the possible arrangements of dice those which are favorable or unfavorable for each one of the players, which is easily accomplished."

This rule is then illustrated by a number of problems. For example: Pierre needs one game to win; Paul, two; and Jacques, three games; Pierre needs five; Paul, six games; and so on.

The fourth part of Montmort's book contains solutions of various problems concerning chances, in particular of the five proposed by Huygens in 1657. Here the problem of the "duration of play" is also included. Concluding this section, Montmort lists the following four problems (these were originally given at the end of the first edition):

1. Determine the odds of the banker in the game of Treize (thirteen).
2. A problem connected with the card game called "*Her*."†
3. A problem connected with the game called "*Ferme*," similar to the card game "*Point*."‡
4. A problem connected with a game (not entirely of pure chance) similar to solitaire.

Next [134, pp. 283–414] the correspondence between Montmort and Nicholas Bernoulli, together with one letter from John Bernoulli to Montmort and a reply from Montmort, are presented. In a letter to Montmort, dated 9 September, 1713 [134, pp. 401–402], N. Bernoulli proposes the following problem: Two players A and B toss a coin under the following conditions: The game is to continue until the first head appears; player B is to give two coins to player A if a head appears on the first toss; four coins if the head appears on the second toss; eight if on the third; and so on. The question is what amount should A pay B before the game starts to make it a fair game?

In order for the game to be fair, A should pay B the number of coins equal to the mathematical expectation of the gain, which is

$$2(1/2) + 4(1/4) + 8(1/8) + \cdots + 2^n(1/2^n) = n.$$

Hence, the mathematical expectation of the gain tends to infinity, and A must pay B an infinite number of coins before starting the game, which clearly does not make sense. Daniel Bernoulli (1700–1782) studied this

†Description of this game is given in Todhunter [176, p. 106].

‡Description of this game is given in Todhunter [176, p. 110] (*Translator's remarks*).

problem and published his investigations in the *Papers of St. Petersburg Imperial Academy of Sciences* [15]. For this reason, this problem is called the Petersburg game or the Petersburg paradox.

De Moivre's (1667–1754) principal works on probability theory are "The Doctrine of Chances" ([133]; first edition, 1718; second, 1740; third (final), 1756)† and "De Mensura Sortis" published in the *Philosophical Transactions of the Royal Society* for the year 1711 [132].

Abraham de Moivre was a member of the Royal Society as of 1697 and was also elected a member of the Paris and Berlin Academies of Science. He is well known for his contribution to the theory of series. He discovered the rules for raising a complex number to a power, as well as extracting the nth root of complex numbers. These rules are now called de Moivre's formulas. De Moivre also obtained, independently of Stirling, the asymptotic expansion $n! = (2\pi n)^{1/2} n^n e^{-n}$ and extensively utilized it in his work.

In his treatise "The Doctrine of Chances" de Moivre discusses problems relating to the duration of play and investigates topics connected with J. Bernoulli's theorem.

The problem of duration of play was first proposed by Huygens. De Moivre repeatedly deals with this problem in both "De Mensura Sortis" [132, p. 227] and "The Doctrine of Chances" [133, p. 52].

We shall consider this problem briefly: Given two players A and B, the probabilities of winning a round are p and $q = 1 - p$ respectively for each of the players. At the beginning of the game A has a coins and B b coins. The loser in each round is to give one coin to his adversary. What is the probability P_a that A will lose all his capital before winning all of B's capital? P_b is defined analogously. It can be shown that the game is finite, i.e. $P_a + P_b = 1$.

De Moivre obtains the following solution:

$$P_b = \frac{(q/p)^a - 1}{(q/p)^{a+b} - 1}.$$

He found also that the expected number of games is equal to

$$E(N) = (bP_b - aP_a)/(p - q).$$

He suggests a general rule for determining $P_{a,n} + P_{b,n}$, where $P_{a,n}$ is the probability of A being ruined in the course of n games and $P_{b,n}$ the corresponding probability for B. He also determined the value of $P_{b,n}$ for the case when a is infinite and $P_{a,n} = 0$.

However, a few months before these results were published (the appearance of the *Philosophical Transactions* for 1711 was delayed substantially),

---

†This posthumous edition was reprinted in 1967 by Chelsea, New York (*Translator's remark*).

Montmort, in 1713, published solutions for the values of $P_{a,n}$ and $P_{b,n}$.

In 1710 Montmort derived a method for computing $P_{a,n}$ and $P_{b,n}$ in the symmetric case $p = q$. He wrote about this discovery to John Bernoulli, who transferred the letter to his nephew Nicholas. In his reply, dated 26 February, 1711, which was published by Montmort in his "Essai d'Analyse," Nicholas Bernoulli presents, without proof, a solution to this problem for the general case $p \neq q$.

The expression for $P_{b,n}$, using modern notation, can be written as follows:

$$P_{b,n} = \sum_t \left[p^{ts+b}q^{ts} \sum_i \binom{n}{i}(p^{n-b-2ts-i}q^i + q^{n-b-2ts-i}p^i) \right]$$
$$- \sum_t \left[p^{ts+s}q^{ts+a} \sum_t \binom{n}{i}p^{n-b-2ts-2a-i}q^i + q^{n-b-2ts-2a-i}p^i) \right],$$

where $s = a + b$. The summation is with respect to those t that yield non-negative exponents.

Having studied mortality tables, de Moivre presents at the end of his book "The Doctrine of Chances" (in the section "The Doctrine of Chances Applied to Valuation of Annuities") a simple equation for the mortality law between the ages of 22 and the limit of longevity ("The Limit of Life") which he considered to be 86 years. This is an equation of a straight line $y = 86 - x$, where x denotes an age between 22 and 86; y represents the number of people who reach age x. Assigning to x the values 23, 24, 25, and so on, we obtain that 63 (percent of) persons reach age 23, 62 reach age 24, 61 age 25, and so on.

As was indicated above, it does not follow from Bernoulli's theorem that as n increases, the fraction m/n will necessarily approach the value p. The number of occurrences of the event, i.e., the number m, is subject to chance, and there are therefore possible large deviations of m/n from p. In his work "Miscellanea Analytica" (1730) de Moivre investigates the problem of the probabilities of various deviations between m/n and p. de Moivre obtains a solution for the particular case $p = \frac{1}{2}$, i.e., he investigated for $p = \frac{1}{2}$ the probabilities of various values of $|(m/n) - p|$.†

Later, Laplace extended de Moivre's theorem to arbitrary values of $p \neq 0, 1$. The de Moivre–Laplace theorem is the second basic limit theorem in probability theory (the first one being Bernoulli's theorem discussed in the previous section).

---

†See Pearson's observation quoted in the footnote on p. 75 (*Translator's remark*).

3   Simpson's contributions to probability theory

Thomas Simpson (1710–1761) was the first to investigate continuous distributions in probability theory. He is, however, known mostly for his contributions to mathematical analysis.

Simpson wrote two books devoted to probability theory: "The Doctrine of Annuities and Reversions" (1742) [168] and "The Nature and Laws of Chance" (1740) [167], and also an article devoted to probabilistic justification of utilizing the arithmetic mean in data analysis, entitled "A Letter to the Right Honourable George Earl of Macclesfield, President of the Royal Society, on the Advantage of Taking the Mean of a Number of Observations in Practical Astronomy" [169].

The second of the above works is concerned mainly with games of chance. The treatise is divided into 30 different problems. The problems and the methods of their solution are very similar to those contained in de Moivre's book "The Doctrine of Chances" [133]. This subject was already of some interest by the middle of the eighteenth century. New problems had arisen in probability theory. One such problem is discussed in Simpson's article [169] quoted above. In Proposition 1 of this paper he states the problem:

> Supposing that the several chances for the different errors that any single observation can admit of, are expressed by the terms of the progression
>
> $$r^{-V}, \ldots, r^{-3}, r^{-2}, r^{-1}; r^0, r^1, r^2, r^3, \ldots, r^V,$$
>
> (III.1)
>
> it is proposed to determine the probability, or odds, that the error, by taking the mean of a given number (n) of observations exceeds not a given quantity m/n. [169]

To be more precise, the terms of the series (III.1) are not the odds (probabilities), since their sum is not equal to 1, but are proportional to these probabilities. We shall comment below on this rather peculiar choice of the distribution, but note at this stage only that the quantity r does not appear in the final formulas and that is probably why Simpson does not explain the meaning of this quantity.

Simpson considers the probability of error m/n in the arithmetic mean equal to a fraction whose denominator is the sum of terms of the sequence (III.1) raised to the nth power and the numerator is that term of the denominator which contains r^m. Here the explanation

$$(1 - r)^{-n} = 1 + nr + \tfrac{1}{2}n(n + 1)r^2 - \cdots$$

is utilized. Simpson notes that the probability of error in the arithmetic mean being contained in the interval $[-m/n, m/n]$ is equal to the sum of probabilities calculated by the above-stated rule for $m = 0, \pm 1, \pm 2, \ldots, \pm m$.

It is doubtful that Simpson attached any direct importance to the result obtained, since the distribution (III.1) is of a rather special nature. The only application to this rule obtained by Simpson is for the case $r = 1$ (where the positive and negative errors are equally probable). This case, as Simpson remarks, corresponds to obtaining the points with n dice where each die has $2V + 1$ faces. The probability of such an event is determined in problem XXII of Simpson's book "The Nature and Laws of Chance." We note that an analogous problem (number XXXVIII) is discussed in de Moivre's "The Doctrine of Chances" [133].

What is quite remarkable in Simpson's paper of 1756 is his approach to the problems of the theory of errors and the analogy observed with the tossing of dice.

Next in this paper Simpson investigates in place of (III.1) another "distribution function" defined by the probabilities (more precisely the number of outcomes):

$$r^{-V}, 2r^{1-V}, 3r^{2-V}, \ldots, (V+1)r^0, \ldots, 3r^{V-2}, 2r^{V-1}, r^V, \qquad \text{(III.2)}$$

and observes that the sum of terms in the sequence (III.2) when raised to the power t (the number of observations) is equal to

$$r^{-Vt}(1 + r + r^2 + \cdots + r^V)^{2t} = r^{-Vt}(1 - r^{V+1})^{2t}/(1-r)^{2t},$$

and hence this new case is reduced to the first one. Following this, Simpson again discusses the particular case $r = 1$.† Unlike the previous discussion the case $r = 1$ is directly utilized and the only objection to Simpson's exposition might be that it would have been desirable to point out the auxiliary role of the series (III.1) and (III.2) from the very beginning.

And so, Simpson selects the triangular distribution:

the chances of several errors $-5, -4, -3, -2, -1, 0, +1, +2, +3, +4, +5$, included within the limits thus assigned, are respectively proportional to the terms of the series 1, 2, 3, 4, 5, 6, 5, 4, 3, 2, 1, which series seems much better adapted than if all the terms were to be equal, since it is highly reasonable to suppose that the chances for the different errors decrease, as the errors themselves increase. [169]

After computing the required probability of the error in the arithmetic mean (this is carried out by means of the sequence (III.2) for $r = 1$), Simpson concludes his paper with the following assertion:

---

†Plackett [148] describes Simpson's results in terms of generating functions. Indeed, Simpson utilizes methods that are actually equivalent to the utilization of these functions.

Upon the whole of which it appears that the taking of the mean of a number of observations greatly dimishes the chances for all the smaller errors, and cuts off almost all possibility of any great ones; which last consideration alone seems sufficient to recommend the use of the method not only to astronomers but to all others concerned in making of experiments of any kind (to which the above reasoning is equally applicable). [169]

This assertion is more precise than the analogous (preliminary) remark presented at the beginning of the paper:

Whatever series is assumed for the chances of the happening of the different errors, the result will turn out greatly in favor of the method now practiced by taking a mean value. [169]

In 1757 this paper was reprinted with substantial additions in a collection of Simpson's papers under the general title: "Miscellaneous Tracts on Some Curious ... Subjects in Mechanics, Physical Astronomy and Speculative Mathematics" [170, pp. 64–75]. In this work Simpson was the pioneer (excluding Bayes, whose results were published somewhat later) in introducing continuous distribution functions. However, strictly speaking, he made only a first step in this direction.

Simpson seeks the limit of the probability of error in the arithmetic mean for the distribution (III.2) as the number of possible errors in observations increases indefinitely and illustrates this case by means of a diagram (Fig. 1).

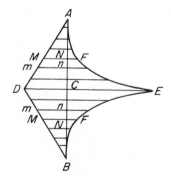

Figure 1

The probability that an error in observation is contained in the very small interval Nn, as Simpson observes, is proportional to the area $MNnm$ (or the length MN). In this diagram Simpson constructs a curve AEB, which represents the error in the arithmetic mean: If, for example, $CN = m/t$, then the area $CNFE$ expresses the probability that the error in the arithmetic mean is less than m/t.

Simpson is not at all concerned about the obvious discrepancy between his calculations and the diagram: It is assumed in his calculations that $|V| \to \infty$, while the diagram is drawn for a finite V, but with continuous changes in the argument. In any case, the calculations are correct, but are not completed.

We note that for a modern reader the graph of the curve is constructed in a somewhat unusual manner (the independent variable is marked vertically).

Simpson does not mention J. Bernoulli at all, possibly because the latter considered only the binomial distribution. However, the limiting approach was introduced by Simpson perhaps merely for aesthetic reasons—he was concerned primarily with the case of a finite number of observations—while the main emphasis in Bernoulli's derivation is in the unlimited increase of the number of experiments.

As far as the properties of random errors in observations utilized in Simpson's arguments, these properties should have been recognized rather easily at an earlier stage. In publications of later years—after Gauss—they were invariably repeated in all texts and manuals. But neither Simpson, nor presumably other authors up to the second half of the nineteenth century, the only exception being Lambert, connected these properties with the realization of the law of large numbers. Moreover, the concept of random errors developed somewhat independently of the gradually evolving notion of a random (chance) variable.

It should be pointed out that Simpson's results were rediscovered, within approximately a decade (around 1770), by Lagrange [94].

We shall now discuss briefly the role of the arithmetic mean in the history of probability theory. The notion of the arithmetic mean appears in many formulas for approximate calculation of areas and volumes known from ancient times. According to Lurie [110], this notion stemmed from practical problems in economics and social activities. This was also the opinion of Leibniz [101, p. 16]. Moreover, the arithmetic mean along with the geometric mean and seven (!) other means were known to the circle of Pythagoreans [118, p. 63].

It is remarkable that in an Indian manuscript of the sixteenth century, the use of the arithmetic mean was very clearly associated with compensating the approximate (nonexact) nature of the formulas employed [8, p. 97], and it was noted that precision increases with the number of observations. This utilization of the arithmetic mean is valid at present also. For example, repeated measurements of angles in triangulation work are intended not only to compensate errors in the observations proper, but also to compensate for the effect of horizontal refraction, since the formula for the rectilinear propagation of light is known to be incorrect.

The other purpose in applying the arithmetic mean to original data by the ancient (Babylonian) mathematicians and astronomers was to compensate for systematic errors in observations. This is mentioned—however without appropriate supportive evidence—in Weiman's book [179, p. 204]. If this opinion is correct, then one may state that (as early as 500–300 B.C.) in Babylon the results of data processing were considered dependent on the conditions surrounding the observations.†

We have not mentioned until now that Simpson's paper was prompted by the fact that "some persons, of considerable note, have been of the opinion, and even publicly maintained, that one single observation taken with due care, was as much to be relied on as the mean of a great number" (the beginning of [169]). Although views and statements of this kind were not predominant (even Simpson himself at the beginning of the revised version of his paper written in 1757 points out that the methods utilized by astronomers in taking the arithmetic mean are almost universally accepted), they are, however, quite indicative.‡

The seventeenth century witnessed the invention of the telescope, vernier, and gauge, while astronomic-geodesical instruments acquired an "almost" modern form. Errors in observations were on a sharp decline. James Bradley's observational errors were measured in seconds, i.e., were 60 times smaller than those encountered by Tycho Brahe [162]. Bradley utilized the arithmetic mean systematically however; he wrote at least on one occasion that among several observations of a star he chose either the mean value, or the observation closest to the mean [162, p. 29].

After all, Bradley should indeed have been allowed this small piece of frivolousness in view of the exceptional precision of his observations. However, this period of "calm" and satisfaction with choosing one observation, even the closest to the mean, could not have continued very long. Bradley was the one who observed that the new discoveries (such as that of aberration in 1725) were leading to new problems (those of mutation) and here "systematic series of observations and experiments" are required.

Moreover, during the same period (the second half of the eighteenth century) relatively precise calibrated measurements were carried out by those who had to achieve the highest possible degree of precision. When reading these accounts, we found no indication that usage of the arithmetic mean had been abandoned.

We now return to Leibniz, who wrote in this connection that "the basis of all these theoretical constructions [in probability theory] is the so-called

---

†Modern objections to undiscriminatory use of the arithmetic mean are voiced in P. Huber's article [*Ann. Math. Statist.* **43**, 1041–1067 (1972)].

‡Plackett [148, pp. 130] however, points out that:

Beyond the fact that the basic parameters in the schemes represent a compromise between observation and the needs of computation, nothing has survived to indicate how they were estimated [by Babylonians] from the original data, which are themselves almost wholly absent (*Translator's remark*).

prostapheresis, i.e. the arithmetic mean between several equally applicable propositions." Leibniz asserts that "this is an axiom: *aequalibus aequalia*— propositions of equal value are to be taken equally into consideration" [100].

In this treatise [101] Leibniz mentions only de Witt by name; however, we may assume that the idea of "considering equally propositions of equal value" (whose consequence is the rule of the arithmetic mean) served as some sort of a premise for the classical definition of probability.

4   Thomas Bayes and his theorem

Bayes' formula appears in all texts on probability theory. The formula's content is as follows: Let event B occur in conjunction with one and only one of n disjoint events $A_1, A_2, \ldots, A_n$, i.e.,

$$B = \sum_{i=1}^{n} BA_i, \qquad P(B) = \sum_{i=1}^{n} P(BA_i).$$

Using the total probability rule we obtain

$$P(B) = \sum_{i=1}^{n} P(A_i) P_{A_i}(B).$$

In view of the multiplication rule, we have for each A_k $(k = 1, \ldots, n)$:

$$P(A_k B) = P(A_k) P_{A_k}(B) = P(B) P_B(A_k).$$

Using the last result we obtain an expression for $P_B(A_k)$:

$$P_B(A_k) = \frac{P(A_k) P_{A_k}(B)}{P(B)} = \frac{P(A_k) P_{A_k}(B)}{\sum_{i=1}^{n} P(A_i) P_{A_i}(B)};$$

and, in view of the total probability rule, we have

$$P_B(A_k) = P(A_k).$$

This is the well-known Bayes theorem (more details are given, e.g., in [69, Section 9]). Assuming a knowledge of the multiplication theorem, Bayes' formula can be obtained directly as an alternative form of the multiplication rule for probabilities. Therefore, the meaning of the following assertion, which very frequently appears in various textbooks, is ambiguous: "This formula was discovered in the middle of the eighteenth century by the English mathematician Bayes" (see, e.g. [28, p. 218]).

What in fact was Bayes' contribution? Biographical data concerning Bayes is scarce and often misleading. In books on the history of mathematics, such as Cantor's or Poggendorff's, no personal details are given, except the fact that he was Fellow of the Royal Society. Even in the "Great Soviet

Encyclopedia" (BSE) there is no mention of his birthdate and the date of his death is given incorrectly as 1763. It is known that Thomas Bayes was born in London in 1702. He was educated privately. There is some speculation that one of his tutors was de Moivre, who earning his living in London during that period by teaching mathematics. Thomas' father, Joshua Bayes, F.R.S., was the minister of a Presbyterian meeting house in London. Thomas Bayes was also ordained and began his ministry helping his father.

In 1720 in Tunbridge Wells (about 35 miles outside London) a religious congregation was organized. Shortly thereafter Bayes moved to minister at the Presbyterian church there. In 1731 he produced a tract entitled "Divine Benevolence or an Attempt to Prove that the Principle End of Divine Providence and Government Is the Happiness of His Creatures." This tract was published by John Noon. In 1736 John Noon published an anonymous tract entitled "An Introduction to the Doctrine of Fluxions and a Defence of Mathematicians against the Objections of the Author of the Analyst." This work was attributed to Bayes by his contemporaries and by many other modern authorities.

In 1742 Bayes was elected a Fellow of the Royal Society. In 1752 he retired and was succeeded in his ministry by William Johnston, who inherited Bayes' library. Bayes continued to live in Tunbridge Wells until his death on 17 April, 1761.

Apart from the tract already noted, Bayes also wrote a letter on asymptotic series to John Canton, which was published in the *Philosophical Transactions of the Royal Society* (pp. 269–271, **53**, 1763).

In these same *Transactions*, for the same year 1763 (see [13]), Bayes' memoirs on probability theory were also published, entitled "An Essay towards solving a Problem in the Doctrine of Chances by the late Rev. Mr Bayes, F.R.S., communicated by Mr Price in a Letter to John Canton, A.M., F.R.S."

This work with the revised title "Thomas Bayes's Essay towards Solving a Problem in the Doctrine of Chances" was reprinted in 1958 in *Biometrika* [13, pp. 293–315].

The paper is preceded by the accompanying letter of Richard Price to Canton, dated 10 November, 1763. The letter starts: "I now send you an essay which I have found among the papers of our deceased friend Mr Bayes and which, in my opinion, has great merit, and well deserves to be preserved" [13, p. 296].

Price then describes the content of Bayes' paper and requests that this essay be communicated to the Royal Society. The essay was indeed read at the meeting of the Society on 23 December, 1763.

Price indicates in his letter that Bayes "had ... the honour of being a member of that illustrious [Royal] Society and was much esteemed by many in it as a very able mathematician" [13, p. 296].

Bayes' essay starts with the formulation of a general problem:

> *Given* the number of times in which an unknown event has happened
> and failed: *Required* the chance that the probability of its happening
> in a single trial lies somewhere between any two degrees of probability
> that can be named. [13, p. 298]

After this general statement of the problem Bayes proceeds with the basic
definitions and propositions of probability theory.

Price's comments on this part of Bayes' essay are as follows:

> Mr Bayes has thought fit to begin his work with a brief demonstra-
> tion of the general laws of chance. His reason for doing this, as he says
> in his introduction, was not merely that his reader might not have the
> trouble of searching elsewhere for the principles on which he has
> argued, but because he did not know whither to refer him for a clear
> demonstration of them. He has also made an apology for the peculiar
> definition he has given of the word *chance* or *probability*. His design
> herein was to cut off all dispute about the meaning of the word, which in
> common language is used in different senses by persons of different
> opinions. [13, p. 298]

We may conclude from these remarks that there existed an introduction
to Bayes' work which was not published. It may be assumed that Price did
not include this introduction with his letter to Canton, otherwise there would
have been no reason to repeat its content in such detail.

Bayes begins with the formulation of seven definitions which are reproduced below in full:

> **Definition 1**   Several events are *inconsistent*, when if one of them
> happens, none of the rest can.
>
> **2**   Two events are *contrary* when one, or other of them must; and
> both together cannot happen.
>
> **3**   An event is said to *fail*, when it cannot happen; or, which comes
> to the same thing, when its contrary has happened.
>
> **4**   An event is said to be determined when it has either happened
> or failed.
>
> **5**   The *probability of any event* is the ratio between the value at
> which an expectation depending on the happening of the event ought
> to be computed, and the value of the thing expected upon its happening.
>
> **6**   By *chance* I mean the same as probability.
>
> **7**   Events are independent when the happening of any one of them
> does neither increase nor abate the probability of the rest. [13, pp.
> 298–299]

Some of these definitions (e.g., 1 and 7) concur almost completely with the definitions used at present. However, Bayes' definition of probability is somewhat unclear.

Next Bayes proves a number of basic theorems which he calls propositions. The first proposition is the addition theorem of probabilities.

Proposition 1   When several events are inconsistent the probability of the happening of one or other of them is the sum of the probabilities of each of them.

The proof of this proposition, as given by Bayes, reveals very clearly the manner in which Bayes comprehends and utilizes the notion of probability. We therefore present this proof in full:

> Suppose there be three such events, and whichever of them happens I am to receive N, and that the probability of the first, second, and third are respectively a/N, b/N, c/N. Then (by the definition of probability) the value of my expectation from the first will be a, from the second b, and from the third c. Wherefore the value of my expectations from all three will be $a + b + c$. But the sum of my expectation from all three is in this case an expectation of receiving N upon the happening of one or other of them. Wherefore (by Definition 5) the probability of one or other of them is $(a + b + c)/N$ or $a/N + b/N + c/N$. The sum of the probabilities of each of them. [13, p. 299]

As a corollary of this theorem Bayes obtains that if the probability of the occurrence of an event is P/N, the probability of its "failure" is $(N - P)/N$. He writes further: "If it be certain that one or other of the three events must happen, then $a + b + c = N$" [13, p. 299].

In the case of equally likely outcomes, the probability, according to Bayes, is the ratio of the mathematical expectation to the sum of all the values of the random quantity. Indeed, if

$$p_1 = p_2 = \cdots = p, \qquad \text{then} \qquad p = \frac{\Sigma x_i p_i}{\Sigma x_i} = \frac{p \Sigma x_i}{\Sigma x_i} = p.$$

Next Bayes proves the multiplication theorem for probabilities:

> The probability that the two subsequent events will both happen is compounded of the probability of the first and the probability of the second on supposition the first happens. [13, p. 299]

The modern formulation of this theorem is $P(A \text{ and } B) = P(A) P_A(B)$.

From the multiplication theorem of probabilities Bayes obtains the following corollary: "If of two subsequent events the probability of the first be a/N, and the probability of both together be P/N, then the probability of the second on the supposition the first happens is P/a" [13, p. 300]. This means that $P(A \text{ and } B) = P(A) P_A(B)$ implies that

$$P_A(B) = \frac{P(A \text{ and } B)}{P(A)} = \frac{P/N}{a/N} = \frac{P}{a}.$$

We now present several other propositions given in Bayes' essay [13, p. 301]:

If there be two subsequent events, the probability of the second b/N and the probability of both together P/N, and it being first discovered that the second event has happened, from hence I guess that the first event has also happened i.e. both events happened, the probability I am in the right is P/b.

Indeed, if it is known that the second event happened, then the probability that both happened is equal to the probability of the first event under the condition that the second happened:

$$P_{II}(I) = \frac{P(I \text{ and } II)}{P(II)} = \frac{P/N}{b/N} = \frac{P}{b}.$$

In the next proposition Bayes discusses the probability of simultaneous occurrences of several independent events. In the corollary to this proposition he writes:

If there be several independent events, and the probability of each one be a, and that of its failing be b, the probability that the first happens and the second fails, and the third fails and the fourth happens will be *abba*. [13, p. 301]

The last (seventh) proposition in Bayes' essay contains a derivation of J. Bernoulli's formula

$$P_{m,n} = C_n{}^m p^m q^{n-m}.$$

If the probability of an event be a, and that of its failure be b in each single trial, the probability of its happening p times, and failing q times in $p + q$ trials is $Ea^p b^q$, if E be the coefficient of the term in which occurs $a^p b^q$ when the binomial $(a + b)^{p+q}$ is expanded. [13, p. 302]

The derivation of this formula concludes Section I of Bayes' essay. Section II begins with a description of a model which plays an important role in the succeeding part of the essay.

Consider a square table (or plane) $ABCD$ (Fig. 2) and a ball W thrown on this table. It is assured that "there shall be the same probability that it rests upon any one equal part of the plane as another, and that it must necessarily rest somewhere upon it" [13, p. 302], i.e., that the ball W may rest at any part

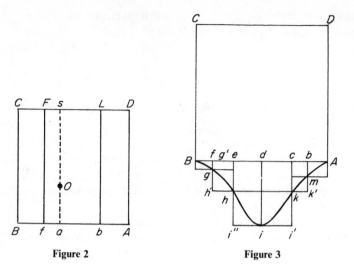

Figure 2   **Figure 3**

of *ABCD* with the same probability. Two arbitrary points *f* and *b* are chosen on *AB* and two lines parallel to *AD* are drawn passing through these points. The lines intersect *CD* at points *F* and *L*, say.

Following this construction, Bayes proves Lemma 1. "The probability that the point 0 will fall between any two points in the line *AB* is the ratio of the distance between the two points to the whole line *AB*" [13, p. 302]. In other words, the probability that a ball, which was thrown randomly on *ABCD*, will rest in the rectangle *bfFL* is equal to *fb/AB*.

The first case considered is the case when the rectangles *fBCF*, *bfFL*, and *AbLD* are commensurable, followed by the case when they are not: "If the rectangles *Cf*, *Fb*, *LA* are not commensurable, yet the last mentioned probability can be neither greater nor less than the ratio of *fb* to **AB**" [13, p. 303].

The proof of this last assertion is carried out using the contradiction method; it is first shown that the required probability cannot be less than *fb/AB* and further that it cannot be greater than this value.

Next, the figure *BghikmA* is constructed on the line *BA* as follows (Fig. 3): Let $y = Ex^p r^q$, where

$$Ab/AB = x; \qquad Bb/AB = r; \qquad bm/AB = y;$$

and *E* is the coefficient of the term $a^p b^q$ in the expansion of $(a + b)^{p+q}$.
Thereafter the following proposition is stated:

> I say that before the ball *W* is thrown, the probability that point *o* should fall between *f* and *b*, any two points named in the line *AB*, and

withall that the event M† should happen p times and fail q in $p + q$ trials, is the ratio of *fghikmb*, the part of the figure *BghikmA* intercepted between the perpendiculars *fg* and *bm* raised upon the line AB to CA the square upon AB. (13, p. 303]

This proposition is also proved by the contradiction method. Let the required probability equal not the ratio of the area of the above-defined figure to the area of $ABCD$, but the ratio of the area of some figure D, greater than $S_{fghikmb}$, to the area of $HBCD$. Choose points e, d, and c, so many and so placed that the sum of the areas of the rectangles *bckk′*, *cdii′*, *dei′i*, *efhh′* constructed by means of these points differs from $S_{fghikmb}$ less than area D does. The points are chosen in such a manner that *di* is the longest line between AB and the curve *BghikmA*.

Let the point o be projected on the point $o′$ which coincides with the point e. Then the probability that the event M will occur is equal to $X = Ae/AB$, and the probability of its nonoccurrence is $r = Be/AB$. Then the probability of p occurrences of event M and q nonoccurrences of this event in $p + q$ trials will be $EX^p r^q$ and in this case $y = eh/AB = EX^p r^q$. If the point $o′$ lies on the interval *ef*, the probability that the event M will occur p times and fail to occur q times in $p + q$ trials may not exceed eh/AB. We thus have two events:

First, the point $o′$ falls in the interval *ef*. The probability of this event is ef/AB. Second, the event M occurs p times and fails to occur q times in $p + q$ trials. The probability of the latter is not greater than eh/AB. The probability of their joint occurrence is not greater than $(ef/AB)(eh/AB) = S_{efh′h}/S_{ABCD}$. Passing in this manner from interval to interval (from *fe* to *ed*, and so on) and summing up the obtained probabilities, we arrive at the conclusion that the required probability will not exceed the ratio of the sum of areas of circumscribed rectangles to S_{ABCD}. But S_D is greater than this sum, therefore the required probability is less than S_D/S_{ABCD}, which contradicts the original assumption. In the same manner the assumption that S_D is less than $S_{fghikmb}$ leads to a contradiction. These contradictions show that the required probability is equal to $S_{fghikmb}/S_{ABCD}$.

The following corollary is then noted:

Before the ball W is thrown, the probability that the point o will lie somewhere between A and B, or somewhere upon the line AB, and withal that the event M will happen p times, and fail q times in $p + q$ trials is the ratio of the whole figure AiB to CA. But it is certain that the point o will lie somewhere upon AB. Wherefore, before the ball W is thrown the probability the event M will happen p times and fail q in $p + q$ trials is the ratio of AiB to CA. [13, p. 305]

†The event M indicates that the ball rests between AD and *os*.

A direct conclusion from the above is the following proposition:

> If before anything is discovered concerning the place of the point
> o, it should appear that the event M had happened p times and failed
> q in $p + q$ trials, and from hence I guess that the point o lies between
> any two points in the line AB, as f and b, and consequently that the
> probability of the event M in a single trial was somewhere between the
> ratio of Ab to AB and that of Af to AB; the probability I am in the right
> is the ratio of that part of the figure AiB described as before which is
> intercepted between perpendiculars erected upon AB at the points
> f and b, to the whole figure AiB. [13, p. 305]

Next, Bayes concludes that considering event M and taking into account
only its occurrences and nonoccurrences in n trials, one can state a proposition concerning its probability and also determine the probability of this
proposition. Bayes assumes that the probability distribution is uniform on
AB.

Since the ratios of areas appear in all the computations, Bayes firstly
cancels all the values of the ordinates by the amount E. He then proceeds to
prove the following Proposition 10:

> If a figure be described upon any base AH having for its equation
> $y = x^p r^q \cdots$, I say then that if an unknown event has happened p times
> and failed q in $p + q$ trials, and in the base AH taking any two points
> as f and t you erect the ordinates fC, tF at right angles with it, the
> chance that the probability of the event lies somewhere between the
> ratio of Af to AH and that of At to AH, is the ratio of $tFCf$, that part
> of the before-described figure which is intercepted between the two
> ordinates, to $ACFH$ the whole figure insisting on the base AH. [13,
> p. 306]

Bayes states that the validity of this proposition follows from Proposition
9 and previous remarks. "Now, in order to reduce the foregoing rule to

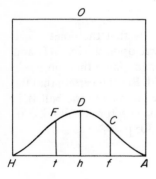

Figure 4

practice, we must find the value of the area of the figure described and several parts of it ..." [13, p. 306].

Let $AH = 1$ and let the rectangle constructed on AH be a square which will likewise have an area of 1 (Fig. 4).

Since $Cf/AH = y$; $Af/AH = x$ and $Hf/AH = r$, then in our case $Cf = y$; $Af = x$; $Hf = r$. From the equation of the curve $y = x^p r^q$ we obtain $y = x^p(1 - x)^q$, since $r + x = 1$. Therefore

$$y = x^p(1 - x)^q = x^p - qx^{p+1} + \frac{q(q - 1)x^{p+2}}{2}$$

$$- \frac{q(q - 1)(q - 2)x^{p+3}}{2 \cdot 3} + \cdots.$$

Next, referring to Newton's proposition, Bayes writes: "Now the abscisse being x and the ordinate x^p, the correspondent area is $x^{p+1}/(p + 1)$ and the ordinate being qx^{p+1} the area is $qx^{p+2}/(p + 2)$; and in like manner of the rest" [13, pp. 306–307].

If the abscissa is x and the ordinate

$$y = x^p - qx^{p+1} + \frac{q(q - 1)x^{p+2}}{2} - \frac{q(q - 1)(q - 2)x^{p+3}}{2 \cdot 3} + \cdots,$$

then the corresponding area is equal to

$$\frac{x^{p+1}}{p + 1} - \frac{qx^{p+2}}{p + 2} + \frac{q(q - 1)x^{p+3}}{2(p + 3)} - \frac{q(q - 1)(q - 2)x^{p+4}}{2 \cdot 3 \cdot (p + 4)} + \cdots. \quad \text{(III.3)}$$

But since $x = Af/AH = Af$ and $y = Cf/AH = Cf$, this is the expression for an arbitrary area $ACf = ACf/HO$. Analogously one obtains that

$$HCf = \frac{HCf}{HO} = \frac{r^{q+1}}{q + 1} - \frac{pr^{q+2}}{q + 2} +$$

$$\frac{p(p - 1)r^{q+3}}{2(q + 3)} - \frac{p(p - 1)(p - 2)r^{q+4}}{2 \cdot 3 \cdot (q + 4)} + \cdots.$$

$$\text{(III.4)}$$

Next, Bayes denotes the ratio ACf/HO in the form

$$\frac{x^{p+1}r^q}{p + 1} + \frac{qx^{p+2}r^{q-1}}{(p + 1)(p + 2)} + \frac{q(q - 1)x^{p+3}r^{q-2}}{(p + 1)(p + 2)(p + 3)}$$

$$+ \frac{q(q - 1)(q - 2)x^{p+4}r^{q-3}}{(p + 1)(p + 2)(p + 3)(p + 4)} + \cdots + \frac{x^{n+1}q(q - 1) \cdots 1}{(n + 1)(p + 1)(p + 2) \cdots n} + \cdots,$$

$$\text{(III.5)}$$

where $n = p + q$.

In the case $r = 1 - x$, Eq. (III.5) reduces to equation (III.6) after some algebraic manipulations. Replacing $x = 1 - r$ in (III.6) and carrying out appropriate manipulations, one obtains (III.4) from (III.6).

Further, Bayes proves the following assertion: "If E be the coefficient of that term of the binomial $(a + b)^{p+q}$ expanded in which occurs $a^p b^q$, the ratio of the whole figure $ACFH$ to HO is $\{(n + 1)E\}^{-1}$, n being $p + q$" [13, p. 308].

Based on previous results (Articles 1 and 4), Bayes now notes the following corollary: "The ratio of ACf to the whole figure $ACFH$ is

$$(n + 1) E \left[\frac{x^{p+1}}{p + 1} - \frac{qx^{p+2}}{p + 2} + \frac{q(q - 1)x^{p+3}}{2(p + 3)} - \cdots \right],$$

and if, as x expresses the ratio of Af to AH, X should express the ratio of At to AH, the ratio of AFt to $ACFH$ would be

$$(n + 1) E \left[\frac{X^{p+1}}{p + 1} - \frac{qX^{p+2}}{p + 2} + \frac{q(q - 1)X^{p+3}}{2(p + 3)} - \cdots \right]" \qquad [13, p. 308].$$

Following this, Bayes formulates his basic rule:

> *Rule* 1   If nothing is known concerning an event but that it has happened p times and failed q in $p + q$ or n trials, and from hence I guess that the probability of its happening in a single trial lies somewhere between any two degrees of probability as X and x, the chance I am in the right in my guess is $(n + 1) E$ multiplied into the difference between the series
>
> $$\frac{X^{p+1}}{p + 1} - \frac{qX^{p+2}}{p + 2} + \frac{q(q - 1)X^{p+3}}{2(p + 3)} - \cdots$$
>
> and the series
>
> $$\frac{x^{p+1}}{p + 1} - \frac{qx^{p+2}}{p + 2} + \frac{q(q - 1)x^{p+3}}{2(p + 3)} - \cdots.$$

E being the coefficient of $a^p b^q$ when $(a + b)^n$ is expanded.

> This is the proper rule to be used when q is a small number; but if q is large and p small, change everywhere in the series here set down p into q and q into p and x into r or $(1 - x)$, and X into $R = (1 - X)$; which will not make any alteration in the difference between the two series. [13, p. 308]

"Thus far is Mr Bayes's essay," writes R. Price immediately after this rule [13, p. 308].

Rule 1 actually gives a method for determining the area of a curvilinear trapezoid by means of a definite integral.

Next Price cites Rule 2 as valid for the case when p and q are large numbers and it is difficult to apply Rule 1 in view of a large number of terms in the series. "Mr Bayes, therefore, by an investigation which it would be too tedious to give here, has deduced from this rule another" [13, p. 308]. He then presents this rule without a proof:

If nothing is known concerning an event but that it has happened p times and failed q in $p + q$ or n trials, and from hence I guess that the probability of its happening in a single trial lies between $(p/n) + z$ and $(p/n) - z$; if $m^2 = n^3/(pq)$, $a = p/n$, $b = q/n$, E the coefficient of the terms in which occurs $a^p b^q$ when $(a + b)^n$ is expanded, and

$$\Sigma = \frac{(n + 1)(2pq)^{1/2}}{nn^{1/2}} Ea^p b^q,$$

multiplied by the series

$$mz - \frac{m^3 z^3}{3} + \frac{(n - 2) m^5 z^5}{2n \cdot 5} - \frac{(n - 2)(n - 4) m^7 z^7}{2n \cdot 3n \cdot 7}$$
$$+ \frac{(n - 2)(n - 4)(n - 6) m^9 z^9}{2n \cdot 3n \cdot 4n \cdot 9} - \cdots,$$

my chance to be in the right is greater than

$$\frac{2\Sigma}{1 + 2Ea^p b^q + 2Ea^p b^q/n}$$

and less than

$$\frac{2\Sigma}{1 - 2Ea^p b^q - 2Ea^p b^q/n}$$

and if $p = q$ my chance is 2Σ exactly. [13, p. 309]

Price observes that there is "a small oversight" in the deduction of the limits for the chances; namely, the third term was omitted $(2Ea^p b^q/n)$. "I have ventured to correct his copy, and to give the rule as I am satisfied it ought to be given."

For the case of large values of mz, Price presents a third rule derived by him. This rule was not available in Bayes' work.

There is an appendix to Bayes' paper, written by Price, where some propositions of Bayes are clarified, and, in particular, the problem of application "of the foregoing rules to some particular cases" is discussed in detail.

The first rule gives a direct and complete solution in all cases. Two other rules are particular approximations of the first rule in the case when the first rule is too cumbersome to apply.

"The first rule may be used in all cases where either p or q are nothing or not large. The second rule may be used in all cases where mz is less than $\sqrt{3}$" [13, p. 311].

Next Price considers several examples. We shall present a few of these: Let us suppose that an event happened once. "What conclusion may we draw from hence with respect to the probability of its happening on a second trial?" In this case $p = 1, q = 0$. Applying the first rule and noting that $n = 1$, we obtain

$$(n + 1)\left(\frac{X^{p+1}}{p + 1} - \frac{x^{p+1}}{p + 1}\right) = X^2 - x^2.$$

Next we put $X = 1$ and $x = \frac{1}{2}$ and obtain that $1^2 - (\frac{1}{2})^2 = \frac{3}{4}$ is the chance that the probability of the occurrence of the event on the second trial in the case of its occurrence on the first trial will lie somewhere between $\frac{1}{2}$ and 1.

If the event occurred twice, then analogously the chance $\frac{7}{8}$ is obtained; and if three times, $\frac{15}{16}$. In general, if the event happened p times in a row, then $1 - (\frac{1}{2})^{p+1}$ is the chance that the probability of its occurrence at the $(p + 1)$th trial will lie between $\frac{1}{2}$ and 1.

Suppose that an event happened ten times without failing. What are the chances that the probability of the given event lies between $\frac{16}{17}$ and $\frac{2}{3}$? *Solution*:

$$p = 10, \quad q = 0, \quad n = 1, \quad X = \tfrac{16}{17}, \quad x = \tfrac{2}{3}.$$
$$X^{p+1} - x^{p+1} = (\tfrac{16}{17})^{11} - (\tfrac{2}{3})^{11} = 0.5013.$$

Using an example with a die, Price carries out the following argument: Because a die must turn on some side even if it is loaded, its first throw does not increase our knowledge about the die and if no information was available before the throw, there would be no information after the first throw as well. Therefore Bayes' rule should be applied starting from the second trial. However, "it should not be imagined that any number of such experiments (where the die falls of the same side without failing) can give sufficient reason for thinking that it would *never* turn on any other side" [13, p. 312].

Price asserts that Bayes' rules are applicable "to the events and appearances of nature." He presents an example with the observations on "the return of the Sun," but immediately warns that "it should be carefully remembered that these deductions suppose a previous total ignorance of nature" [13, p. 313].

It is Price's opinion that these rules assist us in determining a pattern in the regular changes in natural objects and we perceive the Creator's wisdom in this manner.

Next he discusses an example of drawing in a lottery: Let eleven lots be drawn among which there is one prize and ten blanks. It is assumed that the ratio of blanks to prizes in the lottery lies somewhere between $\frac{9}{1}$ and $\frac{11}{1}$. What is the probability of this assumption? Here

$$X = \tfrac{11}{12}, \qquad x = \tfrac{9}{10}, \qquad p = 10, \qquad q = 1, \qquad n = 11, \qquad E = C_{11}^{1} = 11.$$

The corresponding probability is therefore

$$(n + 1)E\left[\left(\frac{X^{p+1}}{p+1} - \frac{qX^{p+2}}{p+2}\right) - \left(\frac{x^{p+1}}{p+1} - \frac{qx^{p+2}}{p+2}\right)\right]$$

$$= 12 \cdot 11 \cdot \left\{\left[\frac{(\frac{11}{12})^{11}}{11} - \frac{(\frac{11}{12})^{12}}{12}\right] - \left[\frac{(\frac{9}{11})^{11}}{11} - \frac{(\frac{9}{10})^{12}}{12}\right]\right\} = 0.07699 \approx \tfrac{1}{12}.$$

Let the same assumption be made in the case of 20 blanks and two prizes. What is the corresponding probability in this case?

$$X = \tfrac{11}{12}, \qquad x = \tfrac{9}{10}, \qquad n = 22, \qquad p = 20, \qquad q = 2, \qquad E = C_{22}^{2} = 231.$$

$$P = (n + 1)E\left\{\left[\frac{X^{p+1}}{p+1} - \frac{qX^{p+2}}{p+2} + \frac{q(q-1)X^{p+3}}{2(p+3)}\right]\right.$$

$$\left. - \left[\frac{x^{p+1}}{p+1} - \frac{qx^{p+2}}{p+2} + \frac{q(q-1)x^{p+3}}{2(p+3)}\right]\right\}$$

$$= 0.10843 \approx \tfrac{1}{9}.$$

If there were 40 blanks and four prizes, the value of P would be 0.1525, and so on.

In the case of 100 blanks and ten prizes the answer, according to Price, may still be found by the first rule, and the probability in this case would be 0.2506.

> From these calculations it appears that, in the circumstances I have supposed, the chance for being right in guessing the proportion of *blanks* to *prizes* to be nearly the same with that of the number of *blanks* drawn in a given time to the number of prizes drawn, is continually increasing as these numbers increase; and that therefore, when they are considerably large, this conclusion may be looked upon as morally certain. [13, p. 314]

Consider now under the same assumptions the case when the blanks have been drawn 1000 times, and prizes 100 times in 1100 trials. In this case the calculations based on the first rule are practically impossible. The second rule should therefore be applied. For this purpose, z should be found which will correspond to our data: $z = \tfrac{1}{100}$. Now our problem will be formulated as follows: Find the chance that the probability of drawing a blank in a single trial would lie somewhere between $\tfrac{10}{11} - \tfrac{1}{110}$ and $\tfrac{10}{11} + \tfrac{1}{110}$. In this case the deviation from the previous assigned limits will be insignificant.

The required probability is greater than

$$\frac{2\Sigma}{1 + 2Ea^p b^q + 2Ea^p b^q/n}$$

and is less than

$$\frac{2\Sigma}{1 - 2Ea^p b^q - 2Ea^p b^q/n}$$

where

$$\Sigma = \frac{(n + 1)(2pq)^{1/2}}{n^{3/2}} Ea^p b^q \left[mz - \frac{m^3 z^3}{3} + \frac{(n - 2) m^5 z^5}{2n \cdot 5} - \cdots \right].$$

In our example we have

$$p = 1000; \quad q = 100; \quad n = 1100; \quad z = \tfrac{1}{100};$$

$$mz = z(n^3/pq)^{1/2} = 1.048808;$$

$$a = p/n = \tfrac{1000}{1100} = \tfrac{10}{11}; \quad b = q/n = \tfrac{100}{1100} = \tfrac{1}{11}.$$

The required probability is greater than 0.7953 and less than 0.9405.

When 10,000 blanks and 1000 prizes are available, the second rule "as the first rule becomes useless, the value of mz being so great as to render it scarcely possible to calculate the series in Σ. In such cases the third rule must be used.

Apart from this essay no other works of Bayes on probability theory are known. The formula given in most texts under the name "Bayes' theorem" (or Bayes' formula) actually appears nowhere in Bayes' writing.† It is Laplace who was responsible for this name.

At the present time the problem investigated by Bayes presents no difficulty. It can be formulated as follows: Let the probability of event A be an unknown quantity p. Find the probability that p lies in the interval (a, b) if in n independent trials this event occurred m times and failed $n - m$ times, assuming that the probability distribution of the quantity p is uniform in the interval $(0, 1)$. The solution can be presented using the following formula:

$$P\{a \leqq p < b | A \text{ occurs } m \text{ times}\} = \frac{\int_a^b x^m (1 - x)^{n-m} \, dx}{\int_0^1 x^m (1 - x)^{n-m} \, dx}.$$

One, however, should not underestimate the value of Bayes' essay. He was the first to derive the binomial distribution "curve" and obtain all its properties. Bayes also established the rule for obtaining the probability that the required probability would lie between given limits.

---

† It appeared in Laplace's essay [98, p. 189], expressed in words only, where Bayes' paper [13] in *Philosophical Transactions* (1763) is also mentioned (*Translator's remark*).

5 Euler's and Daniel Bernoulli's contributions

Toward the middle of the eighteenth century probability theory began to be applied more and more in various areas, first and foremost in demography, insurance, estimation of observational errors, organization of lotteries, etc. A progressively larger circle of mathematicians became involved in the development of this scientific discipline. Among them was Leonhard Euler (1707–1783). Some of Euler's contributions to probability theory and its applications were published during his lifetime, some much later, and some of his works in this field appeared only in 1923 in the seventh volume of his collected works; in this volume all of his papers dealing with probability theory and its applications were assembled [56].

Euler's interest in probability theory apparently stemmed from the following circumstances. In the beginning of the seventeenth century the so-called Génoise Lottery was organized (see also [27]). In the eighteenth century this lottery became very popular in many countries of Western Europe. It was often utilized to replenish the government treasury. The Prussian King Frederick II was constantly in need of money and he readily accepted for consideration various projects to increase government funds. It was Raccolini who proposed the organization of a lottery in Prussia to King Frederick. In his letter to Euler, dated 15 September 1794, King Frederick requested Euler's advice in connection with the setting up of this lottery. In 1763 another lottery proposal was submitted to King Frederick. He then consulted a second time with Euler in a letter dated 17 August 1763. In his replies to Frederick the Great, Euler presents detailed computations of the cost of a ticket for an all-prize lottery. The lottery examined by Euler was designed as follows: Each player picks a number or series of numbers which, he believes, will occur among the five drawn tickets. Moreover, the player also indicates the pay-off he wants to receive if the chosen numbers do appear. Under these conditions Euler carries out all the necessary calculations using numerical examples.

In his memoir entitled "Sur la probabilité des séquences dans la Lotterie Génoise" Euler points out that the computations of probabilities of drawing the numbers given in advance from the five drawn numbers is well known. He therefore considers the probabilities of occurrences of sequences.†

Euler views the problem of obtaining the probability of the occurrence of a sequence as so complex that "enormous difficulties are encountered for its solution."

---

†A sequence is meant to be the occurrence of two or more consecutive numbers among the five tickets "for example, such two numbers 7 and 8 constitute a sequence of two; if there are three such numbers 25, 26, 27, it would be a sequence of three, and similarly for a larger number" [56, p. 113].

He first considers the simplest particular cases and then proceeds to the general problem [56, p. 113]:

First problem: Find the probability of the occurrence of a sequence when two tickets are drawn from n tickets. The total number of outcomes is $C_n^2 = (n - 1)n/1 \cdot 2$. The number of sequences of length two (a sequence of length k will be denoted by (k)) is equal to $n - 1$. Hence

$$P_{(2)} = (n - 1)[2/n(n - 1)] = 2/n.$$

Probability that a sequence does not occur: $P_{(1)} = 1 - 2/n = (n - 2)/2$ and the number of outcomes in which the sequence does not occur is equal to

$$\frac{n(n - 1)}{2} - (n - 1) = \frac{n(n - 1) - 2(n - 1)}{2} = \frac{(n - 1)(n - 2)}{2}.$$

Hence from this argument we also obtain alternatively that

$$P_{(1)} = \frac{(n - 1)(n - 2) \cdot 2}{2n(n - 1)} = \frac{n - 2}{2}.$$

Second problem: Three tickets are drawn. Find the probabilities of (3) and (2).

The total number of outcomes is $C_n^3 = n(n - 1)(n - 2)/1 \cdot 1 \cdot 3$. The number of (3) is $n - 2$. Hence

$$P_{(3)} = \frac{(n - 2) \cdot 2 \cdot 3}{n(n - 1)(n - 2)} = \frac{2 \cdot 3}{n(n - 1)}.$$

The number of (2) $= (n - 2)(n - 3)$, hence

$$P_{(2)} = \frac{(n - 2)(n - 3)n(n - 1) \cdot 2 \cdot 3}{2 \cdot 3n(n - 1)(n - 2)} = \frac{2 \cdot 3(n - 3)}{n(n - 1)}.$$

In the fourth and fifth problems, Euler discusses the probability of obtaining various sequences when four or five tickets are drawn.

The next section of this treatise is entitled "Applications of the Génoise Lottery." Here the following problem is considered: In the Génoise Lottery $n = 90$ and five tickets are drawn. Probabilities of occurrences of various sequences are summarized in the table:

$$\text{(5)} \quad \frac{2 \cdot 3 \cdot 4 \cdot 5}{90 \cdot 89 \cdot 88 \cdot 87} = \frac{1}{511{,}038}$$

$$\text{(4)} \quad \text{(1)†} \quad \frac{2 \cdot 3 \cdot 4 \cdot 5 \cdot 85}{90 \cdot 89 \cdot 88 \cdot 87} = \frac{85}{511{,}038}$$

†The notation $(k)\,(m) \cdots (1)$ denotes $P_{(k)\text{and}(m)\text{and}...\text{and}(1)}$.

$$(3) \quad (2) \qquad \frac{2 \cdot 3 \cdot 4 \cdot 5 \cdot 85}{90 \cdot 89 \cdot 88 \cdot 87} = \frac{85}{511{,}038}$$

$$(3) \quad (1) \quad (1) \qquad \frac{3 \cdot 4 \cdot 5 \cdot 85 \cdot 84}{90 \cdot 89 \cdot 88 \cdot 87} = \frac{3570}{511{,}038}$$

$$(2) \quad (2) \quad (1) \qquad \frac{3 \cdot 4 \cdot 5 \cdot 85 \cdot 84}{90 \cdot 89 \cdot 88 \cdot 87} = \frac{3570}{511{,}038}$$

$$(2) \quad (1) \quad (1) \quad (1) \qquad \frac{4 \cdot 5 \cdot 85 \cdot 84 \cdot 83}{90 \cdot 89 \cdot 88 \cdot 87} = \frac{98{,}770}{511{,}038}$$

$$(1) \quad (1) \quad (1) \quad (1) \qquad \frac{85 \cdot 84 \cdot 83 \cdot 82}{90 \cdot 89 \cdot 88 \cdot 87} = \frac{404{,}957}{511{,}038}$$

For example, the first probability $P_{(5)}$ is obtained as follows: The total number of outcomes is

$$C_{90}^{5} = 90 \cdot 89 \cdot 88 \cdot 87 \cdot 86 / 1 \cdot 2 \cdot 3 \cdot 4 \cdot 5.$$

The number of sequences (5) is equal to $90 - 4 = 86$. Therefore

$$P_{(5)} = 86 \cdot 2 \cdot 3 \cdot 4 \cdot 5 / 90 \cdot 89 \cdot 88 \cdot 87 \cdot 86 = \tfrac{1}{511{,}038}.$$

Using this table, Euler solves a number of related problems. One such problem is to find the probability that at least one sequence (2) is obtained. Next, Euler considers the case of six drawings and only after this discussion does he proceed with the investigation of various sequences in the case when m drawings are taken from n numbers.

In another paper "Reflections of a special type of lottery, the so-called *Génoise Lottery*" [56, p. 466] Euler also deals with problems connected with designing lotteries. This work similarly resulted from plans for organizing a state lottery to replenish treasury funds. Here the probability is determined of obtaining a particular set of tickets, when m tickets are drawn from n tickets. In this paper the problem concerning the price of a ticket is also solved.

Again Euler returned to problems of lottery design in his later paper "Solutio quarundam quaestionum difficiliorum in calculo probabilium" (1785). Here is how he describes this work:

> These problems arose from a popular game in which five tickets are drawn by lot out of 90 tickets numbered consecutively from 1 to 90. Here the following questions may be asked: what is the probability that, in a certain number of drawings, all the ninety numbers will be drawn, or at least 89 or 88 of them or a smaller number? I intend to solve these difficult problems on the basis of the long well-known principles of calculus of probabilities. I am not deterred by the objections of the famous d'Alembert who attempted to cast doubts on this calculus. After this great mathematician abandoned his mathematical

pursuits, it seems that he even declared war on them, since he proceeded to destroy many of the most firmly established basic propositions. Although these objections must have great significance for the ignorant, there is no need to fear that they may cause any damage to "science." [56, p. 408]

It is very interesting to note the appraisal of d'Alembert's role in probability theory given by Euler in the above quotation.

In 1770 Grithaus suggested to Frederick the Great a lottery of the following structure: A man may take the same ticket in five different lotteries each having 1000 prizes to 9000 blanks. Besides his chance for the prizes, he is to have one ducat returned if he wins no prize.

King Frederick commissioned Euler to investigate this lottery. The study resulted in a memoir "Solution d'une question très difficile dans le Calcul des Probabilités" [56, p. 162]. In this work Euler determines the probabilities of all the combinations which may come up in the lottery.

In a number of other papers Euler investigates problems connected with games of chance. One of his most interesting papers on this subject is the memoir entitled "Calcul de la Probabilité dans le Jeu de Rencontre" (1753) [56, p. 11]. He first describes the card game "Rencontre" (this game was discussed by Montmort and N. Bernoulli under the name "Treize"): Two players, each having a complete deck of cards, draw the cards singly until both draw the same card in which case the first of the players wins. If such a match does not occur, then the second player wins. The problem is to find the probability of a win for each of the two players.

Euler replaces the cards by tickets enumerated 1, 2, 3, 4, and so on and assumes that out of two players A and B, only A will draw the tickets in their natural order, 1, 2, 3, and so on.

He notes that the problem depends on the number of tickets and becomes more involved as the number of tickets increases. He thus proceeds with the solution for the cases when the number of tickets is small.

In particular, in the case when A and B each have only one ticket, we obtain $P(A) = 1$ and $P(B) = 0$.

Next the case of two tickets is considered. Clearly in this case the solution is $P(A) = \frac{1}{2}$ and $P(B) = \frac{1}{2}$.

If both players have three tickets each, then as A draws the tickets in their natural order 1, 2, 3, B can draw them in six possible ways. Euler constructs the following table:

A	B					
	1	2	3	4	5	6
1	1	1	2	2	3	3
2	2	3	1	3	2	1
3	3	2	3	1	1	2

All the six possible outcomes are equally probable. It is easily seen that A has four favorable outcomes to win and B two, thus,

$$P(A) = \tfrac{4}{6} = \tfrac{2}{3}; \qquad P(B) = \tfrac{2}{6} = \tfrac{1}{3}.$$

If, however, each player has four cards the corresponding table will be as follows:

A											B													
	1	2	3	4	5	6	7	8	9	10	11	12	13	14	15	16	17	18	19	20	21	22	23	24
1	1	1	1	1	1	1	1	2	2	2	2	2	2	3	3	3	3	3	3	4	4	4	4	4
2	2	2	3	3	4	4	3	3	4	4	1	1	4	4	1	1	2	2	1	1	2	2	3	3
3	3	4	4	2	2	3	4	1	1	3	4	3	1	2	2	4	1	4	2	3	3	1	1	2
4	4	3	2	4	3	2	1	4	3	1	3	4	2	1	4	2	4	1	3	2	1	3	2	1

One can easily compute that in this case $P(A) = \tfrac{15}{24} = \tfrac{5}{8}$ and $P(B) = \tfrac{9}{24} = \tfrac{3}{8}$. A similar table for the case of five cards will contain 120 columns.

Euler remarks that such a table for a large number of cards becomes unfeasible. He therefore asserts that a general rule must be established with which one can deduce the corresponding probabilities for a large number of cards on the basis of the knowledge of these probabilities for a smaller number [56, p. 14].

At the end of this memoir, Euler concludes that as n tends to infinity the chance for A is given by the infinite series

$$1 - \tfrac{1}{2} + \tfrac{1}{6} - \tfrac{1}{24} + \tfrac{1}{120} - \tfrac{1}{720} + \cdots$$

and the chance for B is given by

$$1 - 1 + \tfrac{1}{2} - \tfrac{1}{6} + \tfrac{1}{24} - \tfrac{1}{120} + \tfrac{1}{720} - \cdots.$$

It is well known that the latter series converges to $1/e$. Hence for the case of $n = \infty$, A's chance of winning is equal to $1 - 1/e$ and B's chance is $1/e$.[†]

Finally Euler observes that this[‡] result is "correct," provided the number of cards is larger than 20 and that it is very accurate in the case of the standard game of this type when a deck of 52 cards is used [56, p. 25].

We once again note the characteristic feature of the method used by

---

[†] In modern terminology this is a well-known "secretary problem" (see, e.g., Feller [195]). The result

$$1 - \frac{1}{2} + \frac{1}{3!} - \frac{1}{4!} + \cdots$$

was given by N. Bernoulli (see Todhunter [176, p. 92]), while de Moivre showed the limit of this series to be $1 - 1/e$ [176, p. 157] (*Translator's remark*).

[‡] Asymptotic (*Translator's remark*).

Euler in solving problems. First the simplest cases are considered, then somewhat more involved cases are tackled, and finally the problem in its general form is solved.

In other papers related to games of chance Euler discusses the St. Petersburg paradox, the game *Pharaon*, and other specific card games.

The development of the natural sciences, in particular those of physics and astronomy, posed the important problem of determining the most accurate value based on a large number of observations. In the first half of the eighteenth century the arithmetic mean was considered the most accurate value by the majority of investigators. But with improvements in methods of observation many sources of crude errors were eliminated and, as a result, the principle of the arithmetic mean became subject to criticism, since it was felt that a single well-executed measurement would suffice.†

In 1778 in the memoirs of the Academy of St. Petersburg, *Acta Academiae Petropolitanae* (1777), pp. 3–23, Daniel Bernoulli's memoir "The most probable choice between several discrepant observations and the formation therefrom of the most likely induction" was published.‡ In this work D. Bernoulli writes that the principle of the arithmetic mean may be applied only in the case when it is known that errors in observations are equiprobable. However, in practice this is not usually the case. In Section 2 of his paper Bernoulli remarks: "But is it right to hold that the several observations are of the same weight or moment, or equally prone to any and every error? Are errors of some degrees as easy to make as others of as many minutes? Is there everywhere the same probability?" [18, p. 3].

He compares the application of the arithmetic mean with "a blind archer aiming his arrows at a set mark." He believes that the small chance errors in observations are caused by "innumerable imperfections and other tiny hidden obstacles" [18, p. 5]. In place of the arithmetic mean he proposes an estimate later referred to as the maximum-likelihood estimate. He introduces this estimate under the assumption that if, given the occurrence of a certain event belonging to a system of n mutually exclusive events A_i each with probability p_i and if p_j is the largest of these probabilities, one should assume that the event A_j has occurred.

Bernoulli assumes that a density curve possesses the following two properties: It consists of two symmetric parts, which cross the abscissa axis in a right angle or almost a right angle; the density is zero everywhere but on a finite interval and the probability is maximal in the middle, where the tangent is parallel to the abscissa axis.

---

†Cf. the first footnote on p. 86.

‡This paper, together with a Commentary by Euler, appears in the 7th volume of Euler's collected works [56, p. 262]. The translation of these works appeared in *Biometrika* in 1961 [18].

For his density curve he chooses a semicircle, although he notes that the choice of a semiellipse, arc of a parabola, or some other suitable curve are equally acceptable. The radius of the semicircle is equal to the maximal error or to the distance between the extremal observations.

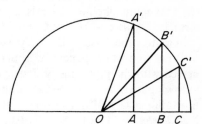

Figure 5

The likelihood function given by Bernoulli is of the form (Fig. 5)

$$[rr - xx]^{1/2} [rr - (x - a)^2]^{1/2} [rr - (x - b)^2]^{1/2} [rr - (x - c)^2]^{1/2} \cdots .$$

He, however, uses the square of this expression

$$[rr - xx] [rr - (x - a)^2] [rr - (x - b)^2] [rr - (x - c)^2] \cdots \qquad [18, \text{p. } 7]$$

where (see Fig. 5)

$$OA' = r ; \qquad\qquad BB' = [r^2 - (x - b)^2]^{1/2}$$
$$OA = x - a ; \qquad\qquad OC = x ;$$
$$AA' = [r^2 - (x - a)^2]^{1/2} \qquad CC' = (r^2 - x^2)^{1/2} .$$
$$OB = x - b ;$$

The likelihood equation even for a small number of observations becomes very cumbersome. For example, Bernoulli constructs the likelihood equation for the case of three observations which reduces to an equation of the fifth degree. He solves this equation for the particular cases when $b = 2a$ and $b = -a$, as well as some others. At the conclusion of his paper Bernoulli remarks that his estimate "gives rise to nothing which is in the least displeasing, still less leads to any absurd result" [18, p. 13].

In connection with this memoir Euler has written a paper "Observations on the foregoing dissertation of Bernoulli," which was published in the same volume of *Acta Academiae Petropolitanae* (1777) and its English translation was printed in *Biometrika* (1961), **48**, pp. 13–18 [18].

In Euler's opinion, the Bernoulli principle is not substantiated. Euler suggests a return to the principle of the arithmetic mean and computes the mean value by using the formula

$$x = \frac{a\alpha + b\beta + \cdots}{\alpha + \beta + \cdots},$$

where $a, b, \ldots$ are the observation values and $\alpha, \beta, \ldots$ are the corresponding weights; moreover, it should be assumed that

$$\alpha^2 = r^2 - (x - a)^2 ; \qquad \beta = r^2 - (x - b)^2 , \ldots .$$

This special combination of both principles of the arithmetic mean and the maximum likelihood leads to an equation of the third degree of the form $nx^3 - nrrx + 3Bx - C = 0$ [18, p. 15].

In many other cases this problem can be reduced to a quadratic equation.†

Other important papers by Euler related to probability theory are connected with the problems of demography and insurance. Among those the following two are of special interest: (1) "Recherches générales sur la mortalité et la multiplication du genre humain" published in *Mém. Acad. Sci., Berlin* **16** (1760), 144–164 (the date of publication was 1767) and (2) "Sur la multiplication du genre humain" in *Adversaria mathematica,* **H. 6** (1750–1755), 328–331.

In these memoirs Euler poses and solves many questions which later became the basis of mathematical demography. Solving problems associated with population growth, on the size of population in forthcoming years, the mortality rate, etc., Euler constructed a complete theory of mortality rates according to age distributions. He also obtained some interesting results concerning the time period for doubling the size of a given population.

Although all the problems and examples considered by Euler are formal in nature, they are of very significant value for the initial development of mathematical demography.

We present some of Euler's results in the field of demography. He was of the opinion that mortality tables cannot be considered universal: Each table is suitable for the location for which it is compiled; mortality in large cities is higher than in small towns, and mortality in the latter is higher than in rural locations. He also notes that mortality of women for certain age groups is lower than that of men, and that the mortality rate of a particular segment of individuals does not reflect the mortality rate of the population as a whole.

The significant attention devoted by Euler to demographic problems is evident from the fact that in the first chapter of his treatise "Introduction to infinitesimal analysis" [57] he presented several problems of a demographic nature. The title of the first chapter is "On exponential logarithmic quantities." Some of these problems are as follows:

The population in a certain region increases yearly in the amount of $\frac{1}{30}$ of its initial number of 100,000 individuals. What will the population

---

†More details and further developments are given in Sheynin's papers [164] and [165] (*Translator's remark*).

of the region be in 100 years? Euler's solution is: A year later the population will be

$$(1 + \tfrac{1}{30}) \, 100{,}000 = (\tfrac{31}{30}) \, 100{,}000 \; ;$$

in 100 years, $(\tfrac{31}{30})^{100} \, 100{,}000$. Taking logarithms we obtain

$$100(\log 31 - \log 30) + \log 100{,}000 = 100(0.014240439) + 5 = 6.4240439.$$

The number which corresponds to this logarithm is 2,654,874. "Thus the population of the region will increase in 100 years more than $26\tfrac{1}{2}$ times" [57, p. 95].

Some other problems discussed in this chapter are:

Let the population be doubled at the end of each century. Find the annual increase in population.

Let the population increase annually in the amount of $\tfrac{1}{100}$ of its initial number. How many years will it take for the population to increase tenfold?

Evidently all these problems were arbitrary, but they served to develop methods and procedures in mathematical demography.

In the paper "Recherches générales sur la mortalité et la multiplication du genre humain" [56, pp. 79–100] Euler introduced several important notions and made a number of significant inferences. The paper begins with a remark that recently annual figures of births and deaths in various localities are being published and that a great discrepancy has been noted among these figures. He then proceeds with the solution of six problems. We shall examine a few of them:

Second Problem: What is the probability that a person of age m years will continue to live another n years?

Let N denote the number of individuals born at the same time, let $(k) \, N$ be the number of individuals still alive after k years. Hence, (k) is the probability of survival. Using this notation, Euler gives the solution of the problem as follows: $P = (m + n)/(n)$, and the probability that an individual will die during the given period is equal to

$$P_1 = 1 - \frac{(m + n)}{(n)}.$$

In order to obtain a solution to the fourth problem Euler introduces the notion of vitality for a person of m years of age. He defines this notion as follows: the vitality is equal to $(m) - (m + z)$, where z is determined by

$$(m + z)/(m) = \tfrac{1}{2}.$$

Euler uses the numerical values of the probabilities (m) from the work of the Dutch demographer M. Kerseboom (1691–1771). Kerseboom's demographic tables were constructed for Holland. Euler notes that the tables which he applied in his paper — as well as all other tables of this kind —

are not universal and can be utilized only for the regions for which they were constructed. He observes that in Kerseboom's tables the difference in mortality between men and women is not taken into account. Euler points out the social factors in demography and especially their influence on infant mortality.

Euler also introduces (for the first time in the theory of mortality) "probable duration of life" and "the order of extinction." These notions are fundamental to the modern theory of mortality tables.

By solving these problems Euler develops the methodology in the application of mortality tables. He actually expounds on a basic part of the mathematical theory of mortality. In the first part of this memoir several problems related to insurance are also solved.

In the second part Euler discusses problems of population increase, solves various problems related to the determination of population size in forthcoming years, the number of deaths, and so on. This led Euler to the construction of mortality tables on the basis of information on the number of survivors and data on the deaths for the corresponding year distributed according to age. Several of Euler's papers are devoted to insurance. They all constitute an important contribution to the theory and practice of actuarial science.

Euler does not, however, touch on the central problems and notions of probability theory in his works on this subject. He mostly confines himself to the solution of specific problems. This characterizes the breadth of his outlook and also his interest in practical applications and needs. In the course of solving these problems Euler often exhibits remarkable skill, especially in refining his methods. The most significant mark in the applications of probability theory Euler made was in the field of demography. He laid the foundations of several basic notions of modern demography and worked out tools for their utilization.

We now proceed to an evaluation of D. Bernoulli's role in the development of probability theory.† Already mentioned was one of D. Bernoulli's memoirs devoted to the study of probability curves. Bernoulli introduced the new idea of applying differential calculus to the problems of probability theory. His main work in this area is the memoir "De usu algorithmi infinitesimalis in arte conjectandi specimen" [16].

Problems in probability, as has been noted at various points in this book, often lead to lengthy and involved calculations utilizing the rather cumbersome methods of combinatorics. Bernoulli recommends substituting for

†Some new light on D. Bernoulli's contributions has been shed recently by O. B. Sheynin [212]. Sheynin analyzes D. Bernoulli's memoir [15] devoted to the normal law and concludes that D. Bernoulli's influence on Laplace, especially concerning applications of probability theory, was comparable to that of de Moivre. The first table of the normal curve was computed by D. Bernoulli and is included in the above-mentioned work (*Translator's remark*).

these methods of solutions the operations of differential calculus. By considering the number 1 to be infinitely small, as compared to large numbers, Bernoulli obtains a series of approximate formulas. These formulas serve as asymptotic expressions which correspond to large values of the parameters occurring in a given problem. Using various examples Bernoulli shows how the application of differential calculus can simplify solutions of problems.

Toward the end of the introductory part of this memoir, Bernoulli remarks:

> Each time the situation changes as a result of a continuous chance process, as for example, cards with different numbers are drawn from an urn in succession and when the laws determining various changes resulting from this operation are investigated, it seems useful to apply infinitesimal calculus, provided each variation can be considered infinitesimally small. This is possible as long as the number of cards remaining in the urn is very large, since in this case unity can be taken as infinitesimally small. The same hypothesis was the basis of the arithmetic of infinitesimal quantities utilized by mathematicians before differential and integral calculi were discovered. However, I realize that this separately posed question requires additional elaboration and I therefore proceed to illustrate this matter by examples in which I first use the usual analysis and then carry on to an application of the algorithm of infinitesimal quantities. [16, Paragraph 1]

In Paragraph 2 Bernoulli formulates the following problem: In a bag there are $2n$ cards; two of them are marked 1, two are marked 2, two are marked 3, and so on. We draw a number of cards; we are required to calculate the probable number of *pairs* remaining in the bag.

Bernoulli first solves this problem using the standard method. Let the total number of cards be $2n$, and hence there are n pairs. After a number of cards have been drawn, let the number of cards remaining in the urn be r, and the number of remaining pairs be denoted by x. There are, therefore, $r - 2x$ "unpaired" cards. Assume now that an additional card is drawn so that $r - 1$ cards remain in the urn. The latter can be chosen either from the paired or from the unpaired (single) cards. The number of favorable cases for the singles is $r - 2x$ and for the pairs is $2x$. In the first case the number of pairs will remain x and in the second it will reduce to $x - 1$. "According to the basic rule of the art of conjecturing, we multiply the above mentioned quantities x and $x - 1$ by the corresponding number of cases and divide the sum of the products by their total number." In other words, the mathematical expectation of the remaining pairs is determined. It is of interest to note that Bernoulli refers to the method of computing mathematical expectation as the basic rule of the art of conjecturing.

The probability that there will remain x pairs is equal to $(r - 2x)/r$, the probability that $x - 1$ pairs will remain is equal to $2x/r$. The mathematical expectation of the remaining pairs is computed as follows:

$$x\frac{r - 2x}{r} + (x - 1)\frac{2x}{r} = \frac{x(r - 2x) + (x - 1)2x}{r} = x - \frac{2x}{r}.$$

"After each new draw of a card, the number of remaining pairs in the urn is reduced by $2x/r$." Actually, a decrease in mathematical expectation is implied here. On the basis of the obtained relationship, we calculate the values of x starting from $r = 2n$ and then proceed to the values $r = 2n - 1, 2n - 2, 2n - 3$, and so on. Before the first draw $x = n$, after the first draw $x = n - 1$. After the second draw x is evaluated by the formula $x = X - 2X/r$, where X is the number of pairs after the previous draw. Thus

$$x = n - 1 - \frac{2(n - 1)}{2n - 1} = \frac{2n^2 - n - 2n + 1 - 2n + 2}{2n - 1}$$

$$= \frac{2n^2 - 5n + 3}{2n - 1} = \frac{4n^2 - 10n + 6}{4n - 2} = \frac{(2n - 3)(2n - 2)}{4n - 2}.$$

This formula, $x = X - 2X/r$, is also applied in the case of further draws. After the third draw we obtain

$$x = \frac{(2n - 3)(2n - 2)}{4n - 2} - \frac{2(2n - 3)(2n - 2)}{(4n - 2)(2n - 2)} = \frac{(2n - 4)(2n - 3)}{4n - 2}.$$

After the fourth draw the value of x will be

$$x = \frac{(2n - 4)(2n - 3)}{4n - 2} - \frac{2(2n - 4)(2n - 3)}{(4n - 2)(2n - 3)} = \frac{(2n - 5)(2n - 4)}{4n - 2}$$

and so on.

Next, D. Bernoulli points out the rule of calculating the value of x after any one of the successive draws. After the $(2n - r)$th draw we have

$$x = \frac{[2n - (2n - r + 1)][2n - (2n - r)]}{4n - 2} = \frac{r(r - 1)}{4n - 2}.$$

This is the final solution of the problem. Bernoulli thus obtains here the mathematical expectation of the number of remaining pairs after the $(2n - r)$th draw.

As an illustration of the method the following example is presented: Two decks of cards—each deck consisting of 52 cards—are shuffled. We obtain 52 pairs of cards and the total number of cards is $n = 104$. 13 cards are then drawn at random and $r = 91$ cards remain. Using the above formula we obtain

$$x = \frac{r(r-1)}{4n-2} = \frac{91 \cdot 90}{4 \cdot 52 - 2} = 39\frac{78}{103}.$$

It thus follows that at first almost as many pairs are broken as the number of cards that have been drawn. After 52 cards are drawn, we obtain

$$x = \frac{52 \cdot 51}{4 \cdot 52 - 2} = 12\frac{90}{103}.$$

It therefore follows that, with some confidence to win, one can bet that there will be at least 12 pairs of cards after 52 cards have been drawn; to bet on 13 pairs would be somewhat risky, however.

D. Bernoulli then remarks that

the honorable theory of the art of conjecturing is at present ignored or underestimated. However, this science helps solve many problems in the province of morality or politics, extracts maximal benefit from human actions and directs them with foresight. [16, Paragraph 4]

Next Bernoulli writes that, as $n \to \infty$, $r \to \infty$ also, and one may disregard numbers 1 and 2 in our formula $x = r(r-1)/(4n-2)$. We thus obtain that $x = r^2/4n$. Concerning the latter equality, D. Bernoulli writes: "I will now show a method by means of which the latter equation $x = r \cdot r/4n$ can be directly and easily obtained by means of the calculus of infinitesimal quantities."

Let r decrease by the amount dr. This decrease is due either to the single cards, $r - 2x$ in total, or to the paired cards, $2x$ in total. In the first case the number of paired cards remains unchanged, i.e. $dx = 0$. In the second case, the decrease dr is due entirely to the decrease in x, i.e. $dr = dx$. "We thus have $r - 2x$ cases, in which $dx = 0$ and $2x$ cases in which the element dx is equal to dr. It thus follows from the basic rule of the art of conjectures that the true value of the element $dx = 2x\,dr/r$, i.e. $dx/x = 2dr/r$." This equation is integrated under the initial condition that if $x = n$, then $r = 2n$. Integrating, we obtain

$$\ln x = 2\ln r + \ln C ; \qquad \ln C = \ln n - 2\ln 2n ;$$

$$\ln x = 2\ln r + \ln n - 2\ln 2n ; \qquad \ln x = \ln r^2 n/4n^2 ; \qquad x = r^2/4n.$$

Q.E.D.

Next Bernoulli solves another problem using the same method. This is a certain generalization of the previous one. The cards in the urn are divided equally into two categories, i.e. black and white, and a white card corresponds to each black card and is marked by the same symbol or number.

Let the initial number of pairs be equal to n and the number of all cards be $2n$ equally divided between white and black. Let us assume, moreover, that the cards belonging to one of the categories are more likely to be drawn from the urn than the cards belonging to the other, in other words, the white and black cards are drawn from the urn in unequal numbers. Finally, let the number of black cards in the urn be s and the white t. What is the probable number of pairs remaining in the urn?

Solving the problem Bernoulli obtains the following differential equation:

$$dx = \frac{x \, ds}{s} + \frac{x \, dt}{t}.$$

The solution of this equation is $x = s/n$ (taking the initial conditions, s and t equal to n at $x = n$, into account).

Next he writes that it is easy to obtain solutions if there are not just two, but three, four, or more categories of cards. He then presents the formulas for these cases constructed analogously to those in the solutions of the previous problem.

The paper concludes with a discussion of the problem for two categories with an additional assumption on the ratio of the likelihood of drawing a black card to that of drawing a white card. The final example deals with the case of this ratio being equal to 2.

It should be noted that in this problem D. Bernoulli does not distinguish clearly enough between the mathematical expectation of a random variable and the actual values of the random variable.

Bernoulli applies the method of infinitesimal quantities in two other papers also (see [140]). In the first, the following problem is considered: In one urn there are n white balls and in the other n black balls. A ball is transferred from the first urn to the second and simultaneously from the second to the first. This operation is performed r times. It is required to determine the mathematical expectation of the number of white balls in the first urn. Solving this problem algebraically, D. Bernoulli obtains the exact answer

$$x = \tfrac{1}{2} n \left[1 + \left(\frac{n-2}{n} \right)^r \right]. \tag{*}$$

Then, using an argument based on infinitesimal analysis he obtains an approximate formula† for large n

$$x = \tfrac{1}{2} n (1 + \exp(-2r/n)). \tag{**}$$

D. Bernoulli wrote a paper on the advantages of inoculation entitled "Essai d'une nouvelle analyse de la mortalité causée par la petite Vérole, et

---

† As $n \to \infty$ the $(*)$ approaches its limiting value given by $(**)$ (*Translator's remark*).

des avantages de l'inoculation pour la prévenir," published in *Histoire de l'Académie de Paris* for 1760 (the date of publication of the volume was 1766). His arguments in this paper are based on probabilistic considerations as well as on utilization of the method of infinitesimal quantities.[†]

However, the most well-known paper of D. Bernoulli related to probability theory is his work "Exposition of a new theory of the measurement of risk"[‡] [15] in which he introduces the notion of moral expectation.

In the beginning of this paper he defines the mathematical expectation in the usual manner by means of the formula $\sum x_i p_i$. Next he introduces the notion of moral expectation (or mean utility) which resulted from the division of "wealth" of α into "wealth" x. This quantity is determined by the formula $y = b \log x/\alpha$ where $b > 0$, or in the differential form: $dy = b\,(dx/x)$. Hence the mean utility resulting from any small increase in wealth is proportional to the infinitesimal increase in wealth and is inversely proportional to the amount of wealth x previously possessed.

Let $C_1, C_2, \ldots, C_m$ be the possible values of the gain and $p_1, p_2, \ldots, p_n$ the corresponding probabilities. The mean utility of the gain is determined by the formula

$$z = p_1 b \log \frac{C_1 + \alpha}{\alpha} + p_2 b \log \frac{C_2 + \alpha}{\alpha} + \cdots + p_m b \log \frac{C_n + \alpha}{\alpha}.$$

This is the mathematical expectation of the quantity $b \log [(x + \alpha)/\alpha]$, where x is the gain and α is the wealth of the winner. We determine x from the equation $z = b \log x/\alpha$; $x = \alpha a^{z/b}$. Substituting here the value of z obtained from the above formula, we get

$$x + \alpha a^\gamma, \quad \text{where} \quad \gamma = p_1 \log \frac{C_1 + \alpha}{\alpha} + p_2 \log \frac{C_1 + \alpha}{\alpha} + p_2 \log \frac{C_2 + \alpha}{\alpha}$$

$$+ \cdots + p_m \log \frac{C_m + \alpha}{\alpha},$$

$$x = \alpha \left(\frac{C_1 + \alpha}{\alpha} \right)^{p_1} \left(\frac{C_2 + \alpha}{\alpha} \right)^{p_2} \cdots \left(\frac{C_m + \alpha}{\alpha} \right)^{p_m}$$

$$= \frac{\alpha}{\alpha^{p_1} \alpha^{p_2} \cdots \alpha^{p_m}} (C_1 + \alpha)^{p_1} (C_2 + \alpha)^{p_2} \cdots (C_m + \alpha)^{p_m}$$

$$= \frac{\alpha}{\alpha^{p_1 + p_2 + \cdots + p_m}} (C_1 + \alpha)^{p_1} (C_2 + \alpha)^{p_2} \cdots (C_m + \alpha)^{p_m}.$$

[†]See further discussion of this paper in Section 7 devoted to d'Alembert (*Translator's remark*).
[‡]Published in St. Petersburg under its original title "Specimen Theoriae Novae de Mensura Sortis" in *Novi Commentarii . . .* **V**, 175–192 (1738); English translation in *Econometrica* **22**, 23–36 (1954).

Noting that $\sum_{i=1}^{m} p_i = 1$, we finally obtain

$$x = (C_1 + \alpha)^{p_1} (C_2 + \alpha)^{p_2} \cdots (C_m + \alpha)^{p_m}.$$

The difference $(C_1 + \alpha)^{p_1} (C_2 + \alpha)^{p_2} \cdots (C_m + \alpha)^{p_m} - \alpha$ is the value of the mean utility of the gain with initial wealth in the amount of α.

We now rewrite this expression in the form

$$\alpha \left(1 + \frac{C_1}{\alpha} \right)^{p_1} \left(1 + \frac{C_2}{\alpha} \right)^{p_2} \cdots \left(1 + \frac{C_m}{\alpha} \right)^{p_m} - \alpha.$$

Expanding it using the binomial theorem under the assumption that α is large and disregarding all the terms starting with the $1/\alpha$ order, we obtain the expression for the mathematical expectation. Indeed,

$$\left(1 + \frac{C_1}{\alpha} \right)^{p_1} = 1 + \frac{p_1 C_1}{\alpha} + \frac{p_1 (p_1 - 1)}{2} \frac{C_1^2}{\alpha^2} + \cdots,$$

$$\left(1 + \frac{C_2}{\alpha} \right)^{p_2} = 1 + \frac{p_2 C_2}{\alpha} + \cdots,$$

$$\vdots$$

$$\left(1 + \frac{C_m}{\alpha} \right)^{p_m} = 1 + \frac{p_m C_m}{\alpha} + \cdots.$$

Hence:

$$\alpha \left(1 + \frac{C_1}{\alpha} \right)^{p_1} \left(1 + \frac{C_2}{\alpha} \right)^{p_2} \left(1 + \frac{C_3}{\alpha} \right)^{p_3} \cdots - \alpha$$

$$= \alpha \left(1 + \frac{p_1 C_1}{\alpha} + \cdots \right) \left(1 + \frac{p_2 C_2}{\alpha} + \cdots \right) \cdots \left(1 + \frac{p_m C_m}{\alpha} + \cdots \right) - \alpha$$

$$= \alpha \left(1 + \frac{p_1 C_1}{\alpha} + \frac{p_2 C_2}{\alpha} + \frac{p_3 C_3}{\alpha} + \cdots \right) - \alpha$$

$$= p_1 C_1 + p_2 C_2 + \cdots = \sum_{i=1}^{m} p_i C_i.$$

D. Bernoulli applies the notion of mean utility (moral expectation) to a problem which became known later as the St. Petersburg's paradox (or game), the paradox in the problem being, as noted previously, the fact that the mathematical expectation of the win in the game is infinite.

If, however, in place of the mathematical expectation, the moral expectation is applied, then a finite value of the win is obtained.

Indeed, let α be the capital of one of the players (A) before the game.

We then obtain the following expression for the moral expectation:

$$(\alpha + 1)^{1/2}(\alpha + 2)^{1/4}(\alpha + 4)^{1/8} \cdots - \alpha = P(\alpha) - \alpha. \qquad \text{(III. 7)}$$

For $\alpha = 0$, the difference (III.7) equals 2; for $\alpha = 10$, it is approximately 3; and for $\alpha = 4$, about $4\frac{1}{3}$; for $\alpha = 1000$, about 6; or, using Bernoulli's terminology: "If Paul owned nothing at all, the value is

$$1^{1/2} \cdot 2^{1/4} \cdot 4^{1/8} \cdot 8^{1/16} \cdots,$$

which amounts to two ducats precisely. If he owned ten ducats, his opportunity would be worth approximately three ducats; it would be worth approximately four if his wealth were one hundred, and six if he possessed one thousand" [15, p. 32].

The amount x which the player should pay for rights on A's position, having wealth α, is determined by the equality

$$P(\alpha - x) - a = 0.$$

In this paper Bernoulli considers another problem also:

Sempronius owns goods at home worth a total of 4000 ducats and in addition possesses 8000 ducats worth of commodities in foreign countries from where they can only be transported by sea. However, our daily experience teaches us that of ten ships one perishes. Under these conditions I maintain that if Sempronius trusted all his 8000 ducats of goods to one ship his expectation of the commodities is worth 6751 ducats. That is

$$(8000 + 4000)^{9/10} \cdot (0 + 4000)^{1/10} - 4000 = (12000^9 \cdot 4000^1)^{1/10} - 4000.$$
$$\text{(III.8)}$$

If, however he were to trust equal portions of these commodities to two ships the value of his expectation would be

$$(8000 + 4000)^{81/100}(4000 + 4000)^{18/100} + (0 + 4000)^{1/100} - 4000$$
$$= (12000^{81} 8000^{18} 4000)^{1/100} - 4000, \qquad \text{(III.9)}$$

i.e., 7033 ducats. Here $\frac{81}{100}$ is the probability that both ships will arrive safely: $\frac{18}{100}$ is the probability that one of the ships will perish, and $\frac{1}{100}$ is the probability that both will perish. [15]

The value of (III.9) is larger than that of (III.8). From here D. Bernoulli concludes that the addition in the number of ships increases the moral expectation.†

Problems connected with moral expectation were further investigated

† See also [171].

by Laplace, Poisson, Lacroix, V. Ya. Bunyakovskiĭ, and others. This notion was criticized on various occasions. For example, Bertrand in his "Calcul des probabilités" ironically remarks:

> The theory of moral expectation became a classical theory, the term classical being used advisedly. This theory was studied, taught, and described in truly famous books. But here its success ended, it was not practically utilized and nothing workable came out of it. [23]

6   G. L. Buffon

In the eighteenth century natural scientists attempted for the first time to apply probability theory to verify certain assertions in their fields. One of the first investigators utilizing this approach was the French scientist G. L. Buffon (1707–1788). His fundamental work "Histoire Naturelle"† which greatly influenced his contemporaries and impressed everyone with its vastness, consisting of 44 volumes, 36 of which he wrote in their entirety.

Buffon utilizes elementary notions of probability theory to justify his hypothesis concerning the origin of the planets. According to this hypothesis‡ the six planets known at that time (Mercury, Venus, Earth, Mars, Jupiter, and Saturn) originated as the result of a collision between the sun and a comet. Buffon observes that these planets all have certain features in common and enumerates these properties:

> The first is that the mutual direction of the six planets is from West to East; the odds are therefore 64 against 1, that the planets will not rotate in the same direction, unless there is a single cause for this type of movement; this assertion is easily deduced from the rules of calculus of probabilities [36, p. 131]

Indeed, if we assume that the movement of any one of the planets around the sun is equiprobable in each direction, then the probability that the six planets will rotate around the sun all in the same direction equals $(\frac{1}{2})^6 = \frac{1}{64}$.

The next feature noted by Buffon is that the inclination of the planets does not exceed $7\frac{1}{2}°$, and the odds are therefore $24:1$ $(7.5° = \frac{1}{24}180°)$ that two planets will be located on planes which intersect in an angle not exceeding 7.5°. And the probability that the other four randomly chosen planes will intersect with the first plane in an angle not exceeding 7.5° is equal to $(\frac{1}{24})^5 = 1/7692624$ (see Fig. 6).

---

†An English translation under the title "Buffon's Natural History" was published in London in 1809 by H. S. Symond (*Translator's remark*).

‡This hypothesis appeared for the first time in Buffon's treatise "Epochs of Nature" in 1740.

Figure 6

Buffon concludes from these observations that there must be a common cause in the origin of the planets and attempts to prove that such a cause could only be a collision between the sun and a comet.

In his works Buffon applies the following type argument: The probability of a certain objectively existing phenomenon (i.e. the probability of the movement of all planets in one direction) is calculated; if this probability is small, the assertion can then be made that this event is not random but subject to a certain regularity, and this regularity should be determined.

This type of argument was applied by natural scientists even at a later period. For example, Kirchhoff (1824–1887) investigated the spectrum of iron consisting of 60 light lines. He discovered that each of these lines in the spectrum of the iron coincides with a certain dark line in the solar spectrum. Kirchhoff then wondered whether these coincidences were due to chance. He found that one cannot distinguish between the lines if the distances between them is less than $\frac{1}{2}$ mm. On the scale on which the spectra were observed the distance between two adjacent lines of the solar spectrum was 2 mm (see Fig. 7). Therefore, if the 60 lines of the iron spectrum are

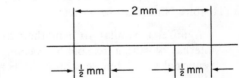

Figure 7

independent of the dark lines of the solar spectrum, then the probability that each one of them is less than $\frac{1}{2}$ mm distant from any one of the lines of the solar spectrum would be equal to $(\frac{1}{2})^{60}$. Thus Kirchhoff concludes that the coincidence of the lines of the iron spectrum and the solar spectrum is due to a specific reason which should satisfactorily explain this phenomenon.

Poincaré considers this type of argument quite plausible. He writes in this connection:

> Let rolling pebbles be left subject to chance on the side of a mountain, and they will all end by falling into the valley. If we find one of them at the foot, it will be a commonplace effect which will teach us nothing about the previous history of the pebble; we will not be able to know its original position on the mountain. But if, by accident, we find a stone near the summit, we can assert that it has always been there,

since, if it had been on the slope, it would have rolled to the very bottom. And we will make this assertion with the greater certainty, the more exceptional the event is and the greater the chances were that the situation would not have occurred. [152, p. 8]

A well-known problem associated with Buffon is "Buffon's needle problem" (1777): A plane is ruled with a series of parallel lines a distance *a* apart. A needle of length *l* is tossed onto the plane. What is the probability that the needle will intersect one of the lines?

If $l < a$, then the required probability is $p = 2l/\pi a$.†

Many investigators (including Buffon)‡ used this result for the experimental determination of the value of π. They carried out a given number of tosses of the needle (n) and calculated the number of intersections with the parallel lines (m). By approximating the probability p by the frequency m/n we can compute the value of π from the relation $p = 2l/\pi a$.

Using this method Wolf, based on 5000 tosses, obtained the value of π to equal 3.1596. In 1855 Ambrose Smith, of Aberdeen, made 3204 trials with a rod three-fifths of the distance between the planks. His experiment resulted in the value of $\pi = 3.1553$. In 1901, Lazzerini carried out 3408 analogous trials and obtained the value of π correct to six digits.

Buffon also advocated the introduction of the notion of moral expectation on a par with mathematical expectation. In his "Essai d'Arithmétique Morale" which appeared in 1777 in the fourth volume of "Supplément à l'Histoire Naturelle" he writes:

> A miser is similar to a mathematician, both value money for its internal worth; a reasonable person does not care about its conventional value, but sees only the benefit that he can derive from it. His argument is more sound than that of a miser and he feels better than a mathematician. A thaler saved by a poor man to pay his legal obligations and a thaler which fills up the bags of a money-lender have the same value in the eyes of a miser and a mathematician: the first adds it to his collection with equal pleasure, the second considers them to be equal units; while a reasonable man appraises the thaler of a poor man as a golden coin and that of the money-lender as a penny's worth. [37]

Buffon assumes that there are circumstances under which moral expectation should be allowed for. Consider the following example. Two persons possess equal fortunes of 100 thalers each and stake one-half of it, i.e. 50 thalers, in a dice game: The winner increases his fortune by one-third, since he accumulates 150 thalers, while the loser decreases his fortune twofold

---

† The derivation is given in [69, p. 49].
‡ Who tried 2048 tosses.

since he remains with 50 thalers. Buffon thus concludes that the game is in essence unfair to the players.

According to Buffon, the measure of the "moral gain should be the ratio of the acquired sum to the total amount of the capital. Let A be the total amount of the capital and a the expected change. The moral gain in case of loss in the amount a is equal to a/A and in case of gain $a/(A + a)$, the difference being

$$\frac{a}{A} - \frac{a}{A + a} = \frac{a^2}{A(A + a)}.$$

Buffon was engaged in various problems connected with probability theory. In practical problems, he suggests that events with probability of occurrence near 1 should be considered certain and those with probability near 0 impossible. For example, he considers probabilities of order 0.0001 as probabilities of impossible events, since a healthy person of 56 years of age is certain to survive the next 24 hours, although statistical data shows that the probability of death for such a person during this period is equal to 0.00001. D. Bernoulli in his letter to Buffon of 19 March 1762, observes that the available statistical data do not distinguish between healthy and sick people in a given age group; therefore, the probability of death in the course of a day for a healthy 56-year-old person is less than 0.00001. Buffon agrees that this probability should possibly have been decreased, but his general conclusion remains unaltered.†

Buffon devoted considerable time and effort to mortality tables, investigating the tables for 15 parishes (12 in France and three specifically in Paris) compiled by M. Dupré of St. Maur. These detailed tables which serve as a source for determination of "probabilities of the duration of life" are presented in Buffon's treatise on natural history [37, p. 153]. Based on these tables he computes the probabilities of the "duration of life" for each age group, interpreting this term as a period of time for which the probability of survival during this period is $\frac{1}{2}$. For example, at the age 40 this period is 22 years and 1 month (see Table I).

Buffon remarks that "the knowledge of probabilities for continuation of life is a very important subject in the natural history of humanity" [37, p. 209]. However, it should be noted that Buffon often manipulates not with probabilities, but with ratios of probabilities. For example, it is shown in a table that the number of newborns in a given year is 23,994 and the number of infant deaths during the first year of their life is 6454. Buffon concludes from these data that "one can assert with probability and wager 17,540 against 6454 or by reducing the numbers, $2\frac{3}{4}$ against 1 that a newborn infant will live one year" [37, p. 217]. Indeed, 17,540 = 23,994 − 6454, which is the number of

---

†Condorcet (see p. 129ff) proposes another numerical estimate of this probability (see, e.g., Todhunter [176, p. 387]) (*Translator's remark*).

TABLE I

TABLE OF THE PROBABILITIES OF THE DURATION OF LIFE[a]

Age	Duration of life		Age	Duration of life		Age	Duration of life	
Years	Years	Months	Years	Years	Months	Years	Years	Months
0	8	0	29	28	6	58	12	3
1	33	0	30	28	0	59	11	8
2	38	0	31	27	6	60	11	1
3	40	0	32	26	11	61	10	6
4	41	0	33	26	3	62	10	0
5	41	6	34	25	7	63	9	6
6	42	0	35	25	0	64	9	0
7	42	3	36	24	5	65	8	6
8	41	6	37	23	10	66	8	0
9	40	10	38	23	3	67	7	6
10	40	2	39	22	8	68	7	0
11	39	6	40	22	1	69	6	7
12	38	9	41	21	6	70	6	2
13	38	1	42	20	11	71	5	8
14	37	5	43	20	4	72	5	4
15	36	9	44	19	9	73	5	0
16	36	0	45	19	3	74	4	9
17	35	4	46	18	9	75	4	6
18	34	8	47	18	2	76	4	3
19	34	0	48	17	8	77	4	1
20	33	5	49	17	2	78	3	11
21	32	11	50	16	7	79	3	9
22	32	4	51	16	0	80	3	7
23	31	10	52	15	6	81	3	5
24	31	3	53	15	0	82	3	3
25	30	9	54	14	6	83	3	2
26	30	2	55	14	0	84	3	1
27	29	7	56	13	5	85	3	0
28	2	0	57	12	10			

[a] From Buffon's "Natural History," Vol. I, p. 99, published in New York by Leavitt and Allen in 1865.

infants surviving up to age 2. Hence, the probability of dying during the first year of life equals $p = 6454/23,994$; the probability of surviving up to the second year is $q = 1 - p = 17,540/23,994$. Dividing q by p we obtain the number given by Buffon:

$$q/p = 17,540/23,994 : 6454/23,994 = 2\tfrac{3}{4}.$$

Analogous tables are calculated for different durations of life. For example, one can wager 15,612 against 8832 or about $\tfrac{3}{4}$ against 1 that a newborn infant will live two years . . . 23,030 against 964 or 24 against 1 that he will not survive 78 years [37, pp. 217, 219], and so on. These tables are computed for each age and occupy more than 100 pages. Some assertions in the commentaries for these tables are in the form of numerical problems. For example, "the odds are 11 to 4 that a newborn infant will survive one year, but will not survive up to 47 years," or "if the father is already 40 years old, then the odds are 3 to 2 that his son, who by then is one year old, will outlive him" [37, pp. 221, 228], and so on.

It is a well-known historical anecdote that to confirm the validity of Bernoulli's theorem, Buffon tried 4040 throws of a coin. In this experiment the heads appeared 1992 times and tails 2048 times.

Buffon serves as an example of a well-known natural scientist in the eighteenth century who began to apply probabilistic concepts in his investigations. However, not all the scientists welcomed probability theory favorably, despite its obvious successes.

7 Opposition from d'Alembert

Jean d'Alembert's (1717–1783) rich scientific legacy includes several works devoted to probability theory [3]. He also touches on topics in probability theory in many of his letters [5].

However, d'Alembert's name appears in the literature on probability theory mainly as an example of the fact that even certain prominent mathematicians sometimes committed errors in solving elementary probabilistic problems.

As an example we present a quotation from a textbook on probability theory.

> A coin is tossed twice. What is the probability that heads will appear at least once? D'Alembert's argument is as follows: Heads may appear either in the first toss, or in the second toss, or in neither toss. There are three cases in total, among those two are favorable to the expected event, therefore the required probability is $\frac{2}{3}$. [28, pp. 35–36]

D'Alembert does not distinguish here between equiprobable and non-equiprobable cases.

Kolman explains the cause of this error by the fact that the basic ideas in probability theory are not natural to classical mathematics and therefore some mathematicians, in particular d'Alembert, committed very elementary errors [89, p. 230]. Other authors on the history of probability do not discuss at all the reasoning behind such an error.†

---

†T. L. Hankins in his new biography of d'Alembert presents the following opinion on this historical anecdote: "It must be recognized that d'Alembert's criticisms, no matter how badly he expressed them, were directed against the unproved assumption that the mathematical laws of probability accurately described physical reality" [198, p. 148]. See also *Mathematical Gazette* (February 1971, pp. 71–72), where a recent case of d'Alembert's type of error is discussed (*Translator's remark*).

D'Alembert, together with Diderot, published and edited the famous "Encyclopédie." The first edition of this monumental work consisted of 35 volumes which appeared during the years 1751–1780. After publication of the seventh volume in 1757, d'Alembert disassociated himself from the "Encyclopédie" (as a result of the widening rift between his and Diderot's views on the foundation of science and scientific methods). However, he had written many articles for the first seven volumes, among them articles on probability theory. The first one chronologically was the article "Croix ou Pile" ("Heads or Tails") published in the fourth volume of the "Encyclopédie" [2, pp. 512–513]. In this article d'Alembert solves the problem of the probability of obtaining heads twice during two tosses of a coin. His solution is incorrect, since he assumes that there are only three cases to be considered, and not four. If a tail occurs on the first throw, then there is no need to throw the coin again, since two heads cannot occur in this case— which is, according to d'Alembert, the first case. The second case is in obtaining heads on both tosses $(+, +)$. The third case is in obtaining a head on the first toss and a tail on the second $(+, -)$. Therefore, according to d'Alembert, the required probability is $\frac{1}{3}$. He misses the point that these three cases are not equally likely. Clearly, there are actually four equiprobable cases: $(+, +)$, $(+, -)$, $(-, +)$ and $(-, -)$, and the required probability is $\frac{1}{4}$.

Next in the same article, d'Alembert computes the probability that in the course of three tosses of a coin, heads appear at least once. Here there are eight equiprobable cases, in which seven are favorable to our event and therefore the required probability is $\frac{7}{8}$. D'Alembert, however, constructs the following table:

(1)   heads;
(2)   tails, heads;
(3)   tails, tails, heads;
(4)   tails, tails, tails.

He assumes that, if on the first or second throws a head appears, then all the remaining tosses are not required. Not realizing that these four cases are not equally probable, d'Alembert erroneously asserts that the required probability is $\frac{3}{4}$ and not $\frac{7}{8}$.

In the same article, d'Alembert deals with the St. Petersburg problem and comments in the following manner on D. Bernoulli's solution of this problem: "We don't know whether his solution is satisfactory; here is a case of a scandal which deserves the attention of mathematicians" [2, p. 512]. However, he himself tends to agree with the possibility of infinite expectations.

In the seventh volume of the "Encyclopedie." d'Alembert presents Necker's objections to his article "Croix ou Pile." Necker (professor of mathematics in Geneva) communicated his objection in a letter; he points out that the three cases considered by d'Alembert are not equally probable. He then presents a correct solution of the problem and remarks at the end of the letter that d'Alembert's opinions on this problem are unacceptable, since they lead to an obvious error.

In another article, "Absent," d'Alembert alludes to an essay by Nicholas Bernoulli. Based on the mortality tables published by Antoine Deparcieux (1703–1768), he concludes that a missing person, of whom no findings have been received, should be considered dead at the age of 75. Next he asserts that, if one accepts Buffon's remark concerning the moral probability of events, a probability less than 0.0001 should be considered physically impossible.

While other articles in the "Encyclopédie" written by d'Alembert, such as "Advantage," "Bassette," "Carreau" (a game of chance), "Die," "Lotterie," "Pari," "Cards," etc., touch upon topics related to probability theory, they are, however, even less significant than those considered above.

In the second volume of his "Opuscules Mathématiques," a paper "Réflexions sur le calcul des Probabilités" [3, pp. 1–25] is presented. At the beginning of the paper, d'Alembert gives the common definition of mathematical expectation and observes that, although this rule has been adapted by all mathematicians, cases come up for which the rule does not hold. An obvious example of a problem for which the notion of mathematical expectation is inapplicable is, in d'Alembert's opinion, the St. Petersburg problem. In this connection he writes: "No one would be willing to pay even a modest sum of money for the right to enter this game, not to mention an infinite amount" [3, p. 2].

Next he discusses the distinction, which should always be considered, between the metaphysically possible and physically possible. To clarify these notions, d'Alembert presents the following example:

It is metaphysically possible to throw two sixes with two dice a hundred times running; but it is physically impossible because it has never happened and never will. This is in d'Alembert's opinion a confirmation of the fact that a very small chance is to be regarded and treated as zero.

D'Alembert proposes to determine the probabilities of events on the basis of an experiment. Here he assumes that if during three tosses of a coin heads appears three times consecutively, it is then more probable that the next toss will result in a tails. He presents these deliberations as an argument against the basic well-established assumptions of probability theory.

He concludes this paper with the following remarks:

> We conclude from these reflections: (1) If the rule I gave in the "Encyclopédie" (not knowing a better one) for the determination of

probabilities in the game with heads and tails is not sufficiently precise, the commonly accepted rule for determination of this probability is even less accurate. (2) In order to arrive at a satisfactory theory of the calculus of probabilities, one must solve several problems which may possibly be insoluble, namely, to establish the probabilities in not equally likely cases; determine when the probability should be considered nonexistent; and, finally, to determine how the expectations or stakes should be estimated depending on the probability being smaller or larger. [3, p. 24]

In spite of his generally negative attitude toward probability theory, d'Alembert posed several questions of basic importance: (1) How to define probability when equally likely cases are not available. (2) What probabilities may be disregarded, the meaning of this operation, and so on.

In a memoir on inoculation entitled "Essai d'une nouvelle analyse de la mortalité causée par la petite Vérole, et des avantages de l'Inoculation pour la prévenir" (1770), D. Bernoulli concludes that inoculation against smallpox increases the average duration of life by three to four years. D'Alembert criticizes Bernoulli's conclusion in his article "Sur l'application du Calcul des Probabilités à l'inoculation de la petite Vérole" [4]. He does not deny the advantages of inoculation, but feels that its advantages and disadvantages were not compared properly by Daniel Bernoulli and that the former is overestimated. In d'Alembert's opinion, Bernoulli regarded the subject as related to interests of the state, and showed that inoculation was to be recommended because it augments the mean duration of life. However, d'Alembert considers the subject as it relates to a private individual, who may consider the advantage of gaining three or four years of his life versus the possible immediate risk due to an inoculation. The relative value of these two alternatives is too indefinite to be estimated. In this connection, d'Alembert presented a paper on 12 November 1760, at the Academy of Sciences in which he denounced D. Bernoulli's ideas. He writes:

I assume that a 30-year-old person is due to live another 30 years and he may rely on nature to live out the remaining 30 years without inoculations. I then assume that after the treatment the mean expectancy of life would be 34 years. But would it not seem necessary—in order to estimate the advantages of inoculations—to compare not only the mean life expectancy of 30 years with that of 34 years, but also the risk in the odds of 1 to 200 of dying in the course of a month due to inoculation with the remote advantage of living four years longer after age 60? [4]

D'Alembert points out the existence of two different methods of estimating the life expectancy for a person of a given age. The first estimate is the mean life expectancy, and the second the probable life expectancy (or the probable duration of life), which is such a duration of life that it has equal chances of not being reached as of being exceeded. D'Alembert does not suggest that we distinguish between these two notions. His idea is that each one of them may serve for the determination of the expected duration of life. This kind of consideration he advances against probability theory, noting that this theory presents two answers to the very same question.

In Volume 1 of d'Alembert's "Oeuvres" [7] (Paris, 1821–1822) his Réflexions sur l'Inoculation" and "Doutes et Questions sur le Calcul des Probabilités" were published. These essays were an alternative exposition of the above-mentioned works adapted for readers less versed in mathematics. Concerning probability theory in general, d'Alembert writes:

> I was the first who dared to express doubts concerning certain basic principles of this theory. Certain great mathematicians found these doubts worthy of attention, other great mathematicians found them to be absurd (why should I soften the expression they used?). The problem is to find out whether they were wrong or not, or by applying these principles whether they were doubly wrong. Their conclusion, which they did not find it necessary to substantiate, added courage to some mediocre mathematicians who hastened to write about this problem and attack me without hearing my arguments. I shall try now to explain as clearly as possible so that my readers will be able to judge me. [7]

It should be noted that the arguments presented in these works against probability theory repeat arguments of the previous papers and actually do not contain new material.

In the fourth volume of "Mathematical Works" ("Opuscules Mathématiques") [5] extracts from d'Alembert's correspondence concerning probability theory are presented. Here such topics as the St. Petersburg game, analysis of specific games, and problems of duration of life, inoculation, and others are discussed. In one of these letters, d'Alembert objects to the notion of mathematical expectation which was already commonly accepted.

> Let someone be asked to choose one combination out of 100. Let among these 100, 99 result in a gain of 100 francs each, while the hundredth results in a gain of 99,000 francs. Who then would be so unreasonable as to prefer the combination which gives 99,000 francs? The expectations in both cases are actually different, although they are the same according to the rules of probability theory. [5]

D'Alembert in his letters repeatedly reiterates his already familiar objections to probability theory: "My objections were formed about thirty years ago, when reading the excellent book of M. Bernoulli."

Returning to the solution of the problem of obtaining two heads in two tossings of a coin, d'Alembert writes:

> If the three possible cases in this game are not equally probable, then it seems to me that this is not due to the fact to which it is usually attributed, that the probability of the first is $\frac{1}{2}$, and the probabilities of the two others are $\frac{1}{2} \cdot \frac{1}{2}$ or $\frac{1}{4}$. The more I think about this, the more convinced I become that mathematically speaking these three tosses are equally probable. [5]

He also asserts that, for purposes of computing probabilities, tossing one coin successively n times is not the same as tossing n coins simultaneously.

In one of his letters d'Alembert considers the following problem: What is the number of successive tosses of a die required in order to be able to lay a bet that a given number of points will occur? He observes that, according to the accepted rules, such a bet can be gainfully laid with four tosses. Indeed, $(\frac{5}{6})^4 < \frac{1}{2}$ and $1 - (\frac{5}{6})^4 > \frac{1}{2}$, but $(\frac{5}{6})^3 > \frac{1}{2}$ and $1 - (\frac{5}{6})^3 < \frac{1}{2}$. But then he asserts that, according to a conclusion reached by some player, the theory here does not correspond to practical reality.

Next, d'Alembert refers to letters he received from three mathematicians. One of them is characterized as a very well-learned author who contributed successfully to the field of mathematics, and who is known for his excellent work in philology. His correspondent writes [5]: "Everything that you have said concerning probility theory is excellent and very obvious; the old calculus of probabilities is now destroyed." D'Alembert remarks in this connection: "I do not claim at all to destroy the calculus of probabilities; I wish it to be amended and improved."

The seventh volume of his collected works, published in 1780, contains d'Alembert's memoir "Sur le calcul des Probabilités" [6]. This treatise is devoted mainly to the St. Petersburg problem. He begins the exposition with the following statement:

> I wish to apologize to the mathematicians for my return to a discussion of this problem. But I must admit that the more I think about it the more convincing become my doubts concerning the principles of the generally accepted theory. I wish these doubts to be clarified and that the theory—whether certain principles of the theory are changed or whether it is retained as it is—will in future be set forth in a manner that will at least eliminate obscurities. [6]

D'Alembert proposes the following completely arbitrary and unjustified solution. He assumes that, if a head occurs on the first toss, then the chances of its occurrence on the second are $(1 + \alpha)/2$, but not $\frac{1}{2}$. If a head occurs on both the first and second tosses, the chances for its occurrence on the third toss are $(1 + \alpha + \beta)/2$, and so on. Here $\alpha, \beta, \ldots$ are positive numbers and their sum is less than 1. He concludes his solution with the following remark:

> This is sufficient to show that all the terms of the stake starting from the third toss decrease. We have proved that the whole stake, which is the sum of these terms, is finite even under the assumption of an infinite number of requirements. Thus the result of the solution of the St. Petersburg problem presented here is not subject to the insurmountable difficulty of ordinary solutions. [6]

It is of interest that Diderot wrote a supplement to d'Alembert's article "Absent,"† published in the supplementary volume of the "Encyclopédie" (1776).

It should be emphasized that d'Alembert was not the only one who objected to the basic principles of probability theory. The whole history of this science is full of struggles—against distortions, against unjustified applications to the laws of social development, for recognition as a bona fide mathematical discipline, and against idealistic falsifications and misinterpretations.

8   Jean Antoine de Caritat, Marquis de Condorcet

Condorcet (1743–1794) was a political leader during the period of the French revolution, well known as a sociologist and economist, but less known as a mathematician. However, he became a member of the Paris Academy of Sciences on the strength of his contributions to mathematics.

Most significant of his works in probability theory is the "Essai sur l'application de l'analyse à la probabilité des décisions rendues à la pluralité des voix" which was published in 1785 in the "History of the Royal Paris Academy of Sciences," Volume IV, for the years 1781–1784. The treatise consists of a "Preliminary Discourse" and a five-part essay.

In the "Preliminary Discourse" Condorcet presents the basic results of his book in a form suitable for laymen and briefly expounds on the basic principles of probability theory.

The content of the first part is described in the first paragraph:

> In the first part we assume that the probability of a decision by each of the voters is known and the probability of a decision reached by the majority of voters under different conditions is to be determined; at first only one session is considered, at which one vote is taken (it is

---

†Dealing with the division of property of a man who has been missing for a long period of time (*Translator's remark*).

assumed that the same session continues to vote until the required majority of votes is obtained); next a decision based on the results of several sessions is considered, assuming that a choice is made between two opposing propositions or one of three propositions; finally, the choice is made from among several people or several objects, whose qualities remain to be determined. [176]

Next, in the first part of the treatise 11 hypotheses are considered, some of which are discussed below.

Condorcet's first hypothesis: "Let there be $2q + 1$ voters who are supposed exactly alike as to judgment; let v be the probability that a voter decides correctly, e the probability that he decides incorrectly, so that $v + e = 1$: it is required to find the probability that there will be a majority in favor of the correct decision of a question submitted to the voters" [176, p. 353]. (We may observe that the letters v and e are chosen as the initial letters of the words *vérité* (truth) and *erreur* (error).)

In the second hypothesis it is required under the same conditions to determine the probability of a given plurality of votes, etc.

Condorcet appears to have considered the ninth hypothesis of primary importance. He writes:

> Up till now only one tribunal was considered. However, in many countries the same case is decided by a series of tribunals or must be decided several times by the same tribunal each time following new rules until a definite number of identical decisions is obtained. This type of situation can be subdivided into several cases which we shall now investigate individually. Indeed, one may require: (1) unanimity of decisions; (2) a given majority, either in the absolute value or in relation to the number of decisions rendered; (3) a given number of consecutively coinciding decisions. [176, p. 36]

In this hypothesis Condorcet examines the probability of the correctness of a decision which has been confirmed in succession by an assigned number of tribunals in a series to which the question has been referred. For this purpose, he solves several problems which are of interest on their own.

Problem: Suppose that the probability of an event in a single trial is v, what is the probability that in r trials the event will occur p times in succession?

This problem was also considered by de Moivre [133, problem 74].

Problem: Find the probability that in r trials the event will occur p times in succession before it fails p times in succession.

Problem: Find the probability that the event which occurred in $m + n$ trials m times, will occur in the succeeding $r + t$ trials r times.

Condorcet obtains the required probability in the latter problem as follows:

$$p = \frac{(m + n + 1)!\,(m + r)!\,(n + t)!}{m!\,n!\,(m + n + r + t + 1)!}.$$

As a corollary we obtain from this result that the probability of an occurrence of an event in the next trial, if it occurred in the preceding m trials, is equal to

$$p = \frac{m + 1}{m + 2}.$$

This type of problem was later studied by Laplace. In particular, he comments about the last result: "Thus we find that an event having occurred successively any number of times, the probability that it will happen again next time is equal to this number increased by unity divided by the same number, increased by two units" [98, p. 19].

The absurdity of this conclusion is obvious. For $m = 1$, we obtain $p = \frac{2}{3}$. If we perform an experiment which results in a certain outcome, it will then follow from the given rule that, in repeating this experiment, the probability of the same outcome is equal to $\frac{2}{3}$. For example, in an urn there is an unknown number of labeled cards. If the number 597 is drawn at the first draw, the probability that the same number will appear in the second draw (after we return the card to the urn) is $\frac{2}{3}$, according to the above rule. The same conclusion is "valid" for any number which was drawn on the first draw.

A prominent contemporary mathematician, G. Pólya, comments about this rule:

> Let us attribute numerical values to m and let us not neglect everyday situations. ... In a foreign city where I scarcely understood the language, I ate in a restaurant with strong misgivings. Yet after ten meals taken there I felt no ill effects and so I went quite confidently to the restaurant the eleventh time. The rule said that the probability that I would not be poisoned by my next meal was $\frac{11}{12}$.
>
> A boy is ten years old today. The rule says that, having lived ten years, he has the probability $\frac{11}{12}$ to live one more year. The boy's grandfather attained 70. The rule says that he has the probability $\frac{71}{72}$ to live one more year.
>
> These applications look silly, yet the rule can beat even this absurdity. Let us apply it to the case $m = 0$: the conclusion is as valid for this case as for any other case. Yet for $m = 0$ the rule asserts that any conjecture without any verification has the probability $\frac{1}{2}$. Anybody can invent examples to show that such an assertion is monstrous. (By the way, it is also self-contradictory.) [156, p. 136]†

---

†From G. Pólya, "Patterns of Plausible Inference" (Vol. II of "Mathematics and Plausible Reasoning"), revised edition with Appendix (copyright © 1968 by Princeton University Press). Reprinted by permission of Princeton University Press.

In the tenth hypothesis Condorcet assumes that the voters may also abstain from voting.

The eleventh hypothesis deals with selecting one of three candidates for office.

In all the problems of the first part three quantities are given: the number of voters, the required majority of votes, and the probability of correctness of each of the votes. In the second part it is assumed that only two of these quantities are known as well as the result of the vote. Under this assumption several problems are solved. As we have already seen, some mathematicians advocated the idea of moral expectation and moral certainty. In particular, Buffon was one who advocated these ideas. In the second part of his treatise, Condorcet criticizes Buffon's proposition on moral certainty. He writes:

> This opinion is incorrect in essence, since it leads to a mixture of two things of a quite different nature—probability and certainty; it is similar to mixing an asymptote to a curve with a tangent at a very distant point. Such propositions cannot be allowed in the exact sciences without destroying their accuracy. [176, p. 376]

In the same part of the thesis, Condorcet also discusses the subject of mathematical expectation, referring to the works of Daniel Bernoulli.

The purpose of the third part of his essay is described as follows:

> It should contain the investigation of two different problems. The first is concerned with the determination based on observations of the probability of a court decision or the vote of each one of the jurors; the second deals with the determination of the degree of required probability in order to be able to proceed under various circumstances either reasonably or justly. But it is evident that an investigation of these two problems first requires the establishment of general principles from which one can determine the probability of a future or unknown event not on the basis of the number of possible combinations realized in this event for the complementary event, but only on the basis of the knowledge of order of familiar or previous events of this kind. This is the subject matter of the problems below. [176, pp. 377–378]

The third part consists of 13 problems. We present some of them:

Problem 1 It is known that event A occurred (previously) m times and event N n times. It is then assumed that one of these two events did occur. Find the probability that this was event A (or event N).†

Problem 5 is unquestionably of interest. It is required, under the assumption of the first problem, to obtain the probability that (1) the probability of A is not less than the given value; (2) this probability will differ from

---

†Note the relation to Bayes' problems (*Translator's remark*).

$m/(m + n)$ by no more than a; (3) this probability will differ from $(m + 1)/(m + n + 2)$ by no more than a. If the probability of A is given, it is required to determine the limit a.

Problem 7 Suppose there are two events A and N, which are mutually exclusive, and that in $m + n$ trials A has occurred m times, and N has occurred n times. What is the probability that in the next $p + q$ trials A will occur p times and N will occur q times?

In the introduction to the fourth part, Condorcet writes:

Up to the present time we have investigated our problem only abstractly and the general propositions were quite far from reality. This part is designed to develop a method which introduces into the computations some basic data to be taken into account in order that the result obtained will be relevant in practice. [176]

Part 4 is subdivided into six questions, devoted to the capabilities and truthfulness of the voters. In these problems Condorcet attaches great importance to the unanimity of the decisions.

In the introduction to the fifth part, Condorcet remarks:

The subject of this final part involves the application of the principles developed above to certain examples. It would have been desirable that this application be made on the basis of real data, but difficulties inherent in obtaining such data—which would be out-of-reach for a private individual—have compelled us to confine ourselves to an application of these theoretical principles to simple propositions, in order to at least demonstrate the procedure to be followed by those able to obtain the actual data for such an application. [176, pp. 390–391]

Next, four examples are considered. The first deals with the structure of the tribunal in civil cases, the second of a criminal court, the third example is concerned with the choice of candidates for office, and the fourth discusses the probability of a correct decision in a large gathering.

In order to understand the method by which the basic notions and principles of probability theory were set up, as well as to indicate the pitfalls and delusions encountered by far from second-rate mathematicians of that period, Condorcet's long memoir on probability theory (Mémoire sur le calcul des probabilités"), published in six parts, is of interest.

In the first part of this memoir, Condorcet discusses the St. Petersburg problem. In the second, short part (consisting of eight pages) the problem of application of probability theory to astronomical observations is investigated. The third part starts with the following passage:

La destruction du Gouvernement féodal a laissé subsister en Europe un grand nombre de droits éventuels, mais on peut les réduire à deux classes principales; les uns se payent lorsque les propriétés viennent à changer par vente, les autres se payent aux mutations par succession, soit directe ou collatérale, soit collatérale seulement. [176, p. 395]

[The destruction of the feudal system in Europe left a large number of possible rights which can be subdivided into two principal classes: those that are acquired when the ownership changes hands through the sale of the property, and those that are acquired by means of inheritance by direct or collateral line, or by collateral line only.]

Condorcet then determines the amount of money to be paid down in order to release the property from such feudal rights over it.

The fifth part of Condorcet's memoir is entitled "Sur la probabilité de faits extraordinaires"; and the sixth, "Application des principes de l'article précédent à quelques questions de critique." The two parts are closely connected.

In the sixth part, Condorcet first discusses the notion of probabilities of events, remarking that one should not understand the notion of "proper probability"† of an event as a ratio of the number of available combinations to the total number of combinations:

> For example, if a card is selected out of ten cards and the witness tells that this is a particular card, then the probability proper of this event which should be compared with the probability originating from the evidence is not the probability of choosing this card—which is equal to $\frac{1}{10}$—but the probability of choosing this card in preference to any other given card and, since the probabilities are all the same, the probability proper will be $\frac{1}{2}$ in this case. This distinction is necessary and is also sufficient to explain the differences of opinion between two groups of philosophers. One group is of the opinion that the very same statement by a witness should produce for an unusual event the same probability as for an ordinary event. If, for example, I trust a reasonable person who tells me that a woman gave birth to a boy, I should equally believe him when he tells me that she gave birth to four boys. Others, on the contrary, are convinced that the evidence does not retain its validity for unusual events or events with small probability, and are amazed that if one ticket out of 100,000 is chosen, and a trustworthy witness asserts that the first prize is won by number 256, then no one will doubt his evidence, although one can wage the odds of 999,999 against 1 that this event will not occur. However, on the basis of the previous remark, it is clear that in the second case the "probabilité propre" will be $\frac{1}{2}$, and the evidence remains valid, while in the first case, since the "probabi-

---

† "Probabilité propre" in his terminology.

lité propre" is very small, the probability of the evidence will be almost negligible. Therefore, I suggest that the "probabilité propre" of an event should be defined as a ratio of the number of combinations which produces this or a similar event to the total number of combinations.

Thus, in the case when one of the ten cards is drawn, the number of combinations corresponding to the draw of a given card is one and the number of combinations corresponding to the draw of some other given card is also one, i.e., the "probabilité propre" is expressed by the number $\frac{1}{2}$. [176, pp. 401–402]

After presenting a number of examples, Condorcet applies these arguments to the credibility of certain statements concerning the duration of the reigns of the seven Emperors of Rome. According to some historians the entire duration of these Emperors' reigns was 257 years. Examining the credibility of this statement (under the assumption that in an elective monarchy an emperor at the date of his election will be between 30 and 60 years old), Condorcet concludes that the "probabilité propre" is about $\frac{1}{4}$.

For another statement in Roman history, namely, that augur Accius Naevius cut a stone with a razor, he computes the "probabilité propre" to be 2/1,000,000.

We thus observe that the notion of "probabilité propre" is not well grounded. In contrast to the usual notion of probability, "probabilité propre" is purely subjective and mathematically unrigorous. This notion did not survive in science. However, the fact that it arose points up the complexities in the course of the development of the notion of probability.

It should be noted that the whole direction of this memoir by Condorcet is unfounded. Not having a clear idea concerning the realm of applicability of probability theory and searching for possible applications, Condorcet stumbled onto the wrong path. All his attempts to apply probability theory to the composition of tribunals, decisions at gatherings, and so on were continued for a period of time after Condorcet (and actually they were not originated by him) and occasionally attracted serious attention from various mathematicians. However, in time, these attempts were finally rejected as being beyond the scope of probability theory. Nevertheless, one should not conclude that everything written on this subject should be completely discarded. We observe from the above survey of Condorcet's contributions that he did occasionally pose specific mathematical problems which constituted a definite contribution to the development of probability theory.

9   Laplace and his contributions to probability theory

Pierre-Simon de Laplace (1749–1827) began his scientific activity in the seventies of the eighteenth century. In 1773 he was elected adjunct of the

Paris Academy of Sciences and he became a full-fledged member in 1785.

This was a difficult period, being on the eve of the French Revolution. The activity of the Academy reflected to a certain degree the complexity of the political situation. In spite of its efforts to remain to a great degree a caste and closed organization, it was compelled from time to time to deal with problems resulting from the revolutionary situation.

On 14 July 1789, the armed masses of Paris stormed and took the Bastille, but the Academy tried not to react at all to the beginning of the revolution. It continued its activities pretending that nothing had happened. Four days after the capture of the Bastille, Laplace presented at the regular meeting of the Academy the results of his investigations on the oscillation of the plane of the Earth's orbit. But, despite these efforts of the Academy to shield itself from the surrounding train of events, it was impossible to continue in this manner any longer.

Within the Academy voices of dissent were heard with increased frequency, requesting reorganization of the statutes of the Academy in a more democratic direction.

In December 1789 a commission was elected to draft a proposal of new statutes for the Academy. The membership of this Commission included Borda, Laplace, Condorcet, and Teal; very soon a proposal was presented for consideration. But the draft was unacceptable even to the rather reactionary oriented Academy who found it unresponsive to the new ideas and the changes in the political situation.

Many members of the Academy began an active participation in the political struggle.

The only mention of Laplace's activity at that period of time is his election on 8 May 1790 as a member of the *Commission on Weights and Measures* (Bureau des Longitudes). But soon Laplace and Lavoisier were recalled from membership on this Commission, in view of lack of "republican virtues and too little hatred of royalty" [9, p. 34].

The revolution was mounting and the danger of remaining in its center —Paris—became more imminent. In the spring of 1793 Laplace therefore moved from Paris to provincial Melun.

This move marked the end of the first period of Laplace's activity, and it seems to us that the opinion of the majority of biographers of Laplace concerning his revolutionary zeal is erroneous.

Laplace offered asylum to Jean Sylvain Bailly, a well-known astronomer and member of the Paris Academy of Sciences. Bailly was wandering across France to avoid punishment for an execution that took place at Champ-de-Mars on 17 July 1791. At that time Bailly, in his capacity as Mayor of Paris, ordered the National Guard to open fire on a Republican demonstration.

When Bailly was on his way to Laplace, Melun was occupied by a division of the revolutionary army. Laplace forewarned Bailly, who had not reached

Melun as yet, about this development. But Bailly arrived in spite of the warning. Soon he was recognized by an eyewitness of the execution and was arrested at Laplace's home and shortly thereafter was guillotined by order of the Revolutionary Tribunal.†

On 20 April 1794 another close friend of Laplace, Bochard de Saron was executed, and on 8 May Lavoisier suffered the same fate.

France was under threat of invasion. Famous scientists, such as Carnot, Gaspard Monge, Jean Baptiste Fourier, Antoine François de Fourcroy, Vandermondt, and many others devoted their energies to the defense of France. Laplace, however, at that time was sitting snug in the provinces.

In the middle of 1794 the reversal in the direction of the Revolution began. On 27 July 1794, (9 Thermidor) Robespierre was executed. Only after that did Laplace return to Paris. He was invited as a professor to l'École Normale which was organized by a decree of the National Convention of 30 October 1794 for the purpose of training teachers. His lectures, given in 1795, were published as an introduction to the second edition of the "Théorie Analytique des Probabilités" [97] (in 1814) under the title "Essai Philosophique sur les Probabilités."‡ (The first edition of the "Théorie analytique des probabilités" appeared in 1812.) But even before its publication—during the period when these lectures were read—they gained widespread popularity all over France and were taken down by special stenographers.

In 1795, in place of the abolished Academy, a National Institute was established. Laplace was appointed a member of this Institute and was soon elected president of the subsection on mathematics. Napoleon Bonaparte was also a member of the Institute and friendly relations developed between them. At the beginning, however, Laplace's relation to Napoleon was restrained. When Napoleon left on his African Campaign, Laplace did not join him, despite the fact that a large group of prominent scientists, including Monge, did accompany him. After the revolt of 18 Brumaire (9 November 1799), when Napoleon discharged the Council of Five Hundred and the Directory, Laplace put his fate in Napoleon's hands, anticipating him as the dictator of France, and extended to him his full support.

In 1802, the third volume of "Celestial Mechanics" was published. Laplace dedicated this volume to "Bonaparte, the Pacificator of Europe to whom France owes her prosperity, her greatness and the most brilliant epoch of her glory."§

---

†Additional details related to this period of Laplace's life are given in F. N. David's article [190, p. 39–40] (*Translator's remark*).

‡The English translation by Truscott and Emory [98] was published by Dover in 1951. (The first English version appeared in 1902.) (*Translator's remark*)

§P. Suppes, *Proceedings of the Conference on Foundations of Probability Theory, London, Ontario, May 1973*, observes the interesting fact that there is in this systematic treatise on the solar system "no detailed analysis of error terms or any application directly of a quantitative theory of errors" (*Translator's remark*).

After Napoleon's coronation and his elevation to Emperor (1804), Laplace sent him special congratulatory greetings. It should be remembered that many scientists at that time were indignant at this transformation of France into an Empire and expressed their protest in some manner (Monge, Arago, and others).

Laplace, however, completely supported Napoleon even in the period of the Empire. Nevertheless, Laplace was the first who deserted Napoleon when his situation became unstable. Laplace was completely inactive during the "100 days" assuming correctly that the restoration of the monarchy was imminent.

As soon as the Bourbons ascended the throne, Laplace offered his oath of allegiance. Louis XVIII showered him with awards. Laplace spoke as a staunch monarchist in the Chamber of Peers. When Charles X introduced censorship, the National Institute protested. Laplace, however, refused to preside at the meeting at which the protest was discussed; in addition, he published a special proclamation to the effect that he was in no way connected with this protest.

Laplace's political views did change but not significantly; he was a monarchist throughout his life—considering the constitutional form of government headed by an enlightened monarch as the best form of government. Toward the end of his life he became an ultraroyalist. It is therefore doubtful that his political views fluctuated in accordance with the various types of government in France from the Convention to the Restoration of 1815 as many of his biographers suggest (see, e.g. [163, p. 824]).

There is a well-known story about an alleged encounter with Napoleon during which Laplace retorted to Napoleon that he had no need of the hypothesis concerning the existence of "the author of the universe." Actually such a conversation did not take place. This conversation during Laplace's lifetime, as well as later, was attributed also to other scientists. It is, however, known that during Laplace's lifetime, when this conversation was to be included in a volume of biographies of scientists, Laplace was the one who protested this action.

We shall now discuss some of Laplace's methodological aims and ideas which perhaps are most clearly expressed in his "Philosophical Essay on Probabilities."

At the beginning of his course of lectures Laplace discusses the relationship between chance and necessity. According to Laplace's views, only necessity prevails in nature: "The curve described by a simple molecule of air or vapor is regulated in a manner just as certain as the planetary orbits; the only difference between them is that which comes from our ignorance" [98, p. 6]. And further: "All events, even those which on account of their insignificance do not seem to follow the great laws of nature, are a result of it just as necessarily as the revolutions of the sun" [98, p. 37]. This idea is also expressed in his other works: "A curve traced by a light atom which is seemingly randomly driven by wind, is actually directed in the same precise manner as the orbits of planets" [95, p. 175].

Here is another more explicit statement:

> Given for one instant an intelligence which could comprehend all the forces by which nature is animated and the respective situation of the beings who compose it—an intelligence sufficiently vast to submit these data to analysis—it would embrace in the same formula the movements of the greatest bodies of the universe and those of the lightest atom; for it, nothing would be uncertain and the future, as the past, would be present to it eyes. [98, p. 4]

This represents a vividly expressed deterministic philosophy which was widely practiced and used at that time. Laplace describes here an "ideal" state of science. But a science which would attempt to explain and to determine the trajectory of a single molecule with all its random deviations would be—as Engels remarks—"not a science but simply a game." Moreover, "chance is not explained here by necessity, but rather conversely necessity is reduced to the outcome of a naked chance" [55, p. 175].

By eliminating the chance completely, Laplace is forced to define probable events without reference to this concept. He considers those events which are not completely known as probable. "Probability is relative in part to (our) ignorance, in part to our knowledge" [98, p. 6]. Laplace introduces the purely subjective criterion of equal possibility of events, considering that two events are equally possible if there is no reason to believe "that one of them will occur rather than the other." This criterion is based not on our knowledge, but on our ignorance of true causes.

Assuming that our knowledge is incomplete concerning many objects and events, Laplace proposes applying probability theory to all problems of the natural sciences and society (moral sciences). In Chapter X of the essay ("Application of the Calculus of Probabilities to the Moral Sciences") he then asserts:

> Let us not offer in the least a useless and often dangerous resistance to the inevitable effects of the progress of knowledge; but let us change only with an extreme circumspection our institutions and the usages to which we have already so long conformed. We should know well by the experience of the past the difficulties which they present; but we are ignorant of the extent of the evils which their change can produce. In this ignorance the theory of probability directs us to avoid all change; especially is it necessary to avoid sudden changes. [98, p. 108]

He also comes out against severe and unjustified sentences: "Analysis confirms what simple common sense teaches us, namely, the correctness of judgments is as much more probable as the judges are more numerous and more enlightened. It is important then that tribunals of appeal should fulfill these two conditions" [98, p. 132]. Next he describes somewhat mockingly a tribunal of a thousand and one judges.

Laplace bases his demand of a large-size tribunal on the following grounds: Assume that n judges are to decide "yes" or "no" on the guilt of an accused. If the decision taken by each one of the judges is independent, then Bernoulli's form of the law of large numbers will be applicable and the problem concerning the guilt or innocence of the defendant will be correctly decided, if the majority of the judges votes properly. Consequently, if the tribunal consists of a sufficiently large number of judges and decides according to majority rule, then practically no errors should be committed by the tribunal.

This type of argument was criticized by various authors on many occasions. S. N. Bernstein writes in this connection as follows: "The judges' decisions are based on the same witness testimony and the same material evidence .... In other words, in the case of a court decision, the condition of independence between the opinions of the judges is violated, which radically changes matters" [22, p. 179].

Laplace unjustifiably corroborates many of his assertions on various aspects of social life using arguments based on probability theory, considering that the important problems in any field "are for the most part only problems of probability" [98, p. 7].

Laplace admitted the existence of an external material world outside and independent of our intellect. The environment, according to Laplace, is perceived by our sense organs, and the criterion of knowledge is in conformity with observations.

Laplace believed that phenomena and the actual nature of things do not coincide, and the purpose of science is to correct "the illusions and deceptions of our senses, by perceiving true objects in their deceptive appearance and manifestation" [95, p. 2]. Nature should be approached by comparing various factors; the phenomena should be examined from various points of view in their development; a collection of facts is not sufficient; one should compare and experiment. "If humanity would confine itself to a mere collection of facts, science would be a fruitless nomenclature" [95, p. 54]. However, it is Laplace's opinion that one cannot perceive the actual nature of all phenomena.

Laplace considered that the law of universal gravitation not only "descends to the complete explanation of all celestial phenomena in their minutest detail" [96, p. 1], but also explains "almost all natural phenomena, solidity, crystallization, refraction of air, elevation and depression of liquids in capillary spaces, and in general that all chemical compounds result from forces" [96, p. 2] which Laplace calls particular gravitations, being particular cases of the law of universal gravitation. Concerning this law he remarks:

> We shall see that this great law of nature represents all the celestial phenomena even in their minutest details, that there is not one single inequality of their motions, which is not derived from it, with the most admirable precision, and that frequently anticipated observations by revealing the cause of several singular motions, just perceived by astronomers, and which were either too complicated or too slow to be determined by observations alone, except after a lapse of ages. By means of it, empiricism has been entirely banished from astronomy, which is now a great problem of mechanics. [95, p. 733]

Laplace undertook the assignment of demonstrating that within the limits of the solar system nature obeys the single law of universal gravitation.

According to Laplace, not all of nature can be perceived. We perceive only the relations between specific phenomena and reduce the causes of all the phenomena to a small number of terminal causes which are completely unrecognizable. He emphasizes this idea on various occasions:

> We can, by a train of inductions judiciously managed, arrive at the general phenomena from whence these particular facts arise. It is to discover these grand phenomena and to reduce them to the least possible number, that all our efforts should be directed; for the first causes and intimate nature of beings will be forever unknown. [96, p. 7]

"The generality of laws represented by celestial movements seems to point to the existence of a singular principle" [96, p. 174]. This "singular principle" is unknown. One may, of course, interpret this "unknown singular principle" in various ways, but any interpretation cannot be far from a recognition of a Divine existence, which became Laplace's conviction after the Restoration.

Laplace transfers the law of mechanics to psychology and reduces human feelings to mechanical processes. He reduces even thought itself to simple mechanical oscillations. Laplace asserts that complex ideas are formed from simple ones as the tides in oceans are formed from partial tides caused by the sun and the moon [98].

Laplace's principal scientific contribution is in the field of celestial mechanics. However, he also wrote some fundamental works in mathematics and physics.

His first papers in probability theory are dated at the middle seventies of the eighteenth century.† In 1810 Laplace obtained his most important result in probability theory presently known as Laplace's theorem. The essence of this theorem is that the binomial probability distribution under suitable normalization and unlimited increase in the number of trials approaches the normal probability distribution.‡ After this Laplace published his classical treatise "Théorie analytique des Probabilités" in 1812 [97]. This work sustained three editions (1812, 1814, and 1820) during Laplace's lifetime.

In this volume Laplace presents all his basic results in probability theory. He systematized the previous, often uncoordinated results, improved the methods of proofs, laid the foundations for study of various statistical regularities, successfully applied probability theory to estimations of errors in observations, and so on.

Laplace was a very prominent scientist whose contributions to the development of probability theory were invaluable.

One of the applications of probability theory of special interest to Laplace is in the field of demography. Not only does he discuss these problems in his basic works on probability theory, but also wrote special memoirs on this subject.

Laplace expresses his views on the changes in composition of the population, discusses methods of indirect population counts, and estimates of precision in such counts; he also develops the theory of sampling census and other problems. In his "Philosophical Essay on Probabilities" [98] Laplace describes a method for constructing mortality tables. Next he introduces the notion of mean duration of life, assuming that those who die before age one, live on the average six months, those who die before age two, live on the average one and one-half years, and so on. "A table of mortality is then a table of the probability of human life. The ratio of the individuals inscribed at the side of each year to the number of births is the probability that a new birth will attain this year" [98, p. 142].

We note that the method of calculating probabilities suggested above does not correspond to the classical definition of probability, given by Laplace as the ratio of equally possible cases.

Laplace considers the problem of computing population size using mortality tables. He asserts that mortality tables are valid only at a given time and in a given place.

Discussing various causes which affect mortality, Laplace writes: "The

---

†His first work on probability theory "Mémoire sur la Probabilité des causes par les évènements" appeared in 1774.

‡The investigations by Eggenberger (1894) [193] and Pearson (1925) [207] reveal that de Moivre (1730) was the first to derive the normal law. (See also an article by Sheynin [213] "On the history of the limit theorems of de Moivre and Laplace" in *Istoriya i Metodologiya Estestvennykh Nauk* **9**, 199–211 (1970)) (*Translator's remark*).

greater or less salubrity of the soil, its elevation, its temperature, the customs of the inhabitants, and the operations of governments have a considerable influence upon mortality" [98, p. 142].

Proceeding to the influence of diseases on mortality, Laplace discusses problems of inoculation. Relating the argument between D. Bernoulli and d'Alembert on this subject, Laplace supports Bernoulli's point of view. To determine the mean life in the case when a certain cause of death is eliminated, Laplace argues as follows [97]:

Let N be the total number of births; x the given age; U the number of children who survived after x years, under the condition that one cause of death is eliminated; u the number of children who reached age x with the existence of this cause; and let B denote the cause.

Let $z\Delta x$ be the probability that a person of age x dies as a result of B during the small period of time Δx, $uz\Delta x$ is the number of people who died as a result of B during the time Δx, provided u is large. $y\Delta x$ is the probability that a person of age x dies as a result of other causes during the period Δx; $uy\Delta x$ is the number of people out of u who will die as a result of other causes during Δx.

The total decrease in the number of people during Δx will be $-\Delta u = u(y+z)\Delta x$.

Analogously,

$$-\Delta U = Uy\,\Delta x; \qquad -\frac{\Delta U}{U} = y\,\Delta x;$$

$$\Delta u = uy\,\Delta x + uz\,\Delta x; \qquad y\,\Delta x = -\frac{\Delta u}{u} - z\,\Delta x;$$

$$-\frac{\Delta U}{U} = -\frac{\Delta u}{u} - z\,\Delta x; \qquad \frac{\Delta U}{U} = \frac{\Delta u}{u} + z\,\Delta x.$$

If x is small, then the decrement can be replaced by differentials

$$\frac{dU}{U} = \frac{du}{u} + z\,dx; \qquad \ln U = \ln u + \int_0^x z\,dx;$$

$$U = u\exp\int_0^x z\,dx, \qquad \text{for} \quad x = 0, \qquad U = u = N.$$

If the functional relation between z and x is known, then the relation between U and u can be obtained. In this case it is easy to transform one mortality table into another, assuming that a certain cause of death has been eliminated. But the function (z of x) is not defined. Therefore, $\int_0^x z\,dx$ is calculated approximately as follows: $uz\,\Delta x$ is the number of people of age x dying as a result of B during Δx. Assume Δx is equal to one year,

z is then equal to a ratio in which the denominator is equal to the number of children out of N, who survive in the middle of the same year. In such a manner u is calculated for every age group.

Concerning the increase in the mean duration of life by three years due to inoculation, Laplace writes:

> An increase so considerable would produce a very great increase in the population if the latter, for other reasons, were not restrained by the relative diminution of subsistences. [98, p. 146]

> If easy clearings of the forest can furnish abundant nourishment for new generations, the certainty of being able to support a large family encourages marriages and renders them more productive. Upon the same soil the population and the births ought to increase simultaneously in geometric progression. [98, p. 147]

As of 1745, the sex of newborns was noted in the registers of births in Paris. From the beginning of 1745 to the end of 1784, 393,386 boys and 377,555 girls were baptized. This ratio is $393,386/377,555 \approx 25/24$.

Discussing the ratio of births of boys to that of girls in various countries in Europe, Laplace concludes that "the ratio of births of boys to that of girls in Paris is that of 22 to 21 (no note being taken of foundlings) [98, p. 68]. Taking into account the data collected by de Humboldt in America, Laplace writes: "He has found in the tropics the same ratio of births as we observe in Paris; this ought to make us regard the greater number of masculine births as a general law of the human race" [98, p. 65].

Laplace assumes that the ratio 25/24 obtained for Paris differs from the ratio 22/21 because the Paris birth-registers include foundlings, while "parents in the country and provinces, finding some advantage in keeping boys at home, sent fewer boys, relative to the number of girls to the Hospital for Foundlings in Paris, according to the ratio of births of the two sexes."

Next Laplace relates about a sampling population census of France which he conducted in 1802. Accurate data in 30 departments (districts) scattered over the whole of France were collected. The census of 23 September 1802, resulted in 2,037,615 inhabitants. During the period of three years (1800, 1801, and 1802) there were 110,312 births of boys, 105,287 births of girls, i.e. 215,599 in total; deaths of males were 103,659 in number; deaths of females, 99,443 in number; marriages, 46,037. We have $110,312/105,287 \approx 22/21$; $46,037/215,559 \approx 3/14$. The ratio of the population to the number of yearly births is $(2,037,615)(3)/215,599 \approx 28.352845$. Assuming that the number of yearly births in France is equal to 1,000,000, Laplace concluded that the whole population of France contains 28,352,845 inhabitants.

It can be seen from the above that Laplace played a significant role in the development of statistics, in particular he contributed greatly to the application of probability theory to demography.

In "Théorie analytique" [97] the so-called classical definition of probability theory is given: the probability $P(A)$ of event A is equal to the ratio of the number of possible outcomes of a trial which are favorable to event A to the number of all possible outcomes of the trial. It is assumed in this definition that the individual possible outcomes of the trial are equiprobable.

Laplace attaches to this definition a subjective meaning introducing the principle of insufficiency. The principle asserts that, if the probability of an event is unknown, we assign a certain number to it, which seems reasonable. If there are several events that form a complete system, but the individual probabilities are unknown, we assume that all these events are equally probable.

For example, if a coin is not symmetrical, but we do not know on which side it is biased, then the probability of a head in a first toss is still one half; in view of our ignorance of the favorable side of the coin, the probability of a simple event is increased as much—if the asymmetry is favorable for this simple event—as it is decreased—if the asymmetry is unfavorable.

Concerning this passage, Gnedenko remarks as follows:

> Clearly one should not say that in the first, or second, or any toss of the coin the probability of a certain side is $\frac{1}{2}$, it is simply unknown. Its determination or estimation should be made not by doubtful means which deprive the notion of probability of its objective role as a numerical characteristic of definite real phenomena. [77, p. 105]

Laplace solves the St. Petersburg problem from the point of view of moral expectation. Discussing the question of moral expectation, he asserts that a person who has no property nevertheless possesses something equivalent to a certain amount of capital. For, he would not agree to accept an insignificant amount of money granted only once, under the condition that, after he spends this, he will be refused any means of subsistence.

Since the amount of the moral expectation depends on the available capital, it is still unclear even after Laplace's explanation how one can utilize this notion. Although Laplace did not intend to discredit the notion of moral expectation, its inconsistency became obvious nevertheless.

If we consider the expansion $(x + x^2 + x^3 + x^4 + x^5 + x^6)^n$, then the value of the coefficient of x^s is equal to the number of outcomes with n dice, giving the sum of points equal to s. Laplace generalizes this method of calculation to the method of generating functions widely used at present.

A function

$$f(t) = \sum_{n=0}^{\infty} f_n t^n = f_0 + f_1 t + f_2 t^2 + \cdots + f_n t^n + \cdots$$

is called the generating function of the sequence $f_0, f_1, \ldots, f_n, \ldots$. It is assumed that the power series converges for at least one value of $t \neq 0$.

The sequence $f_0, f_1, \ldots, f_n, \ldots$ can be a sequence of numbers or a sequence of functions. In the latter case the generating function will depend not only on t but also on the arguments of the functions f_n.†

Laplace solves Buffon's needle problem and several other problems. We shall discuss a number of them.

Problem Find the number m of drawings in the French lottery such that the probability of the appearance of all 90 numbers would be equal to $\frac{1}{2}$.

Laplace attains that in this case $m = 85.53$. He then proceeds to obtain a general formula for the number of drawings required so that the probability of the appearance of all the numbers in the lottery will be equal to a given number. He also solves the general problem. A lottery consists of s numbers, and in each drawing n numbers appear. What is the probability p that in m drawings of the lottery all s numbers will appear?

Another problem: Assume that with $m + n$ observations, event A occurred m times, and the complementary event B n times ($m > n$). What is the probability that the probability of occurrence of the event A is greater than the probability of the occurrence of event B? For this probability Laplace obtains the following expression:

$$p = \frac{(\mu - m)(\mu - m + 1)\cdots\mu}{1 \cdot 2 \cdots m}\left[\frac{(\frac{1}{2})^\mu}{\mu} + \frac{m}{\mu}\frac{(\frac{1}{2})^{\mu-1}}{\mu - 1}\right.$$

$$\left. + \frac{m}{\mu}\frac{m-1}{\mu-1}\frac{(\frac{1}{2})^{\mu-2}}{\mu-2} + \cdots + \frac{m}{\mu}\frac{m-1}{\mu-1}\cdots\frac{1}{\mu-m+1}\frac{(\frac{1}{2})^{\mu-m}}{\mu-m}\right],$$

where $\mu = m + n + 1$.

Next the following problem is considered: Find the probability p that an event, which was observed m times in a row, will be repeated in the next k observations. (The answer is $p = (m + 1)/(m + k + 1)$.)‡

We have already pointed out above the inconsistency of this type of calculation.

Laplace also discusses again the stake-division problem (problem of points) [97, p. 22]. A and B agree to play until s wins. The game is interrupted after A has won p games and B q games. How should the stakes be divided? Naturally, the stakes should be divided proportionally to the probabilities of A and B winning the set of s games.

---

†Generating functions are used not only in probability theory, they are also applied in algebra and other branches of mathematics.

‡This problem is the famous *law of succession of Laplace* (see, e.g., Feller [195, p. 113] for a comprehensive discussion of this rule) (*Translator's remark*).

Laplace obtains the probability of winning for A as a function of $V = s - p$ and $W = s - q$, as follows:

$$f(V, W) = \left(\frac{1}{2}\right)^V \left[1 + V \cdot \frac{1}{2} + \frac{V(V+1)}{1 \cdot 2} \left(\frac{1}{2}\right)^2 + \cdots \right.$$
$$\left. + \frac{V(V+1)\cdots(V+W-2)}{(W-1)!} \left(\frac{1}{2}\right)^{W-1} \right].$$

The corresponding probability for B is $1 - f(V, W)$.

Another problem is posed by Laplace: Find the best combination of observations for the determination of an unknown quantity under the condition that positive and negative errors are equally probable and the number of observations is indefinitely large. Without any assumptions on the distribution law of the errors, Laplace obtains that the method of least squares yields the best possible combination of observations.

Laplace also gave a new proof for James Bernoulli's theorem. He obtains the asymptotic formula for the probability of the sum of independent random variables each one of which admits only all the integer values between $-a$ and $+a$. In his derivation he actually employs the basic ideas of the theory of characteristic functions.

This fundamental treatise of Laplace contains a large number of various problems. All these problems and applications are very significant in probability theory. But the focal point of the book, as well as of the whole activity of Laplace in the field, was the proof of one of probability theory's most important limit theorems. This theorem deals with the distribution of deviations of the frequency of occurrence of an event in a sequence of independent trials from its probability. This result is referred to nowadays as Laplace's theorem (or the de Moivre–Laplace theorem, since the particular case of $p = \frac{1}{2}$ was obtained by de Moivre).† Here is the statement of Laplace's theorem: Let the probability of the occurrence of a given event E in n independent trials be equal to p ($0 < p < 1$) and let m be the number of trials in which event E actually occurred; then the probability of the inequality

$$z_1 < \frac{m - np}{(npq)^{1/2}} < z_2 \qquad (q = 1 - p)$$

differs by an arbitrarily small amount from

$$(2\pi)^{-1/2} \int_{z_2}^{z_1} \exp(-z^2/2)\, dz,$$

provided n is sufficiently large. Another name for this assertion is the integral theorem of Laplace (or the global Laplace theorem).

†See also the first footnote on p. 142 (*Translator's remark*).

The probability of exactly m occurrences of event E in n trials is approximately equal to

$$(2\pi npq)^{-1/2}\exp(-z^2/2),$$

where $z = (m - np)/(npq)^{1/2}$. This last assertion is sometimes called the local Laplace theorem.

Laplace's theorem is utilized in practice for values of n of the order of a score and onward. The errors obtained in such cases are sufficiently small.

Laplace attributed great importance to his theorem. He believed that his law explained completely the behavior of random mass schemes to which, according to Laplace, the majority of real-world phenomena belong. It was his contention that the model based on his law is almost universal. This scheme can be described as follows: A large number of quantities which vary randomly are given, but the laws governing these variations are unknown. It, however, turns out that the resultant of these random quantities—in its variation around the mean value—obeys a certain probability law, as given above. This law was termed the normal law by H. Poincaré.

Only after this work of Laplace did the the widespread applications of probability theory become feasible as a scientifically justified method.

Occasionally, however, Laplace misinterprets the far-reaching conclusions of his contributions to probability theory. His basic error in this connection is that he considers the history of human society as a field governed by pure chance and therefore assumes that probability theory is the science capable of rendering a complete analysis and explanation of this history—so that the analysis of social phenomena falls within the realm of probability theory. It is Laplace's view that all the regularities of any field of mass phenomena may possibly be reducible to the unique normal law, as the celestial phenomena are reduced to the unique law of universal gravitation. Based on this point of view, he attempts to apply probability theory to court procedures, decisions at gatherings, and so on. Such an unfounded and erroneous extension of the applications of probability theory had a negative effect on the development of this science.

K. Biermann, the contemporary German mathematical historian, observes that Laplace overshadowed all of his predecessors in probability theory. "However, a certain overestimation of his contributions is noticeable in the literature" [25].

10   Distribution of random errors

Random errors occur in the course of observations of any kind. The problem of how to avoid them or at least to cope with them has attracted the attention of scientists since early times. However, this can be solved satisfactorily only by means of probabilistic methods. This problem was considered in detail in the early nineteenth century.

Two mathematicians independently and almost simultaneously obtained the basic result, the derivation of the normal "law for the distribution of random errors" (the so-called "law of errors"). One was the famous German scientist Carl Friedrich Gauss (1777–1855) and the other a rather obscure American mathematician Robert Adrian (1775–1843). They arrived at their result through different paths. Adrian was solving a particular problem and, when generalizing it, obtained the distribution law of random errors. Gauss, on the other hand, was investigating the general theory of errors in observations and the normal distribution of random errors became a necessary and most important part of this theory. The derivation of the distribution of random errors as given by Gauss served as a basis for further development of the theory of errors. Although Adrian published his analysis somewhat earlier than Gauss, the impact of his derivation was less noticeable than that of Gauss. Adrian's paper was published in 1808 in *The Analyst, or Mathematical Museum* 1 [1, pp. 83–87, 93–109]. The most interesting parts of this paper are two derivations of the normal law for the distribution of random errors in observations.

Let *AB* be the true value of any quantity, for example of a certain distance. The measure of this quantity (by observation or experiment) is *Ab*, the error being *bB* (Fig. 8).

Figure 8 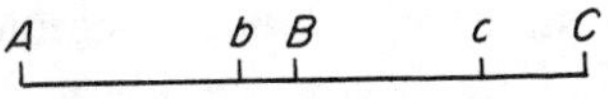   **Figure 9**

"Let *AB*, *BC*, ... be several successive distances of which the values by measure are *Ab*, *bc*, ..., the whole error being *Cc*; now suppose the measures *Ab*, *bc*, are given and also the whole error *Cc*" (Fig. 9) [1, p. 94]. Adrian assumes "as an evident principle" that the errors in measurements of *AB*, *BC* are proportional to their lengths (or to their measured values *Ab*, *bc*). Introducing the notation $Ab = a$, $bc = b$, $Cc = c$, and denoting the errors of the measures *Ab*, *bc* by x, y, respectively, we obtain for the "greatest probability" the equation $x/a = y/b$. Let X and Y be the probabilities that "the errors x, y happen in the distances a, b." The probability of the mutual occurrence of these errors is equal to XY. It is required to find X and Y under the condition that the probability XY will be maximal.

Introducing the notation $f(x) = \ln X$, $\varphi(y) = \ln Y$. Then the maximum of XY corresponds to the relation $f(x) + \varphi(y) = \max$. Differentiating the last relation, we obtain† :

---

† Adrian does not indicate with respect to which argument the derivative is taken. He does not even mention the arguments of the function. All this represents, of course, a certain defect in his work.

$$f'(x)x' + \varphi'(y)y' = 0; \qquad f'(x)x' = -\varphi'(y)y'.$$

But for the maximal probability $x + y = $ const and $x' + y' = 0; x' = -y'$. Dividing the equations we obtain $f'(x) = \varphi'(y)$. Now this equation ought to be equivalent to $x/a = y/b$. This is satisfied in the simplest form if

$$f'(x) = mx/a \qquad \text{and} \qquad \varphi'(y) = my/a.$$

Consider the first relation:

$$f'(x) = mx/a;$$

$$df(x) = (mx/a)\,dx; \qquad \int df(x) = \int (mx/a)\,dx; \qquad f(x) = a_1 + mx^2/2a;$$

$$f(x) = \ln X = a_1 + mx^2/2a; \qquad X = \exp(a_1 + mx^2/2a).$$

The function $U = \exp(a_1 + mx^2/2a)$ is called by Adrian "the general equation of the curve of probability" [1, p. 94]. Next he proves that $m < 0$ [1, p. 95].

In the same paper Adrian presents a second derivation of the distribution law for random errors in observations. In this derivation he considers the measurements of a segment AB with equally probable errors in the length and in the azimuth. Adrian assumes that the locus of the equal probability of the location of point B, determined by the measurements of the length and azimuth of AB, is the simplest curve, i.e. a circle with the center at point B. Under these conditions he obtains that the probabilities of errors are $\exp(c + \frac{1}{2}nx^2)$ and $\exp(c + \frac{1}{2}ny^2)$ correspondingly, where x and y are the errors, $Bm = x$, $mn = y$, and $c = $ const† (Fig. 10).

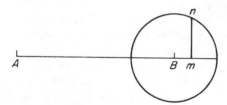

Figure 10

In the same article by Adrian the derivation of the least-squares principle is given as well as the derivation of the principle of the arithmetic mean, and a method "to correct the dead reckoning at sea by an observation of the latitude" are presented.

Gauss published his derivation of the normal law of distribution of random errors in observations in 1809 in his famous work "Theoria motus corporum coelestium."

† More details are given in Sheynin's papers [164, 166.]

His occupation with problems of astronomy and geodesy led him to develop methods of processing results of observations. The common procedure in astronomy and geodesy is for various observers to perform numerous observations at various locations, using different instruments. The results of these observations are subject to errors. Therefore the problem of determining the most probable value of the observed quantity arises. These problems led Gauss to develop the theory of errors, which is directly connected with the ideas and notions of probability theory.

I. M. Vinogradov states that

the extensive approximate calculations, which were carried out by Gauss personally in solving problems related to astronomy and geodesy led him to a deeper development and investigation of the method of least squares and to a clarification of the central position assumed by the normal distribution curve in problems connected with probability theory. [64, p. 9]

Along with the unusually wide scope of Gauss's activities, a characteristic feature of his investigations is a deep interrelation between theoretical and applied problems. He often discovered general mathematical ideas as a result of solving specific problems. This is particularly relevant to his work in the field of probability theory.

Although Gauss's contributions to probability theory are connected with applications, they were not devoted solely to applied problems. His works have had significant impact on the development of various branches of probability theory. For example, his theory of errors prompted an investigations of conditions under which the normal distribution law is applicable. Gauss's contributions raised the problem of estimating the parameters of the normal distribution. He also substantiated the least-squares method using probability theory, taking as an axiom the principle of the arithmetic mean.

Gauss's first work related to probability theory was the famous "Theoria motus corporum coelestium in sectionibus conicis solem ambientium" (Perthes and Besser, Hamburg, 1809; English translation by C. H. Davis, Boston, 1857; French translation by Edm. Dubois, 1864; German translation by Carl Haase, Hanover, 1865). In the last part of this work Gauss for the first time presented his theory of errors in observations.

Two other papers are related to this topic: "Disquisitio de elementis ellipticis Palladis," Comment. (Göttingen) I (1808–1811) and "Bestimmung der Genauigkeit der Beobachtungen" (Z. für Astron. (1816), 185–197).

These works were generalized and supplemented in his treatise "Theoria combinationis observationum erroribus minimis obnoxiae," Comment. (Göttingen) V (1819–1822), which appeared in 1823. In 1828 a supplement to this work: "Supplementum theoriae combinationis observationum erroribus minimis obnoxiae," *Comment. (Göttingen)* **VI** (1823–1827) was published.

In 1845–1851 Gauss wrote "Application of probability theory for determination of the balances of widows' pension funds" and also computed "Tables for determination of the time periods for various types of obligatory incomes for survivors."†

Gauss's notes and letters are also of great value and interest from the point of view of probability theory.

In the eighteenth century the need for an optimal way of combining observations to obtain the most reliable results became evident.

Tycho Brahe as early as the eighties of the sixteenth century, measured the same object under varied conditions for the purpose of elimination of errors. Combining these observations, Brahe strived to eliminate random errors. This problem was always of interest to investigators.

Adrien-Marie Legendre in his treatise "New methods for the determination of the orbits of comets"‡ developed the method of least squares in the appendix entitled "On the method of least squares". He writes: "Among all the principles, which may be suggested for this purpose, there is none simpler than the one we utilized in the previous discussion: the method is to minimize the sum of squares of the errors" [65, pp. 7–8]. Legendre formulates this principle in a clear manner and observes that it should be very useful in various problems of physics and astronomy, where the derivation of the most precise results possible from observations is required.

Gauss presented his method for the first time in 1809. However, he observes that "Our principle, which we have made use of since the year 1795, has lately been published by *Legendre* in the work *Nouvelles méthodes pour la détermination des orbites des comètes*, Paris, 1806" [65, p. 104]. He also gives this date in his letter to Laplace of 30 January 1812.

In a letter to H. M. W. Olbers in 1802 he mentions, however, that "starting from 1794, I have been utilizing the method ... which has also been applied in Legendre's work ...." [65, p. 8]. Gauss mentions two dates, 1794 and 1795. Contemporary authorities are inclined to accept 1794 as the correct one. §

In the beginning of the third section of the second book of "Theoria motus ..." Gauss presents several interesting remarks.††

Since all our measurements and observations are nothing more

---

† See in this connection Gauss's letters to C. L. Gerling reproduced in Dunnington's biography of Gauss [192, pp. 233–234] (*Translator's remark*).

‡ Nouvelles méthodes pour la détermination des orbites des comètes, Paris, 1806.

§ See, for example, E. T. Bell [186, p. 259] for more details on this controversy between Legendre and Gauss (*Translator's remark*).

†† Quoted here from C. H. Davis' translation ("The Computation of an Orbit from Three Complete Observations. From the Theoria Motus of C. F. Gauss," pp. 249–250. Metcalf, Cambridge, Massachusetts, 1857) (*Translator's remark*).

than approximations to the truth, the same must be true of all calculations resting upon them, and the highest aim of all computations
made concerning concrete phenomena must be to approximate, as nearly as practicable, to the truth. But this can be accomplished in no other
way than by a suitable combination of more observations than the
number absolutely requisite for the determination of the unknown
quantities. [65, p. 89]

Next he observes that the results should be derived not from single
observations, but from combined ones, so that "the accidental errors might
as far as possible mutually destroy each other" [65, p. 90].

Further he considers the following problem: Given equidistant observations of a certain quantity, let the random errors possess the "differential"
density of probability distribution $\varphi(\Delta)$. It is required to determine $\varphi(\Delta)$
under the assumption that the most probable value of the quantity under
consideration is equal to the arithmetic mean of the observed values.
In Gauss's words:

Let us suppose ... that there is no reason why we should suspect
one [observation] to be less exact than another, or that we are bound
to regard errors of the same magnitude as equally probable in all.
Accordingly the probability to be assigned to each error Δ will be expressed by a function of Δ which we shall denote by $\varphi(\Delta)$. ... We can
(at least) affirm that its value should be a maximum for $\Delta = 0$, equal,
generally, for equal opposite values of Δ, and should vanish if, for Δ
is taken the greatest error, or a value greater than the greatest error.
$\varphi(\Delta)$, therefore, would appropriately be referred to the class of dis-
continuous functions, and if we undertake to substitute any analytical
function in the place of it for practical purposes, this must be of such
a form that it may converge to zero on both sides, asymptotically,
as it were, from $\Delta = 0$. [65, p. 93]

Under these assumptions, Gauss obtains that the function $\varphi(\Delta)$ is given by
$(h/\sqrt{\pi})\exp(-h^2\Delta^2)$. Gauss denotes the value h as "the measure of precision of the observations." As a corollary, Gauss derives the assertion
that the probability density of a given population of observations attains
its maximum when the sum of squares of deviations of the observed values
from the true value of the observed quantity is minimized. This principle
is then extended for the case of not equally distant observations. According
to Gauss, this principle should be taken as an axiom, and the arithmetic
mean of a number of observations on the same quantity should be postulated
as the most probable value of this quantity.

After deriving the "normal law of the distribution of random errors"
$[\varphi(\Delta) = h\pi^{-1/2}\exp(-h^2\Delta^2)]$, Gauss points out a certain defect in this
law. According to this law, errors of any magnitude are possible.

The function just found cannot, it is true, express rigorously the probabilities of the errors; for, since the possible errors are in all cases confined within certain limits, the probability of errors exceeding those limits ought always to be zero, while our formula always gives some value. [65, p. 96]

It should be emphasized that in deriving the normal distribution Gauss made extensive use of the principle of the arithmetic mean which he formulates in the following manner:

It has been customary certainly to regard as an axiom the hypothesis that, if any quantity has been determined by several direct observations, made under the same circumstances and with equal care, the arithmetic mean of the observed values affords the most probable value, if not rigorously, yet very nearly at least, so that it is always safe to adhere to it. [65, p. 96]

This axiom attracted widespread attention among investigators. Many authors tried to prove it by reducing it to, from their points of view, simpler propositions. A survey and discussion of these attempts is given in Whittaker and Robinson's volume [180].

The normal distribution was considered a universal law for a long period of time. This state of affairs resulted in a delay in the development of quantitative methods for discarding some observations, since the normal law admits the possibility of errors of any magnitude. Hence it was assumed that all the observations should be retained. Only in the middle of the nineteenth century did the first probabilistic criteria for rejecting observations begin to appear.

In his memoir "Bestimmung der Genauigkeit der Beobachtungen" (1816) Gauss investigates the estimation of h based on results of observations.

Gauss introduces the (error) function $\theta(t) = 2\pi^{-1/2} \int_0^t \exp(-t^2)\,dt$ and presents a small table of the values of this function. The value of the argument ρ such that $\theta(\rho) = 0.5$ is specially singled out. This value is $\rho = 0.4769363$. The quantity $\rho/h = r$ is called the probable error for the function $\theta(ht)$.†

In paragraph 3 of this paper, Gauss presents the statement of the problem: "Assume now that in performing m observations the errors α, β, γ, δ, ..., occurred and let us investigate what conclusions can be drawn concerning the values of h and r" [65, p. 122]. Assuming that the observations are independent, Gauss obtains that the probability density of the system

---

† As it was mentioned above, the first table of a *normal curve* appeared in D. Bernoulli's treatise of 1771 (*Translator's remark*).

of observations obtained from the experiment is equal to

$$y = C_m h^m \exp[-h^2(\alpha^2 + \beta^2 + \gamma^2 + \cdots)],$$

where $C_m = \text{const.}$

Next he obtains that value of h which corresponds to the maximum of function y. Differentiating,

$$\begin{aligned}
y' &= C_m\{mh^{m-1}\exp[-h^2(\alpha^2 + \beta^2 + \gamma^2 + \cdots)] \\
&\quad + h^m \exp[-h^2(\alpha^2 + \beta^2 + \gamma^2 + \cdots)](-2h)(\alpha^2 + \beta^2 + \gamma^2 + \cdots)\} \\
&= C_m h^{m-1}\exp[-h^2(\alpha^2 + \beta^2 + \gamma^2 + \cdots)] \\
&\quad \times [m - 2h^2(\alpha^2 + \beta^2 + \gamma^2 + \cdots)].
\end{aligned}$$

For $y' = 0$, $h = 0$; thus we get $y = 0$ and $m - 2h^2(\alpha^2 + \beta^2 + \gamma^2 + \cdots) = 0$, hence

$$h = [m/2(\alpha^2 + \beta^2 + \gamma^2 + \cdots)]^{1/2}.$$

(A simple verification then shows that this is the value of h corresponding to the maximum of y.)

Hence the most probable value of r is

$$r = \frac{\rho}{h} = \frac{\rho}{[m/2(\alpha^2 + \beta^2 + \gamma^2 + \cdots)]^{1/2}}$$

$$= \rho[2(\alpha^2 + \beta^2 + \gamma^2 + \cdots)/m]^{1/2}.\dagger$$

Next, Gauss calculates the most probable values of the sum of the nth powers of the observed values and estimates the unknown values of h and r in terms of these sums.‡

The most complete exposition of the theory of errors is contained in Gauss's paper "Theoria combinationis observationum erroribus minimis obnoxiae." Gauss writes in this memoir that no matter how carefully the observations are carried out, errors are unavoidable. Some errors may be random, others may be predicted and evaluated since these are either constant or vary in a regular manner. The latter type of error is referred to as a systematic error. Gauss, however, points out that such a division of errors into two kinds is relative, and in many cases depends upon the problem at hand.

---

†The modern notation used in this book is different from the original Gauss's notation (*Translator's remark*).

‡ An early application of the method of moment estimation (*Translator's remark*).

This paper of Gauss's is devoted to the study of laws governing the distribution of random errors. Gauss proceeds with the most general assumptions concerning the probability density of the errors $\varphi(x)$. Since positive and negative errors appear equally often, $\varphi(-x) = \varphi(x)$. Next small errors occur more often than the large ones, hence the value of $\varphi(x)$ will be maximal at $x = 0$ and diminishes constantly with the increase of x. Clearly, the value of the integral $\int \varphi(x)\,dx$ in the limits from $x = -\infty$ up to $x = +\infty$ always equals 1. The basic problem considered by Gauss is actually as follows: Let the variables y, x_1, x_2, $\ldots x_\sigma$ be linearly related, i.e.,

$$y = a_1x_1 + a_2x_2 + \cdots + a_\sigma x_\sigma \qquad \text{or} \qquad y = \sum_{s=1}^{\sigma} a_s x_s,$$

while the a_s are unknown.

To determine these unknowns, the values of

$$y_1 = a_1x_{11} + a_2x_{21} + \cdots + a_\sigma x_{\sigma 1},$$

$$y_2 = a_1x_{12} + a_2x_{22} + \cdots + a_\sigma x_{\sigma 2},$$

$$\vdots$$

$$y_N = a_1x_{1N} + a_2x_{2N} + \cdots + a_\sigma x_{\sigma N},$$

or compactly

$$y_r = \sum_{s=1}^{\sigma} a_s x_{sr}, \qquad r = 1, 2, \ldots, N,$$

are obtained from the experimental data.

The experimental determination of y_r is subject to error. Consequently we actually obtain, instead of the value y_r, the value $\eta_r = y_r + \Delta_r$. Given x_{sr} and the obtained values of η_r, it is required to determine the best possible approximate values α_s of the quantities a_s.

According to Gauss, these α_s ought to be determined from the condition

$$\left[\eta_1 - (\alpha_1x_{11} + \alpha_2x_{21} + \ldots + \alpha_\sigma x_{\sigma 1})\right]^2$$

$$+ \left[\eta_2 - (\alpha_1x_{12} + \alpha_2x_{22} + \cdots + \alpha_\sigma x_{\sigma 2})\right]^2 + \cdots$$

$$+ \left[\eta_N - (\alpha_1x_{1N} + \alpha_2x_{2N} + \cdots + \alpha_\sigma x_{\sigma N})\right] = \min.$$

Or, using the $\sum$ notation,

$$\sum_{r=1}^{N} \left(\eta_r - \sum_{a=1}^{\sigma} \alpha_s x_{sr}\right)^2 = \min. \tag{III.10}$$

α_s are then uniquely determined from the system of equations derived from condition (III.10).

These equations are called normal equations and are of the form:

$$\alpha_1(x_{11}x_{11} + x_{12}x_{12} + x_{13}x_{13} + \cdots + x_{1N}x_{1N})$$
$$+ \alpha_2(x_{21}x_{11} + x_{22}x_{12} + \cdots + x_{2N}x_{1N}) + \cdots$$
$$+ \alpha_\sigma(x_{\sigma 1}x_{11} + x_{\sigma 2}x_{12} + \cdots + x_{\sigma N}x_{1N}) = \eta_1 x_{11} + \eta_2 x_{12} + \cdots + \eta_N x_{1N};$$

$$\alpha_1(x_{11}x_{21} + x_{12}x_{22} + \cdots + x_{1N}x_{2N})$$
$$+ \alpha_2(x_{21}x_{21} + x_{22}x_{22} + \cdots + x_{2N}x_{2N}) + \cdots$$
$$+ \alpha_\sigma(x_{\sigma 1}x_{21} + x_{\sigma 2}x_{22} + \cdots + x_{\sigma N}x_{2N}) = \eta_1 x_{21} + \eta_2 x_{22} + \cdots + \eta_N x_{2N};$$

$$\vdots$$

$$\alpha_1(x_{11}x_{\sigma 1} + x_{12}x_{\sigma 2} + \cdots + x_{1N}x_{\sigma N})$$
$$+ \alpha_2(x_{21}x_{\sigma 1} + x_{22}x_{\sigma 2} + \cdots + x_{2N}x_{\sigma N}) + \cdots$$
$$+ \alpha_\sigma(x_{\sigma 1}x_{\sigma 1} + x_{\sigma 2}x_{\sigma 2} + \cdots + x_{\sigma N}x_{\sigma N}) = \eta_1 x_{\sigma 1} + \eta_2 x_{\sigma 2} + \cdots + \eta_N x_{\sigma N}.$$

Or, using the $\sum$ notation:

$$\sum_{s=1}^{\sigma} \alpha_s \sum_{r=1}^{N} x_{sr}x_{ir} = \sum_{r=1}^{N} \eta_r x_{ir}, \qquad i = 1, 2, \ldots, \sigma.$$

Indeed, (III.10) is minimized (and equals zero) if the expression in each of the square brackets vanishes, i.e.,

$$\alpha_1 x_{11} + \alpha_2 x_{21} + \cdots + \alpha_\sigma x_{\sigma 1} = \eta_1,$$

$$\alpha_1 x_{12} + \alpha_2 x_{22} + \cdots + \alpha_\sigma x_{\sigma 2} = \eta_2,$$

$$\vdots$$

$$\alpha_1 x_{1N} + \alpha_2 x_{2N} + \cdots + \alpha_\sigma x_{\sigma N} = \eta_N.$$

The above values of η_r $(r = 1, \ldots, N)$ satisfy the system of the normal equations.

The obtained approximations α_s of the values a_s are free from systematic bias (unbiased), i.e. the mathematical expectation of α_i is equal to a_i.

Gauss's results in the theory of errors are presented almost without modification even nowadays in many textbooks on statistics.

In the fourth volume of Gauss's collected works, incidental notes concerning probability theory are included. We shall only mention one problem suggested by Gauss in his letter of 30 January 1812 to Laplace. This problem served as a starting point for the metric theory of numbers.

Let M be an unknown quantity between 0 and 1, and possible values (between 0 and 1) be either equiprobable, or obey a certain (other) distribution. Assume that M is expanded into a continued fraction:

$$M = \cfrac{1}{a' + \cfrac{1}{a'' + \cdots}}.$$

What is the probability that, after a finite number of terms up to $a^{(n)}$ in this expansion are discarded, the fraction

$$\cfrac{1}{a^{(n+1)} + \cfrac{1}{a^{(n+2)} + \cdots}}$$

will be confined between 0 and x (where $0 < x < 1$)?

In conclusion we note that, despite the fact that probability theory was not the focal point of Gauss's broad scientific interests, he nevertheless contributed greatly to this branch of mathematics, enriching it with numerous significant results. Moreover, Gauss's ideas were applied and further developed in various areas of probability theory (see [64]).

11 The state of probability theory in Europe prior to the advent of the Russian (St. Petersburg) school

Probability theory in the first part of the nineteenth century developed in the midst of strong conflict and contradictions. The creative work of Laplace may serve as a distinct example of this situation, as do the scientific works of Poisson.

Simeon Denis Poisson's (1781–1840) mathematical legacy is vast and voluminous. The following of his works are related to probability theory: "On the probability of mean results of observations"† (1827), "Continuation of the memoir on mean results of observations"‡ (1832), "Sur l'avantage du Banquier au jeu de Pharaon," "On the probability of births of boys and girls," and several others.

All these papers were included in various forms in Poisson's main work on probability theory, "Recherches sur la probabilité des jugements en matière criminelle et en matière civile" [155], published in 1837.§ His well-known theorem is also contained in this volume.

In his book Poisson first presents a brief survey of previous results in probability theory. He pays special attention to Laplace's and Condorcet's works on moral probability.

---

†"Sur la probabilité des résultats moyens des observations," *Conn. des Temps*, pp. 273–302, 1827.

‡"Suite du mémoire sur la probabilité des résultats moyens des observations," *Conn. des Temps*, pp. 3–22, 1832.

§F. Haight in his "Handbook of the Poisson Distribution" (Wiley, New York, 1966) observes that certain authors give the date 1832 for Poisson's "Recherches . . . ." The "Catalogue Générale des Livres Imprimés de la Bibliothèque Nationale" (Vol. **139**, p. 982) gives only 1837, and the volume dated 1837 does not refer to any earlier edition (*Translator's remark*).

Poisson asserts that the analytic theory of probabilities is applicable to the evaluation of the correctness of court decisions. He poses the problem of determining probabilities of errors in court decisions and suggests that Bernoulli's theorem does not present a mathematical foundation for such an application. To create a basis for this application Poisson investigates limiting theorems of probability theory. His investigations resulted in a famous theorem which he designated "the law of large numbers."

Poisson's theorem deals with the following: If n independent trials are performed, resulting in the occurrence or non-occurrence of an event A, and the probability of occurrences of events is not the same in each one of the trials, then, with probability as close to unity as desired (in other words, as close to certainty as desired), one can assert that the frequency m/n of the occurrence of event A will deviate arbitrarily little from the *arithmetic mean $\tilde{p}$* of probabilities of occurrences of events in the individual trials. This theorem is formulated in modern notation as follows:

$$\lim_{n \to \infty} (|m/n - \tilde{p}| < \varepsilon) = 1.$$

If, however, the probability of the occurrences of events remains constant from trial to trial, then $\tilde{p} = p$, and Poisson's theorem in this case reduces to Bernoulli's theorem.

According to Poisson, all events of a moral as well as of a physical nature are subject to this universal law. Poisson viewed this theorem not only as mathematical fact, but also as a philosophical truism. It served as grounds for his investigations concerning the correctness of court decisions and phenomena of a moral nature. He believes that by means of this principle the probability of any human decision may be determined regardless of the reasons for these decisions. Formulas are presented in his work, based on the analysis of a very large number of past court decisions, which give the "exact probability" for each citizen to be accused, convicted, or acquitted (provided legislation remains unaltered).

In his book Poisson also derived the so-called "law of small numbers":

As the deviation of the value of p from the value $\frac{1}{2}$ increases, the asymptotic representation of $P_{m,n}$† in the form $(2\pi)^{-1/2} \exp(-x^2/2)$ becomes less and less accurate. In order for Laplace's theorem to give a reasonably accurate approximation to $P_{m,n}$, the number of observations must be substantially increased, which is not always convenient or even possible. Therefore, the problem arises of obtaining an asymptotic formula which will be particularly suitable for small p (or small $q = 1 - p$, since these cases are reduced to each other). This problem was solved in Poisson's book "Recherches sur la probabilité des jugements . . ." [155]. Poisson obtains that, as $p_n \to 0$ with

---

†$P_{m,n}$ is the probability of m occurrences of A in n trials (*Translator's remark*).

$n \to \infty$, the probability that an event will occur m times approaches

$$P_{m,n} = \frac{\lambda^m}{m!} e^{-\lambda},$$

where $\lambda = np_n$.

This formula of Poisson's can be utilized as an approximating expression for $P_{m,n}$ for a fixed but small p and large n.†

The Polish statistician L. Bortkiewicz (1868–1931) renamed the Poisson distribution the law of small numbers.‡ He also applied this distribution to rare events, such as deaths by horse-kick in the Prussian Army, births of triplets, and so on.

The ideas and conclusions in Poisson's book were supported in the first place by mathematicians who considered it natural to apply probability theory to problems of legislation, jurisprudence, and the political and economical sciences.

However, this opinion was not unanimous. A number of mathematicians were very much against such an application of probability theory. They accused Poisson and his followers of compromising mathematical science. In the course of the argument statements were made that Poisson's results were false. This criticism was extended to Laplace as well.

Commenting on the state of probability theory at that period of time, Gnedenko writes that "in spite of the fact that Laplace and Poisson concluded an important and fruitful initial period in the development of probability theory, a period of philosophical cementation of the basis of this science, this period resulted in an indifferent attitude toward probability theory in the West and in a definite rejection of the possibilities of utilizing its methods in studying natural phenomena. This led to the beginning of a long period of stagnation in probability theory in the West" [158, p. 394].

In this connection the scientific works of a famous Belgian statistician, Adolphe Quetelet (1794–1874), are indicative. Residing in Paris in 1823–1824, Quetelet met Laplace and attended his lectures on probability theory. Many of his future works were influenced by Laplace's philosophy.

Quetelet was a prominent statistician whose vigorous activities in the field raised Belgian statistics to a very high level. His theoretical works contained a number of very valuable contributions. But the basic idea propagated by Quetelet is completely unsubstantiated. Quetelet's thesis is that the rules of probability theory are those that govern and direct the activities

---

†F. N. David [49] and other investigators give the credit for the discovery of "Poisson's binomial exponential limit" to de Moivre (1718); cf. p. 168 of her book (*Translator's remark*).

‡"Das Gesetz der kleinen Zahlen" (1898) is the title of Bortkiewicz's book in which the Poisson distribution "was revived after a half century of nearly perfect vacuum" (F. Haight, "Handbook of the Poisson Distribution," Wiley, New York, 1966) (*Translator's remark*).

of the human society. The degrees of inclination to crime, marriage, etc., are, according to Quetelet, nothing but mathematical probabilities. For example, he writes: "This probability (0.0884) can be considered a measure of the apparent inclination to marriage of a city dwelling Belgian" [160, p. 79]. Quetelet set forth as the main problem of statistical investigations the determination of the nature of the average man. According to Quetelet, the average man is the absolute perfect type, while separate individuals are a distorted representation of this type. "If the average man were completely determined, we might, as I have already observed, consider him as the type of perfection; and everything differing from his proportions or condition, would constitute deformity and disease; everything found dissimilar, not only as regarded proportion and form, but as exceeding the observed limits, would constitute a monstrosity."† Quetelet described the average man as everlasting and invariable [160, p. 38]. Striving to discover laws for the preservation of "the average" man and subsequently laws for preservation of the society as a whole, Quetelet actually tried to prove the eternal nature of the existing capitalistic society. "The absolute noninterference into private matters is a supreme principle" [161, p. 81] was one of Quetelet's basic propositions.

K. Marx wrote about Quetelet in 1869:

> He has done a great service in the past: he proved that even seemingly chance phenomena in social life as a result of their periodicity and the periodicity of average figures possess an internal necessity. But he never succeeded in proving this necessity. He did not move forward, but merely continued to extend the scope of his observations and calculations. [126, p. 7]

The number of papers and books devoted to unjustified applications of probability theory to social problems was on a constant increase. As a result, probability theory reached an impasse toward the middle of the nineteenth century. Since the areas of application of probability theory were not clearly defined or well established, the viewpoint that no relation exists between natural science and mathematics on the one hand and probability theory on the other became prevalent.

12 Probability theory in Russia prior to the advent of the St. Petersburg school

If we do not take into account the research activities carried out within the precincts of the St. Petersburg Academy in the eighteenth century by

---

†A. J. Quetelet's "A Treatise on Man" (p. 99) (originally published in Edinburgh (1842); reprinted in 1968, Burt Franklin, New York) (*Translator's remark*).

Euler and D. Bernoulli, we might then conclude that interest in probability theory in Russia started to develop only in the third decade of the last century.

It seems that the problem of teaching probability as a part of higher mathematical education in Russia first came up at *Vilnus University.* At the beginning of the nineteenth century great influence on teaching mathematics in this University was exerted by Prof. I.A. Snyadezkiĭ. In 1808 he presented a keynote address at the scientific session of the University concerning prospects for the development of mathematics at the University. Touching on the subject of probability theory, he noted that when the mathematical sciences at the University were sufficiently developed, it would be necessary to consider the problem of organizing a Department of Probability Theory.

Snyadezkiĭ spoke out on the methodological problems of mathematics on various occasions. At the scientific session in 1817 he devoted his lecture to probability theory. From the fragmented and sparse documents available we can conclude that in 1829–30 the introduction of an additional (elective) course in probability theory was in line with the positive attitude toward this science at Vilnus University.†

In a report to the Ministry of Education concerning this course the authorities at the University wrote as follows:

> The Department of Physics and Mathematics has noted that probability calculations, which constitute a wide field of mathematical sciences have not been taught at the University as yet; however, these calculations are very important and of significant use since the activities of actuarial societies are based on these calculations. The renowned Laplace has applied them to geodesic and similar activities and for this reason out Department considers that it would be useful to familiarize the students of the mathematical sciences with these calculations. [24, p. 62]

Next it is stated in the report that master of philosophy, Sygmund Revkovskiĭ (1807–1893) would be an appropriate candidate for reading this course. A short scientific appraisal of Revkovskiĭ is attached noting that Revkovskiĭ "on his own initiative has been working for the last two years on a treatise dealing with probability calculations" [24, p. 62]. Unfortunately Revkovskiĭ's treatise has not survived.

Thus in 1829–1830 for the first time in Russia Revkovskiĭ began lecturing in probability theory. The outline of his course is available (see [24]).

---

†Yushkevich [217] points out that the first note on probability theory in Russian was given by A. F. Pavlovsky (1788–1856), Professor of Mathematics at Kharkov (a city in N. E. Ukraine) University, in a speech entitled "On probability," delivered in 1821 (*Translator's remark*).

The outline was written in Polish and is in the manuscript section of the Vilnus University Library.

The outline begins with an explanatory note. Revkovskiĭ was a proponent of the deterministic philosophy in the spirit of Bernoulli and Laplace. However, his interpretation of this problem is to a certain extent original and is of definite interest. He writes:

> Every phenomenon in nature is the result of forces which act according to a certain law. If these forces are known, then it is possible in certain cases to compute and therefore to predict beforehand the phenomenon which should take place, in the same manner as astronomers predict the movements of celestial bodies and the times of eclipses.
>
> However, when the forces which cause a particular phenomenon and the laws governing these forces are either completely unknown or are so diversified and complex that they cannot be subject to computations, the determination of such a phenomenon becomes problematic and doubtful and is usually ascribed to fate. Thus fate owes its existence to our ignorance or unawareness: the more knowledge we acquire concerning the laws and forces in nature, the fewer the phenomena that will depend on fate. The *hope* that an event which depends on fate will occur may be smaller or larger. [24, p. 67]

The measure of this hope is, according to Revkovskiĭ, the probability which he defines in the classical manner.

> The science which provides the methods of precisely calculating the "hope" or, in other words, "probability" of a certain event, is a special branch of applied analysis, which will be called the calculus of probabilities. [24, p. 67]

Further he presents the actual outline of the course itself. We reproduce here in full the section concerning probability theory proper.

> The general theory of calculus of probabilities, namely:
>
> 1. How to denote probabilities of compound events when the probabilities of simple events are constant and known.
>
> 2. How to denote probabilities of compound events when the probabilities of simple events are known but are not constant and vary continuously according to certain (probability) laws.
>
> 3. James Bernoulli's law and its corollaries.
>
> 4. How to denote the probabilities of compound future events when the probabilities of simple events are unknown, but constant.
>
> 5. How to express in the best possible manner probabilities of future events based on observations of past events.

6. How to denote a fortiori probabilities of future events when probabilities of simple events are unknown and vary continuously according to some (probability) law. [24, p. 67]

This completes the outline of probability theory proper. Next, various "applications" are described:

1. To natural philosophy. Since natural philosophy is based on observations, the applications are as follows:

What is the probability that a certain function of errors in observations is contained in given limits? What is the general method of obtaining from conditional equations the values of unknown elements closest to the true values?

Due to the imperfection of the analysis, this method can be applied only in certain special cases. The method of least squares gives results which are closest to the truth.

How by means of observations can one detect the presence of a force or a cause in nature, if its action is very small or seldom observed; here the cause may be attributed to errors of observation as well as to fate. This problem is an important topic in the calculus of probabilities and will be explained by several particular examples. [24, pp. 67–68]

Next applications to geodesy and to "moral and political sciences" are presented, where the following problem is considered:

The extent to which the rationale and honesty of the witnesses affect the probability of the event about which they are testifying, and so on [24, p. 68].

According to this program, the next topic of study is mortality tables, followed by applications to various kinds of games and insurance matters. Among other problems discussed is the notion of "moral hope." The program concludes with problems connected with various types of insurance applications.

This outline was submitted in 1830 to M.V. Ostrogradskiĭ for approval. Commenting in a number of instances on the order of topics presented in this program, Ostrogradskiĭ gave it a generally favorable recommendation.

In his review Ostrogradskiĭ also writes concerning the desirability of introducing instruction in probability theory in all universities, as well as in high schools.

I think that the Academy would render a very useful service worthy of this top scientific institution in the State, if it would utilize all its efforts to introduce instruction in the calculus of probabilities to all native universities and even high schools, in order that the foundations of this science would be imprinted in the minds of the students in good time. [144, pp. 277–278]

The digression in Ostrogradskiĭ's review in which he discussed probability theory is of interest: The science of probabilities is one of the most important applications of mathematical analysis: the philosophy of nature owes a debt to this science, which provides many methods of determining grounds out of a large number of observations on which the important astronomical theories are based; it served as a stimulus for the development of worthwhile public institutions such as insurance companies; using this science we determine the existence of causes, whose effect is smaller than the errors encountered in observations. The influence of this branch of analysis is increasing daily and is now being applied to the political and moral sciences [144, p. 277].

Ostrogradskiĭ's review was discussed at the conference of the Academy of Sciences on 2 June 1830, and the following was included in the minutes of this meeting:

> Ostrogradskiĭ observes that it would be a worthwhile action on the part of the Academy of Sciences to do everything in its power to introduce the teaching of probability theory in all universities of the (Russian) Empire. The Academy, approving Mr. Ostrogradskiĭ's review and especially his advice, learned with satisfaction that such a proposal was previously presented to the Minister of Education by Messrs. Vishnevskiĭ and Kollins, whose views were sought by the Commission for reorganization of schools and universities during the compilation of a plan for teaching mathematics. [144, pp. 276–277]

All this indicates that the time had arrived for the teaching of probability theory. In spite of this, however, the introduction of courses in probability theory in Russian universities proceeded at a very slow pace.

In 1830 in Vilnus University a department of probability theory was established and Revkovskiĭ was appointed professor in this Department. Revkovskiĭ's later life is of interest. The tremor of the Polish revolt of 1830–1831 was felt in Vilnus. The University was closed and Revkovskiĭ was sentenced to death, which was later commuted to a life sentence at hard labor. After several years in prison in the Caucasus, he was released and as a private (soldier) was engaged in topography. Later he rose to the rank of captain and, finally, received the certificate of a transportation engineer. After his retirement, he returned to his native Vilnus and was engaged in political economy, publishing several books on this subject.

His works dealing with the application of mathematics to various production processes began to appear starting from 1866.

In Moscow University the first courses in probability theory were given starting from 1850 by A. Yu. Davidov (1823–1885).† In the archives of

---

†See a recent paper by Kh. O. Ondar in *Istor. i Metodol. Estestv. Nauk* (No. 11, Moscow Univ. Publ., 1971, pp. 98–108), where further details concerning Davidov are described (*Translator's remark*).

Moscow University an outline of a course on probability theory is preserved, which was compiled by A. Yu. Davidov for the academic year 1851–1852, designated for third-year students. In view of the obvious interest of this program—being the first program in probability theory at Moscow University—we shall present it in full:

A program of mathematical probability theory for students of the Department of Physics and Mathematics for the academic year 1851–52.

A priori definition of probabilities:
Relative (conditional) probability
Probability of compound events
J. Bernoulli's theorem
On mathematical and moral expectations.

A posteriori definition of probabilities:
Definition of probabilities of causes
Definition of probability of the future event based on observations.

Application of probability theory to statistics:
Construction of mortality tables and their application to the determination of the probable duration of life, average duration of life and measures of longevity.

On insurance institutions:
On life annuities and funds for widows
Determination of the most probable results from observations.

Literature:
Bunyakovskii's *Mathematical Theory of Probabilities*; Poisson's *Probability Theory*; Laplace's *Probability Theory*.

A. Davidov, 11 August 1851 [11]

Probability theory was Davidov's favorite subject and for many years he read courses in probability theory. In the years 1854–1857, he published several papers on this subject. In the paper "An application of probability theory to statistics" [51] he investigates the probability that the value of a given function is contained within prescribed limits.

In the paper "An application of probability theory to medicine" Davidov discusses the application of statistical methods to the determination of symptoms of ailments, diagnostic, and other problems. An extensive table is appended to this work, with which he solves various problems. For example, in 200 cases of an illness, a particular symptom appears 130 times. Can this symptom be considered a characteristic of this particular disease? From the table the bounds on the probability of repetition of this symptom are determined as follows: $\frac{56}{100}$ and $\frac{74}{100}$; since both of these numbers are greater than $\frac{1}{2}$, the conclusion can be drawn that this symptom is indeed a characteristic of the disease.

It should be noted that Davidov was one of the first to come out against the theory of the average man proposed by Quetelet. He writes: "The average man not only cannot represent a type in humanity, but could not possibly exist" [52, p. 16].

The first course in probability theory at the University of St. Petersburg was offered in 1837 by V.A. Ankudovich. He lectured in this course up until 1850. During the decade 1850–1860 lectures on probability theory were given by V. Ya. Bunyakovskiĭ. In the sixties of the last century very prominent mathemathicians began to read courses in probability theory at various universities: at St. Petersburg's University, Pafnutiĭ Lvovich Chebyshev; in Berlin University, Ernst Eduard Kummer (1810–1893); etc.

Among the first important works on probability theory in Russian were the works of Nikolaĭ Ivanovich Lobachevskiĭ.† The following considerations led him to an investigation of problems in probability theory. Lobachevskiĭ stated on various occasions that only by experiment can one determine the properties of the surrounding space. Moreover, he even undertook an experimental verification, the results of which were published in 1829 in his paper "On the foundations of geometry."‡ He computed that the sum of the angles in the triangle Earth, Sun, and Sirius (Dog Star) deviates from $180°$ by the amount less than $0''.000372''$§ [107, p. 209]. Since this deviation was extremely small, a problem concerning the estimation of errors in observations naturally arose, which required for its solution some knowledge of the theory of errors.

In this connection, in his famous work "New elements of geometry with a complete theory of parallels"†† Lobachevskiĭ, in Chapters XII and XIII, considers the solution of rectilinear and spherical triangles taking into account that the initial data is obtained with a certain degree of accuracy. Lobachevskiĭ writes:

> The errors in their combinations may be opposite one another and hence may be partially eliminated. The more numbers are added, the sooner it is bound to occur, therefore, it rarely occurs that the inaccuracy in such situations will be the largest possible. Thus, the solution will be fully adequate if and only if in addition to the precision

†All spellings of Lobachevskiĭ's name in Latin or Germanic languages are phonetic. The translator has seen eight or ten variations (*Translator's remark*).

‡Published in the Kazan *Messenger* (*Translator's remark*).

§Lobachevskiĭ committed an error in this computation. The correct value is $0''.00000372$ (see [107, p. 286]).

‡Published in separate parts in *Scientific Bulletins of the University of Kazan* for the years 1836, 1837, and 1838 (*Translator's remark*).

of the calculations we also present the *probability* of the occurrences of the errors. [108, pp. 397–398]

Thus Lobachevskiĭ arrived at the problem of determining the distribution law of a sum of a given number of mutually independent identically distributed random variables. The problem is solved in the "New elements of geometry," as well as in the paper "Probability of the average results obtained from repeated observations [109], published in 1842 in *Crelle's Journal* and which is a revision of the corresponding sections of "New elements of geometry."

Lobachevskiĭ solves the following problem: find the distribution of the sum $\mu_r = \lambda_1 + \lambda_2 + \cdots + \lambda_r$, where λ_r are mutually independent and each of them admits a value between $-a$ and $+a$ with the probability $1/(2a + 1)$. The sum μ_1 may admit only integral values $-ra \leqq m \leqq ra$.

Lobachevskiĭ calculates "the number of chances" to obtain $\mu_r = m$ which he denotes by $C_r(m)$. The formula he derives is as follows:

$$C_r(m) = \sum (-1)^\lambda C_r{}^\lambda C_{(r-2\lambda)a+m+r-1-\lambda}^{r-1},$$

where the summation is taken with respect to

$$0 \leqq \lambda < \frac{ra + m + 1}{2a + 1}.$$

The second problem considered by Lobachevskiĭ is that of obtaining the distribution of the arithmetic mean $\xi_r = (\delta_1 + \delta_2 + \cdots + \delta_r)/r$, where δ_i are mutually independent and uniformly distributed on the interval $[-1, +1]$.

For $P_r(x) = P\{|\xi_r| \leqq x\}$, Lobachevskiĭ derives the formula

$$P_r(x) = 1 - \frac{1}{r! \, 2^{r-1}} \sum_{0 \leq \lambda \leq (r-rx)/2} (-1)^\lambda C_r^\lambda (r - rx - 2\lambda)^r.$$

He thus successfully solves the problem of obtaining exact formulas for any number of observations.

Lobachevskiĭ presents the computations of $P_r(x)$ for $r = 2, 3, 10$ and gives short tables of $P_r(x)$ for some values of x.

He defines probability in accordance with the definition given by Laplace: "The word probability indicates the content of the number of favorable cases to the number of cases altogether" [108, p. 398].

Lobachevskiĭ's ideas concerning the errors in observations are also of interest. He writes that the errors in observations

may occur due to inaccuracies in the measuring instruments themselves, as well as inaccuracies in the setting up of the instruments. Both these sources of errors are diminished by repeated observations.... Dividing the quantity by the number of repetitions [Lobachevskiĭ probably means "dividing the sum of the quantities obtained by the number of repetitions"], we obtain the average observation.... The method of repetition is especially advantageous with small size instruments, since the fine subdivisions in the scales of large instruments, without diminishing the human error, can only increase the probability in the direction of the required accuracy. [108, pp. 406–407]

It is also interesting to observe that Lobachevskiĭ utilizes letters of the Russian alphabet for his notation of concepts in probability theory. For example, he denotes probability by the letter B, rather than P, and random events by C.†

Although Lobachevskiĭ dealt with problems of probability theory only occasionally and incidentally, nevertheless, his contributions in this field of mathematics are at the level of the best works of his time. Posing specific problems in the theory of errors, he rigorously derives accurate and convenient practical formulas which are relevant even up to the present time (see, e.g. [91]).

The first works on probability theory carried out in Moscow University were N.D. Brashman's paper "Solutions of problems in the calculus of probabilities" (1835) and N.E. Zernov's long memoir.

On 19 June, 1843, Zernov, professor of mathematics at Moscow University, addressed a ceremonial meeting of the faculty with a lecture on "The theory of probability with special applications to mortality and insurance." In the same year, a substantially extended version of this lecture was published in booklet form. In the literature devoted to the history of mathematics this work of Zernov is not discussed at any length. There is only a brief mention of it in two different volumes, devoted to Russian mathematicians of the eighteenth and nineteenth century [78, 159].

After an introduction in which Zernov asserts that probability theory fits the modern trend of increased interest in learning, he presents a brief historical sketch of the development of probability theory.

The origin and initial development of probability theory are described in the traditional manner overestimating the role of games of chance and Chevalier de Méré. Among the works on probability theory of that time he mentions the work of V. Ya. Bunyakovskiĭ, which was not published as yet. He writes that Bunyakovskiĭ authored a treatise "Discourse on mathematical probability theory, the first part of which is completed" [183, p. 6]. However, one can infer from the text of Zernov's book that he did not read Bunyakovskiĭ's work.

---

†B is the first letter of the word вероятность (probability); C is the first letter of the word случайный (random) (*Translator's remark*).

Next Zernov states some general propositions; these characterize him as a representative of the deterministic approach which was prevalent at that time.

Zernov's statements on the area of applications of probability theory are of interest.

> In the natural sciences probability theory proposes rules of mortality and population, through insurance it protects the citizen from disaster, among uncoordinated observations it chooses the suitable one, it guides the astronomer in determining deviations in the movements of celestial bodies, which are very important to the science, but which, in view of their minuteness could be lost within the inevitable errors. Probability theory penetrates to the temple of Themis, providing the measure of truth and mercy to judges, it can guide lawmakers in making rules for social order, estimating them by the observed results; it may shed light for the historian, weighing the authenticity of legends. Finally, the science of probability clarifies logic itself. [183, p. 7]

In this passage the valuable and scientifically justified applications of probability theory (such as statistics, demography, and the theory of errors) are mixed with the illusory and unjustified applications (such as legal procedures, the authenticity of legends, legislation, and court procedures).

Next comes the definition of probability ($p = m/n$) and some elementary properties of probability, such as $p + q = 1$; the probability of certainty $= 1$, and so on, are presented. The addition rule is formulated by the statement: "The probability of the sort is equal to the sum of probabilities of kinds" [183, p. 91]. The multiplication rule is formulated in the usual manner.

The notion of "relative probability" is then introduced—relative probability being the probability of events considered as if no other events take place.

An example follows this definition: An urn contains three red, one black, and two white balls. The probability of choosing a red ball is $P_{red} = \frac{3}{6}$; $P_{white} = \frac{2}{6}$, $P_{black} = \frac{1}{6}$. These are "actual probabilities." One then bets on the appearance of a white or a black ball, disregarding the red ones. The probability of winning this bet on the white ball is $\frac{2}{3}$, on the black ball $\frac{1}{3}$. These are "relative probabilities." They satisfy the relation $\frac{2}{3} = \frac{2}{6}:(\frac{2}{6} + \frac{1}{6})$; $\frac{1}{3} = \frac{1}{6}:(\frac{2}{6} + \frac{1}{6})$. Even this example indicates that the notion of relative probability is superfluous. It should be noted that Zernov always illustrates his exposition with examples.

Next Zernov explains the binomial probability distribution and formulates Bernoulli's theorem. He then remarks: "The importance of this theorem is that it provides the means for adapting the applications of probability theory from insignificant problems with urns containing balls, or dice tossing and card-playing to natural phenomena and political and moral areas" (183, p. 15].

He then investigates the organization of various lotteries and concludes that their structure greatly profits the proprietors.

Next Zernov discusses the notion of moral expectation. He points out that it is necessary to represent the actual value of a given amount of money based not on its face value but in relation to the capital of the person under consideration [183, p. 21]. If the capital of a certain individual is zero, then any amount for him should be considered infinite. To avoid this contradiction, Zernov argues that capital may be expressed not only in terms of money or property. Using this interpretation, almost every person possesses certain property. To justify these considerations he presents the quotation from D. Bernoulli:

> A man who is able to acquire ten ducats yearly by begging will scarcely be willing to accept a sum of fifty ducats on condition that he henceforth refrain from begging or otherwise trying to earn money. For he would have to live on this amount, and, after he had spent it, his existence must also come to an end. I doubt whether even those who do not possess a farthing and are burdened with financial obligations would be willing to free themselves of their debts or even to accept a still greater gift on such a condition. But if the beggar were to refuse such a contract unless immediately paid no less than one hundred ducats and the man pressed by creditors similarly demanded one thousand ducats, we might say that the former is possessed of wealth worth one hundred, and the latter of one thousand ducats, though in common parlance the former owns nothing and the latter less than nothing. [18]

Bernoulli concludes from this observation that "There is then nobody who can be said to possess nothing at all, unless he starves to death." Zernov adds in this connection that "Physical strength, any type of ability, and even the opportunity to beg or acquire debts is capital or in itself property" [183, p. 22].

All further exposition of the theory of moral expectation follows according to D. Bernoulli. D. Bernoulli's formula is presented and several examples are considered, including one dealing with the allocation of goods on several ships.

According to Zernov, need for introducing moral expectation in addition to mathematical expectation is evident from the following example in particular: A person has 100 rubles; let the probability of his acquiring or losing 50 rubles be equal to $\frac{1}{2}$. The ratio of the acquired sum to the future capital is $50/150 = \frac{1}{3}$, and the lost sum to the present capital is $50/100 = \frac{1}{2}$. "It follows from here that the loss of 50 rubles results in a greater tangible loss, although from the point of view of mathematical expectation the game is fair" [183, p. 25]. Proceeding from the notion of moral expectation, Zernov discusses the solution of the St. Petersburg problem. Concluding his exposition of the notion of moral expectation, Zernov observes that mathematical expectation corresponds to simple percentages, while moral to compound ones.

Next Zernov turns to the main area, in his opinion, of application of probability theory—demographic statistics and insurance. First he presents numerous data derived from various countries concerning the ratio of the number of males born to the number of females born. Some other constant ratios, e.g. mortality ratios, are also presented. This data when related to Russia points up the terrible state of existence of the people who lived under merciless deprivation and exploitation.

In Russia, in 1834, 657,822 Russian-Orthodox males died, among them 339,079 up to the age of 5. In general, out of 1,000,000 newborns in Russia, 540,762 (i.e., over 54%) died before the age of 5. In the Moscow Home for Children during the period 1764–1796, only 5360 children survived out of 40,669; during the period 1797–1828 only 26,352 survived out of 116,752. Commenting on this data, Zernov writes: "These numbers clearly reveal the state of the masses. One cannot help observing from this data the results of many ignorant and superstitious customs which are to be found among the common folk pertaining to mothers and newborns." [183, p. 38]. And on another occasion: "The ignorance surrounding the cradle of a Russian peasant child is much more harmful for the State than all of its enemies; this ignorance is, in any case, the most powerful enemy of Russia" [183, p. 49].

In this section of the book Zernov discusses a large number of demographic notions, such as the probable duration of life, the mean duration of life, the measure of longevity, and so on. The basic principles of construction of mortality tables are also considered. Various forms of insurance are described in detail. Malthus' and Quetelet's ideas are briefly discussed.

Further Zernov proceeds to other applications of probability theory: the theory of errors and legal procedures. In the section on the theory of errors, Zernov, after a few general remarks, presents the method of least squares. The most questionable part of Zernov's book is the last part which deals with juridical problems. In conclusion Zernov remarks on the lack of statistical data for scientifically substantiated inferences. The book concludes with the assertion that hardly any other science can be found, except for probability theory, which bears a "direct relation" to so many and such diversified disciplines [183, p. 83].

In his book Zernov attempted to exhibit all the applications of probability theory known to him. However, as a rule, he limited his discussions to a restatement of the known propositions and applications of probability theory and hardly ever expressed his own opinion concerning the topics at hand. This resulted in a certain unevenness in various parts of the book— the important applications of probability theory to errors in measurement and demography are presented on the same level as the applications to judicial problems which are devoid of any scientific or practical importance.

Clearly, such works as Zernov's book may have attracted and indeed did attract attention to probability theory, but did not provide any stimulating ideas to aid its further development.

The prominent Russian mathematician V. Ya. Bunyakovskiĭ (1804–1889) played a significant role in the spurring of interest toward probability theory in Russia and contributed to its propagation.

One of his first works related to probability theory is the paper "Some thoughts on the misconceptions of certain notions related to society, particularly lotteries and games" [129]. Bunyakovskiĭ writes: "One can point out many useful and important truths unknown to the common people . . ., many problems which were solved long ago and which are still considered debatable and even sometimes, due to deep-seated prejudices, are wrongly solved" [129, p. 80]. As examples, Bunyakovskiĭ cites: inoculation, which increases the mean duration of life; the usefulness of insurance companies; and the harm of gaming, etc.

The insurance business, which had been in existence a long time by then, was not widespread in Russia. The general public was suspicious toward insurance. Bunyakovskiĭ raised his voice against this type of prejudice.

The paper basically concerned itself with an investigation of the role of moral advantage and moral expectation. Using these notions, Bunyakovskiĭ concludes that: "Any game is disadvantageous, even if the players are completely honest. Any lottery is disadvantageous for those who buy tickets. It is better to endanger your property in part, than to do it in one shot."

This interesting paper was included in Bunyakovskiĭ's fundamental work entitled "Foundations of the Mathematical Theory of Probabilities" [38].

In the introduction to this book he notes the difficulties encountered during its compilation, in view of the absence of proper Russian terminology.

> This book is the first Russian work containing a detailed exposition of the mathematical foundations of probability theory, as well as its most important applications. Since up to now no single work or even translation of a book devoted to mathematical probability theory has been available, I have undertaken to write a book on this subject for which no established terminology exists. [38, pp. II–III]

Bunyakovskiĭ performed this task brilliantly. The terminology he introduced is valid up to the present almost without alteration.

Bunyakovskiĭ notes that in writing his book he utilized extensively Laplace's "Théorie Analytique des Probabilités."

Next Bunyakovskii proceeds to a discussion of certain general methodological problems. He is a proponent of the deterministic philosophy which flatly rejects the possibility of the existence of randomness.

After quoting Laplace [98] concerning the omnipotence of the intellect which asserts: "Given for one instant an intelligence which could comprehend all the forces by which Nature is animated . . . for it, nothing would be uncertain and the future, as the past, would be the present in its eyes," Bunyakovskii presents the following explanatory remark:

> If all the data on which an event depends were known and if, moreover, we were gifted with a mind so penetrating as to embrace and grasp the mutual relations in this data, we would then be able to solve problems unerringly and predict the occurrence or nonoccurrence of an event. [38, p. 2]

Bunyakovskii's views on the subject of probability theory are rather diffused. He classifies probability theory among topics of applied mathematics, while he comments about its general problems as follows: "The Analysis of Probabilities examines and evaluates numerically events which depend on causes not only completely unknown to us, but also those whith due to our ignorance, are not subject to any suppositions" [38, p. 1]. At another point he writes: "The likelihood, under different circumstances, may be smaller or larger, hence it, as any mathematical quantity, can be measured. This measure in the mathematical sense is called probability and its calculus which deals with its precise determination is the Analysis of Probabilities" [38, p. 3].

Next he discusses the notion of moral certainty. Here the ambiguity and diffusion of the arguments are quite understandable, since the subject itself has never fallen within well-defined confines. Bunyakovskii writes that moral certainty "takes place in those cases, when our mind accepts a certain fact with complete inner conviction, but cannot, however, prove its existence by means of indisputable arguments. For a sensible man, morally certain truths must be as valid as propositions affirmed with mathematical certainty" [38, p. 4].

After this general introduction, Bunyakovskii proceeds to an exposition of probability theory proper. The measure of probability is defined as the ratio of equiprobable cases. Equiprobable events are defined as those "whose existence is in a strict sense equally undecided" [38, p. 4].

It should be noted that the exposition in the book is accompanied by a large number of well-chosen problems and various types of examples. The following two problems are of interest.

Problem 1:   Find how many times a die must be thrown in order that the probability of the appearance of a given number of points, e.g. six, will be equal to a given number, say $\frac{1}{2}$. Bunyakovskiĭ solves this problem correctly and concludes that "in a fourfold throw of the die a single occurrence of six is more probable than its nonoccurrence" [38, p. 23].

Problem 2:   Find the required number m of throws with two dice, that the probability of occurrence of six on both dice will equal $\frac{1}{2}$. Obtaining the solution that $m = 24.6$, Bunyakovskiĭ writes: "With a 24-fold throw of two dice, the occurrence of 12 points is less probable than the opposite event, while in a 25-fold throw the occurrence of 12 points becomes more probable than the nonoccurrence" [38, p. 24].

Next Bunyakovskiĭ digresses to relate the historical anecdote concerning de Méré and Pascal and the above problems. Pascal solved these problems; however, de Méré objected to the solution of the second problem, claiming that, in order to obtain a probability greater than $\frac{1}{2}$ (for the occurrence of 12 points), 24 throws are sufficient, and not 25.

Bunyakovskiĭ describes in detail the problem of the most probable number of occurrences of a given event in a series of trials. He then proceeds to a discussion of Bernoulli's theorem. Concerning this theorem he writes:

Statisticians... base almost all of their conclusions on this theorem. Their results on population in general, on shifts in population, on the number of criminals, fertility of the soil, import and export of goods, etc. are founded on this theorem. The natural sciences, medicine, legal procedures, in other words all fields of knowledge are concerned, to a certain extent, with this principle. [38, p. 35]

Bunyakovskiĭ states Bernoulli's theorem as follows:

If the number of repetitions of a trial which leads to one of two simple events A or B is increased indefinitely, the ratio between the number of occurrences of this event steadily approaches the ratio of their simple probabilities and, finally, with an appropriate number of trials, differs from this ratio arbitrarily little. [38, p. 36]

To clarify the essence of the matter, Bunyakovskiĭ formulates Bernoulli's theorem in the form of the following problem:

A large series of trials is performed, each of which leads to one of the two events A or B; the simple probabilities for A and B are assumed to to be constant; express them by

$$p = a/(a + b) \qquad \text{and} \qquad q = 1 - p = b/(a + b)$$

correspondingly. If we denote by m the number of trials, then the most probable compound event will be $A^x B^{x'}$ for which the ratio x/x' is either equal to the fraction a/b or differs very little from it, and, moreover, $x + x' = 1$. Now the following two questions may arise: (1) How large is the probability P such that in m trials event A will occur not less than $x - 1$, and not more than $x + 1$ times, and, consequently, event B will occur not less than $x' - 1$, and not more than $x' + 1$ times, assuming that 1 is by far the smaller number than x and x'. (2) Assuming that the probability p of event A is unknown, but knowing how many times it did occur in m trials, determine the probability P' that p is contained within the given limits. [38, pp. 39–40]

Bunyakovskiĭ's solution to the first problem is as follows:

$$P = \frac{2}{\pi^{1/2}} \int_0^t \exp(-t^2)\, dt + \frac{m^{1/2}}{m^{1/2}(2xx')^{1/2}} \exp(-t^2),$$

where $t = lm^{1/2}/(2xx')^{1/2}$. $\hspace{3cm}$ (III.11)

The value obtained of P denotes the probability that, after a very large number of m trials is performed, the number of repetitions of event A is contained between the limits $x + l$ and $x - l$, and that of B between $x' - l$ and $x' + l$, assuming that x and x' are integers whose sum is m, and the ratio x/x' is the closest to ratio $p/(1 - p)$ of the simple probabilities of events A and B; l denotes a number of order not higher than $\sqrt{m}$. [38, p. 44]

Next Bunyakovskiĭ computes the integral $\int_0^t \exp(-t^2)\, dt$:

$$\int_0^t \exp(-t^2)\, dt = \frac{1}{2}\pi^{1/2} - \frac{\exp(-t^2)}{2t}\left[1 - \frac{1}{2t^2} + \frac{1 \cdot 3}{(2t^2)^2} - \frac{1 \cdot 3 \cdot 5}{(2t^2)^3} + \cdots \right].$$

Substituting this value in (III.11) and letting t approach ∞, we obtain that $P \to 1$.

It should be concluded from the above, that, as the number of repetitions of the trials increases indefinitely, the ratio of the number of occurrences of event A to the number of occurrences of B steadily approaches the ratio of the simple probabilities of the events A and B, and, finally, differs from this ratio as little as desired. This regularity in the repetition of randomness which is observed in a large series of trials is expressed in Jacques Bernoulli's remarkable theorem. [38, p. 48]

Solving the second problem posed above, Bunyakovskiĭ obtains the following result: Let i be the number of occurrences of an event A in m trials. Then the chance that the probability of the occurrence of event A

is contained in the limits

$$\frac{i}{m} + \frac{t\,[2i\,(m-i)]^{1/2}}{m^{3/2}}\ , \qquad \text{where} \quad t = \frac{lm^{1/2}}{[2i\,(m-i)]^{1/2}}\ ,$$

is given by

$$P' = \frac{2}{\pi^{1/2}} \int_0^t \exp\,(-t^2)\,dt + \frac{m^{1/2}}{\pi^{1/2}\,[2i\,(m-i)]^{1/2}} \exp\,(-t^2).$$

Next Bunyakovskiĭ proceeds to the urn scheme of selecting balls without replacement.

Chapter III is devoted to the topic of mathematical expectation, which he introduces after his discussion of fair games. The problem of fair division of stakes, as well as others, is considered, including the organization of various lotteries, in particular the French. Computing the fair value of the win for a given combination of numbers, Bunyakovskiĭ concludes that, in the existing lotteries, the purchasers of the tickets are substantially underpaid.

Chapter IV is devoted to the topic of moral expectation. Here Bunyakovskiĭ first presents Buffon's and D. Bernoulli's views of this subject and then proceeds to the discussion of the St. Petersburg game.

In succeeding chapters various problems are considered, among them Buffon's needle problem and some of its generalizations.

A long chapter "On the probabilities of human life" is devoted to demographic problems. Here Bunyakovskiĭ also discusses the subject of infant mortality.

Investigating the construction and utilization of mortality tables, he presents as an example a mortality table constructed on the basis of the 1842 statistics for Moscow. Commenting on the fact that out of 9276 newborns, only 5815 of them survived up to age 5, he writes:

> We observe that the probability of survival in villages is significantly higher as compared with large cities; the reason for this discrepancy becomes quite understandable, if we take into consideration the harmful influence of city life on public health, in particular for the lower classes who more than others are subject to diseases, poverty, overcrowded living quarters, and so on. [38, p. 179]

Next a large number of demographic problems is discussed, such as the measure of longevity, mean duration of life, measures of fertility and mortality, the increase in life span due to elimination of certain diseases, etc.

After considering problems connected with population, Bunyakovskiĭ proceeds in the next chapter to the structure of various benefit funds, insurance institutions, and so on. Here many and various problems are solved.

The next chapter entitled "On the most favorable results of observations" is devoted to estimation of errors in observations. Bunyakovskiĭ states that errors in observations are inevitable. However, the problem remains of how to derive the most probable result from the observations. He concludes that the most probable result in the case of equally probable errors is determined by the arithmetic mean of the observations, provided their number is very large. "This rule is also valid in the case when the errors are not assumed to be equally probable, but obey with some restrictions a certain [probability] law" [38, p. 264].

Bunyakovskiĭ then derives the result that random errors in observations are distributed according to a normal law.

The book concludes with a short historical sketch of the development of probability theory.

The comprehensiveness of the contents and the clarity of the exposition makes this work of Bunyakovskiĭ's a distinguished contribution to probability theory not only in Russia, but in contemporary world literature as well. However, in spite of the definite merits of the book, one cannot ignore certain basic errors and misjudgments committed by Bunyakovskiĭ, first and foremost in the application of probability theory to witness testimonies, court decisions, and other unjustified applications to the problems of social life, which are discussed in Chapter XI entitled "Application of probability analysis to testimonies, legends, various kinds of choices between candidates and opinions and to court sentences by majority rulings."

These applications and opinions were repeatedly subjected to criticism. In particular, A. A. Markov dwells on this point in his book "Calculus of probabilities." [121] He examines the following problem presented in Bunyakovskiĭ's book: Six letters are selected at random from the Russian alphabet, and are placed one after another in order of their selection. Two eye-witnesses assert that the selected letters form the word MOCKBA (Moscow). What is the probability that the testimony of these witnesses is valid?" [38, p. 314]. Here it is assumed that the Russian alphabet consists of 36 letters and that the inclination of the witness to tell the truth is expressed by the fraction $\frac{9}{10}$. In the course of the solution of this problem Bunyakovskiĭ makes several completely unjustified assumptions. Markov's comments in this connection are as follows:

> The above cited example, in our opinion, clearly points up the inevitability of various arbitrary assumptions when dealing with this type of essentially undetermined problem. The problem becomes even more arbitrary if we allow that the witnesses may err and if we eliminate the assumption that the testimonies are independent.

... First of all, if the event is impossible, then no testimony can make it even slightly probable. Finally, the information about an event may reach us not from eyewitnesses, but through a chain of witnesses who in turn relay what they have heard from others. As the chain of witnesses increases, the event becomes more and more obscure. Regardless of the mathematical formulas, which we are not discussing here, since they are of little relevance, in our opinion, it is clear that stories on improbable events which allegedly occurred in ancient times should be regarded with extreme skepticism. [121, p. 320]

Bunyakovskiĭ attempts to guard religious traditions against various attacks. He writes:

Every conclusion deduced from an analytic formula is merely the product of the initial premise on which the formula was based. If the premise is false, then the conclusions of the analysis will be in error. One should therefore justly investigate in detail the proposition which serves as the point of origin. When this analysis leads us to the conclusion that there are facts in the spiritual world, which are not subject to physical laws, all the malicious philosophizing of the pseudo-philosophers will be wiped out. [38, p. 326]

Concerning this statement Markov observes:

We can in no way agree with Bunyakovskiĭ that it is reprehensible to doubt the validity of a certain type of [religious] legend. My disagreement with Bunyakovskiĭ in this case is outside the realm of mathematics and concerns the shaky ground of personal desires and interests. [121, p. 320]

In addition to his basic book on probability theory, Bunyakovskiĭ wrote several articles related to probability theory and its applications. He also compiled detailed tables for retirement funds of the military establishment.†

A number of Bunyakovskiĭ's papers are related to population mobility, mortality tables, etc. He also computed estimations of armed forces contingents for various years. Most of his works on problems connected with probability theory originated from specific practical problems. For example, concerning his paper "On the approximate summation of numerical tables" he observes: "The problem posed in this paper consists of determining in the simplest possible manner the approximate sum of entries in numerical tables, or in general, approximate sums of a large number of terms" [39, p. 1].

This problem is reduced to the following: Let a large number of summands be randomly chosen with an arbitrary number of digits. Without adding these numbers, but by counting the number of summands classified ac-

---

†A number of pension plans in Russia were based on these calculations (see [215], p. 301).

cording to different orders, it is required to determine: (1) the most probable or the average sum of these summands and (2) the probability that the deviation of this sum from its true value is contained within the given limits.

This paper bears a direct relation to the calculator devised by Bunyakovskiĭ in 1867 submitted to the Physical and Mathematical Section of the Academy of Sciences. This device is known as "the Russian self-calculator." Bunyakovskiĭ's self-calculators were suitable for summing up a large number of relatively small summands.† His paper also shows that the advent of these calculators was preceded by a probabilistic analysis of problems connected with summation of a large number of terms.

Thus Bunyakovskiĭ's main interests in the field of probability theory were in the area of its applications. This was also prompted by the fact that in 1858 he was appointed chief government adviser on statistics and insurance matters. His major work written during that period was "An essay on mortality laws and the distribution of the orthodox population according to age" (1865).‡ In this paper he developed a new method of constructing mortality tables. Closely related to this essay is the paper "Anthropobiological investigations and their application to the male population in Russia" (1873–1874). In this work Bunyakovskiĭ determines the current distribution of the male population according to age on the basis of comparisons of the register data for previous years.

It is noteworthy that in one of his papers Bunyakovskiĭ proposes the following possible application of probability theory: "The new application is related to grammatical and etymological investigations of a language as well as in the area of comparative philology" [173, p. 48].

A prominent representative of the Russian school of probability theory was M. V. Ostrogradskiĭ (1801–1862).§ His contributions to probability theory were also prompted mainly by practical considerations.

In 1856, after their defeat in the Crimean war, Russia lost its right to operate a fleet in the Black Sea. Mass dismissals of sailors, as well as civil employees, were imminent. To improve their financial lot it was decided to establish a retirement pension fund, which would begin operation in 1859. Both Bunyakovskiĭ and Ostrogradskiĭ were invited to participate in the Commission to set up the rules governing this pension fund. Ostrogradskiĭ's note entitled "A Note on the Retirement Fund" was published in the volume "A Proposal for Establishing a Retirement Pension Fund in the Navy Department" (1868). In this paper Ostrogradskiĭ points out that retirement funds should be constructed on the basis of actuarial computations.

---

†Bunyakovskiĭ's paper on self-calculators prompted Chebyshev to construct his arithmometer (*Translator's remark*).

‡The members of the Greek Orthodox Church, the official church of the Russian Empire, constituted the great majority of the population in Russia (*Translator's remark*).

§Many papers deal with M. V. Ostrogradskiĭ's activities. We refer the reader to [71] and [74].

In the introduction Ostrogradskiĭ describes his paper:

> We present the solution of the problem of the smallest pension and reduce it to such a degree of simplicity that even a person little versed in arithmetic would find no difficulty—only a knowledge of addition is required. The three enclosed tables provide these solutions. [143, p. 298]

Next follow the above-mentioned tables, using which it is easy to compute the amounts of pension for various cases.

Numerical examples supplemented with detailed discussions are included in this work, which was utilized for a long time in practical applications of insurance in Russia.

In almost all of Ostrogradskiĭ's works on probability theory the influence of Laplace's book ("Théorie Analytique des Probabilités") is strongly felt. Ostrogradskiĭ considered probability theory an important tool for the study of regularities which arise in mass phenomena. However, he often committed errors of a philosophical and methodological nature. In the spirit of Laplace, he was a proponent of the law of insufficient reason and applied probability theory to moral problems, in particular to legal proceedings.

Ostrogradskiĭ's first paper devoted to probability theory, "Extract from a memoir on the probability of judicial errors," was presented at a session of the Academy of Sciences on 12 June 1834, and was published in 1838 (pp. XIX–XXV). In this paper Ostrogradskiĭ considers a type of tribunal in which the judges each have a different degree of correct judgment. Assuming that the limits of the veracity for each of the judges are known, Ostrogradskiĭ derives formulas for the probability of an erroneous verdict in a court consisting of a given number of judges.

Solving these types of essentially unsubstantiated problems, Ostrogradskiĭ makes some interesting remarks in conclusion. He finds that if all the judges are capable of arriving at a correct decision with the same probability, then the probability of the court committing an error will depend only on the majority of votes and not on the total number of the judges. This result contradicts Laplace's opinion, who believed that the probability of accepting a wrong decision in the case when it is voted unanimously by 12 judges is different from the probability of accepting this decision by a majority of 12 votes in a court consisting of 212 judges.

In this connection Ostrogradskiĭ observes:

> We have utmost confidence in an impartial court consisting of 12 people, if a decision has been made unanimously: if, however, the court consists of 212 judges and it is known that only 12 judges are of the same opinion, one should wait until the opinion of the majority

becomes known. However, barring the knowledge of the opinion of the two hundred judges, we arrive again at the case in which a court consisting of 12 individuals has unanimously reached a decision. Where then does this great discrepancy in our confidence come from? ... There is actually no difference and we erred by not delving sufficiently into the problem. [143, p. 66]

In the paper "On a problem concerning probabilities" reported on 23 October 1846, at a session of the St. Petersburg Academy of Sciences and published in 1848 (Vol. VI, pp. 321–346), Ostrogradskiï investigates a problem in quality control and studies the following example:

An urn contains white and black balls, whose total number is known, but the number of balls of each color is unknown. A certain number of balls is drawn from the urn and, after counting the number of white and black balls among them, the balls are returned to the urn. It is required to determine the probability that the total number of white balls is contained within specified limits. Or, in other words, we are seeking the relation between this probability and the above mentioned limits. [143, p. 215]

This example was presented not as an exercise in deriving analytical expressions, but as a practical problem. He writes:

To understand the importance of this problem, let us visualize the situation that a person is to receive a large number of objects subject to certain requirements and a certain amount of time is required to check these requirements.

Military suppliers often face this type of problem. The balls in the urn represent the objects supplied, the white balls may, for example, correspond to the acceptable ones, satisfying certain requirements, and the black corresponding to the unacceptable ones. The draw of a certain number of objects to determine their colors corresponds to checking a part of the received order to determine their quality. This part is five, six, or seven percent of the total number of objects taken at random. After drawing this part and computing the number of acceptable in it, we find the probability that the total number of the acceptable objects will stay within the limits set up in advance. These calculations are performed in the same manner as the calculations of white and black balls in an urn. By a suitable choice of the limits as well as of the number of objects to be checked, the required probability may approach certainty as closely as desired.

Hence, if we solve the problem stated above, a supplier may utilize it to reduce almost twentyfold the very tiresome mechanical labor, as for example in the case of sacks of flour or bolts of cloth. [143, p. 215]

The solution of the problem begins with a discussion of the following auxiliary question: An urn contains a given number of white and black balls, but the proportion of these balls is unknown. Let l balls be drawn from the urn. Find the probability that among these l balls there are n white and m black.

Here is the argument as given by Ostrogradskiĭ:

If l balls are drawn, there can be

$$0, 1, 2, 3, \ldots, n, \ldots, l - 1, l$$

white balls and, hence correspondingly

$$l, l - 1, l - 2, l - 3, \ldots m, \ldots, 1, 0$$

black balls.

Since all these different hypotheses, whose total number is $l + 1$, are equally probable, the probability of each one of them, including the one we have in mind, is $1/(l + 1)$. [143, p. 216]

Ostrogradskiĭ does not indicate that the probabilities of various *outcomes* are not equiprobable. However, to obtain computational formulas he uses the following notation:

$$[z]^k = z(z - 1)(z - 2) \ldots (z - k + 1).$$

Further, Ostrogradskiĭ makes the following assumption: "Let there be x white balls and y black balls in the urn. Find the probability that among the l drawn balls there will be n white and m black ones" [143, p. 216]. This probability is

$$[l]^l [x]^n [y]^m/[n]^n [m]^m [s]^l,$$

where s is the total number of balls.

Assume now that l balls were drawn from the urn and n of them turned out to be white, and m of them black. What is the probability that among $s - l$ balls still remaining in the urn there are x white and y black?

This probability is equal to

$$[l + 1]^{l+1} [x]^n [y]^m/[n]^n [m]^m [s + 1]^{l+1}, \tag{III.12}$$

or, in more common notation:

$$C_x{}^n C_y{}^m/C_s{}^{l+1}.$$

Next Ostrogradskiĭ presents some computational formulas and then discusses the question of the variation in the value of the probability given by (III.12) with variations in the values of x and y correspondingly.

Ostrogradskiĭ obtains that under the conditions of the problem the probability that the number of white balls does not exceed q or the number of black balls is less than $b = s - q$ is given by

$$\frac{[l + 1]^{l+1}}{[m]^m [s + 1]^{l+1}} \sum_{k=0}^{k=m} \frac{[m]^k [b]^{m-k} [q + 1]^{n+k-1}}{[n + k + 1]^{n+k+1}}.$$

At the conclusion of the paper a numerical example is discussed in detail. Various tables to aid computation are also presented in this paper.

Ostrogradskiĭ's paper "On insurance" was published in 1847 in a journal *Finskiĭ Vestnik* [*Finnish Herald*].† This paper is of great interest since it is devoted in the main to philosophical and methodological problems. The paper was not completed.

In this paper some critical remarks on the recently published book by Bunyakovskiĭ "Foundations of the Mathematical Theory of Probabilities" (1846) [38] are given. However, the author's name is not mentioned.

The paper commences with the following observations:

> The theory of insurance cannot be presented without the use of probability calculus, as it is based on this subject. We shall therefore attempt firstly to give a clear explanation of the concept of probability. [143, p. 238]

Next he proceeds to his critical remarks addressed to Bunyakovskiĭ:

> The author desires to show that the probability, which he calls likelihood, is indeed a (numerical) quantity. To prove this, he tries to prove that likelihoods can be larger or smaller than one another. Next, when all the arguments in this connection are presented which are, in his opinion, sufficient for our complete satisfaction, the author concludes that likelihood, like any other mathematical quantity, can be measured.
>
> Thus, according to the learned author's view, likelihood is a mathematical quantity only because some likelihoods may be larger or smaller than others.
>
> This opinion is not entirely correct. Indeed, are we not saying, and do we not clearly understand in saying it, that a certain scientist is more conscientious than another, that a Frenchman is braver than a German, that a reader is wiser than a writer, and so on? Thus the conscientiousness, bravery, wisdom, etc., may be larger or smaller and hence they are mathematical quantities, which can be measured, expressed numerically, and be subject to various operations. Arguing in this manner, we increase significantly the realm of the mathematical sciences and may give rise to the foundations of mathematical theories of dishonesty, absurdity, etc. [143, p. 238]

†Vol. 13, No. 1, pp. 29–34 and No. 2, pp. 40–44.

Ostrogradskiĭ interprets the notion of probability from a subjective point of view as the degree of certitude of the observer.

He writes:

> There is no probability in Nature. Everything that occurs in the universe is inevitable and beyond doubt. Probability is a consequence of human weakness; it relates to us, exists for us and can be only for us. A consideration of probability is an important and even necessary addition to those few truths which we know with relative certainty.
>
> Natural events take place in a completely determined sequence. Superior beings could have discovered and proved this sequence of events. We, however, knowing neither the origin nor the mutual relation of these phenomena and observing only an insignificant part of those in the vicinity of our planet, are unable to predict the sequence of these phenomena and even are, for the most part, forever ignorant concerning their existence. As far as we are concerned, the events which we know about from observations are only probable to various degrees. [143, p. 240]

Ostrogradskiĭ discusses in detail the ideas that probability is the measure of our ignorance and that it is a subjective notion, that the universe is deterministic in nature and no chance element is present in it. Those objects or phenomena which we do not know about or which have not been perceived are called random phenomena.

> If the phenomenon is totally dependent on certain other phenomena or events, some of which may cause its occurrence while others prevent it, and if all these cases are such that, *as far as we are concerned*, there is no reason to prefer one of them to the others, then the probability of the expected event is measured by a fraction whose numerator is the number of cases leading to this event, while the denominator is the number of all the cases. [143, pp. 240–241]

This assertion coincides with the so-called classical definition of probability due to Laplace in which the equiprobability is interpreted as the insufficiency of grounds to prefer one event over another.

An example is then examined. Five balls are in an urn, among them three white and two black. A ball is drawn from the urn. What is the probability that it is white? Commenting on this example, Ostrogradskiĭ writes:

> There is no reason to assume that one ball will fall in our hands faster than another. By saying "no reason," we mean that *we* do not have a reason—the reason, however, exists, but is unknown to us. . . . If someone knew the allocation of the balls in the urn and could compute the movement of the hand which draws the balls, we would be able

to tell in advance which ball is going to appear, there would be no probability in this case. . . .

We repeat that probability and equal possibility of the outcomes, as well as the measure of probability exist only in our minds. For those who have total knowledge of all the phenomena, probability not only has no measure, but simply does not make sense. [143, p. 241]

These assertions are typical of the philosophy of mechanistic determinism which prevailed at that period of time in probability theory.

Next Ostrogradskii proceeds in his paper to a discussion of problems connected with insurance and lotteries. First he explains the notion of mathematical expectation. After examining a numerical example of computing an insurance premium, Ostrogradskii advises the reader: "Pay no more, on the contrary try to pay less, in order to have some advantage. Don't worry about the insurance company—it will never lose money" [143, p. 244]. He also criticizes lotteries which bring big profits to the organizers.

The paper concludes on this note. The continuation of the paper, which was intended, never appeared.

In Ostrogradskii's paper "Game of dice" (1847)† certain games of dice are examined and elementary computations of stakes in fair games are carried out.

These two papers are in fact directed against games of chance and lotteries which had at that time acquired widespread popularity. The games and lotteries were often based on the principle that the cost of a ticket or of the stake substantially exceeded the mathematical expectation of the win. This fact, however, was carefully concealed.

There is some evidence that in 1858 Ostrogradskii was reading an elective course on probability theory at the Michailov Ordnance College. There were 20 lectures altogether. The first three lectures were apparently printed lithographically. However, copies of this edition have not been found.

Gnedenko in one of his papers devoted to the history of probability theory [77] presents the text of a brief historical sketch of probability theory which is kept in the manuscript section of the State Public Library of the Ukrainian SSR and conjectures that this is possibly a draft of Ostrogradskii's introductory lecture presented in Michailov College.

One may thus conclude from the discussion above that Ostrogradskii's work on probability theory was on a par with the level of this science at that period.

We are in full agreement with the appraisal given by Gnedenko of Ostrogradskii's contributions to probability theory:

†*Finskii Vestnik*, **13,** No. 3, pp. 29–32.

In spite of the fact that in his definition of probability Ostrogradskiĭ committed methodological errors, slipping toward a philosophy of subjectivism, the general direction of his creative work in probability theory should be evaluated as instinctively materialistic. For Ostrogradskiĭ, probability theory was valuable only as a tool for cognition of the material world. The philosophical vaccilations connected with his definition of probability and related notions cannot overshadow this conclusion. Indeed, the subject matter of his papers in the field is closely connected with practical problems. [77, p. 123]

In concluding this section we point out that both Bunyakovskiĭ and Ostrogradskiĭ contributed greatly to the dissemination of probability theory in Russia.† However, their works were mostly in the mainstream of the old traditional themes and they did not touch upon the central problems of probability theory. Solutions to these central problems were imperative for finding a way out of the impasse reached at that time.

In order to stimulate anew the development of probability, a materialistic approach to the basic problems were required, and new ideas were needed.

These were introduced by Pafnutiĭ Lvovich Chebyshev.

†In 1875 Bunyakovskiĭ at a special celebration was hailed as the innovator of probability theory in Russia (see, e.g., Yushkevich [217]) *Translator's remark*).

IV

Probability Theory in the Second Half
of the Nineteenth Century

1 Chebyshev—The originator of the Russian school of probability theory

The creative works of Ostrogradskiĭ and Bunyakovskiĭ promoted mathematical education and learning in Russia. Their contributions stimulated interest in mathematics, and probability theory in particular, among the younger generation. All this paved the way for the establishment of the St. Petersburg School.

The creator and ideological leader of this largest prerevolutionary mathematical school in Russia was Pafnutiĭ Lvovich Chebyshev (1821–1894). Chebyshev, as well as many other mathematicians, was influenced by the works of Ostrogradskiĭ and Bunyakovskiĭ.

Chebyshev's contributions were prominent in the development of many branches of mathematics. His investigations span the theory of approximating functions by polynomials, theory of numbers, theory of mechanisms, probability theory, and many other areas. In each of these areas, Chebyshev originated basic new general methods; his ideas had a decisive effect on their further development. The development of mathematics was greatly influenced not only by Chebyshev's own contributions, but also by his activities in guiding and supervising the work of many young scientists.

The mathematical school guided by Chebyshev was of paramount importance in the development of mathematics in Russia. The most prominent representatives of this school were A. N. Korkin (1837–1908), E. I. Zolotarev (1847–1878), A. A. Markov (1856–1922), G. F. Voronoi (1868–1908), A. M. Lyapunov (1857–1918), D. A. Grave (1863–1939), V. A. Steklov (1864–1926), and others. The school was united not only by common interests and problems, by the method of discussion of problems and formulation of inquiries, but also by its materialistic approach to science in general and mathematics in particular.

Markov, Lyapunov, and Steklov are known as prominent representatives of the materialistic philosophy in the sciences. Chebyshev succeeded in exerting so great an influence on the development of the science, mainly because of his materialistic approach to the laws of science and its regularities.

As we have seen, toward the middle of the nineteenth century probability theory reached an impasse. Chebyshev was the one who designated the path to follow in its development, breathed new life into it, solved the basic fundamental problems, and attracted the most talented mathematicians into the field by presenting them with challenging questions—all these activities led the way out of the impasse and prompted further rapid development of probability theory.

In a discussion with A. V. Vasiliev† a short time before his death, Chebyshev pointed out half jokingly that mathematics had survived two periods. During the first period problems were posed by the gods (the Delian problem of doubling the cube) and in the second by the demigods (Pascal and Fermat). In the current third period problems were posed for practical considerations. Chebyshev strongly believed that the harder the problem, the more productive the methods for its solution and the wider the scope of its possible applications. The close relationship between theory and practice was the determining factor in his mathematical activities. His views on this subject were expressed pointedly in his address presented at the commencement ceremonies of the University of St. Petersburg in 1856, entitled "The drawing of geographical maps." He writes in this address:

> From ancient times the mathematical sciences have attracted special attention; at present even more so, due to their influence on the arts and industry. The rapprochement of theory with practice yields very

---

†Vasiliev (1853–1929) was a well-known popularizer of mathematical education and ideas in Russia. His works include a biography of Lobachevskiĭ and books on the history of number theory and analysis. See also a brochure by Vasiliev entitled "P. L. Tchebychef et son Oeuvre Scientifique" (Turin, 1898) (*Translator's remark*).

> beneficial results, and it is not practice alone that gains thereby; the sciences themselves develop under its influence: new subjects are opened up for investigation and new aspects of subjects already known are discovered. In spite of the very high level of development of the mathematical sciences by the great geometers of the last three centuries, practice showed up certain inadequacies; it poses new problems for the sciences and in this manner encourages research of totally new methods for their solution. While theory benefits greatly from new applications of further extensions of an old method, theory benefits still more by the discovery of new methods, and in this case science finds a reliable guide in practice. [45, p. 249]

Chebyshev strove toward a rigorous and effective solution of problems, to the construction of algorithms, which would permit clear-cut numerical answers or at least lead to an approximate useable solution. According to Chebyshev an approximate solution is accurate if it is possible to determine bounds for the errors. In his works, Chebyshev often used relatively simple mathematical tools. He applied only functions of real variables, extensively utilized algebra, and, in particular, the methods of continued fractions. A significant role in the establishment of the Russian mathematical school was played by the pedagogical activities of Chebyshev at the St. Petersburg University. He was a remarkable lecturer and instructor. Eminent students of Chebyshev were Lyapunov and Markov. Their scientific activity was under the constant and direct guidance of Chebyshev. A penetrating characterization of Chebyshev's approach to mathematical problems was given by Lyapunov.

> Chebyshev and his disciples always remained on solid ground, guided by the opinion that only those investigations initiated by applications (scientific or practical) are of value and only those theories which arise from the consideration of particular cases are actually useful.
> The majority of P. L. Chebyshev's works and that of his followers tended toward a detailed investigation of problems important from the point of view of applications and which at the same time present special theoretical difficulties and require the construction of new methods, the results being extended into a general theory. [42, p. 20]

However, despite the progressive attitude that prevailed in the St. Petersburg's school, a certain narrowness in approach was one of its characteristics. Chebyshev and many of his students were often cold and skeptical toward various important achievements in Western European mathematics.

In the two-hundred-year development period in probability theory its main achievements were the limiting theorems: the law of large numbers and the de Moivre–Laplace theorem. But the confines of applicability of these theorems and their further refinements and generalizations were not satisfactorily determined. Often basic errors were committed in applying these theorems. Even Laplace and Poisson did not escape these misconceptions. For instance, applying probability theory to an evaluation of the likelihoods of evidence and testimony, the correctness of court decisions, and other problems of public life was inappropriate. This type of unjustified and erroneous application was embarrassing to probability theory and undermined its effectiveness. Gnedenko characterizes this period in the development of probability theory as follows:

> The enchantment with probability theory in the first quarter of the past century is connected with the names of Laplace and Poisson resulting in a large number of papers devoted to applications in various branches of the natural sciences and public life. Many of these had so little validity that they were considered "mathematically scandalous affairs." Disenchantment followed and among Western European mathematicians probability theory began to be thought of as some kind of mathematical entertainment hardly deserving serious attention. [158, p. 390]

Because of this state of affairs, further substantiation and more accurate specification of the basic postulates of probability theory were required. The subject matter of probability theory and the scope of its applications had to be clarified. The special methods and techniques at hand had to be investigated and reinforced. A major contribution in this direction was made by Chebyshev.

Chebyshev's interest in probability theory was permanent and unfaltering. His Master's dissertation at Moscow University in 1846 was entitled "An essay on elementary analysis of probability theory." When Bunyakovskiĭ resigned from the University of St. Petersburg in 1860, Chebyshev replaced him for the academic year 1860–1861 and read a course on probability theory. Many courses were given by Chebyshev throughout his academic career, but the two favorites were theory of numbers and probability theory. Each of these courses he gave 31 times.

Special attention was devoted by Chebyshev in his investigations in probability theory to limit theorems. He wrote only four papers in probability theory, but their influence on the future development of this science was immense. In his first work, which constituted his Master's dissertation, he introduced and utilized his basic premise of deriving accurate bounds on approximating expressions. In his thesis he proved Bernoulli's theorem and also presented corresponding bounds on obtaining approximations. The purpose of his dissertation is stated as follows:

> To prove without using transcendental analysis the basic theorems of the calculus of probabilities and its main applications which serve as a foundation for all sciences based on observations and testimonies ... Up till now elementary courses in probability theory have been restricted to an exposition of results obtained using advanced calculus. To obtain proofs of these conclusions using rigorous but elementary analysis accessible to a majority of students is indeed an important contribution to the elementary exposition of probability theory. [45, pp. 27–28]

Chebyshev begins his first chapter by introducing the notion of probability. For this purpose, he defines equipossible events:

> If out of a given number of different events under certain conditions one event should take place and there is no particular reason to expect that any one of the given events is preferred over the others, we designate these events as *equipossible*. [45, p. 28]

One must admit that this definition is not completely accurate.

Given n cases of which m yield a certain event, then the ratio m/n is taken as the measure of probability of this event (which is called a probable event). In other words: "the ratio of the number of equipossible cases favorable for the given event to the number of all the equiprobable cases" represents the probability of the event [45, p. 28].

Next Chebyshev defines the subject matter of probability theory:

> A science on probabilities, known under the name of probability theory, deals with the determination of probabilities of an event based on the given connection of this event with events whose probabilities are known. [45, p. 29]

It follows from this definition that probability theory is a mathematical science concerned with determining certain probabilities by means of known probabilities. Chebyshev does not discuss how these initial probabilities are obtained.

Next Chebyshev dwells on the theorems that he calls the basic theorems of probability theory. These are the addition rule, multiplication rule, and the following theorem: If F can occur only in conjunction with one of μ events $E_1, E_2, \ldots, E_\mu$ and the probabilities of F after the occurrence of these events are $P_1, P_2, \ldots, P_\mu$, respectively, while the probabilities of $E_1, E_2, \ldots, E_\mu$ are $p_1, p_2, \ldots, p_\mu$, then the probability that E_1 occurs together with F is

$$\frac{P_1 p_1}{P_1 p_1 + P_2 p_2 + \cdots + P_\mu p_\mu}.\dagger$$

The next chapter is devoted to "repetition of events." Here Bernoulli's formula is derived:

$$P_{\mu,m} = \frac{\mu!}{m!(\mu - m)!} p^m (1 - p)^{\mu - m}.$$

Several other formulas are established, for example:

$$\frac{\mu! \, p_1^{m_1} p_2^{m_2} \cdots p_{\sigma-1}^{m_{\sigma}-1} (1 - p_1 - p_2 - \cdots - p_{\sigma-1})^{\mu - m_1 - m_2 - \cdots - m_{\sigma-1}}}{m_1! \, m_2! \cdots m_{\sigma-1}! (\mu - m_1 - \cdots - m_{\sigma-1})!}$$

is the probability that in μ cases event $E_1{}'$ occurs m_1 times; $E_2{}'$, m_2 times; $\ldots$; $E'_{\sigma-1}, m_{\sigma-1}$ times, and the event opposite to $E_1{}', E_2{}', \ldots, E'_{\sigma-1}$ occurs $\mu - m_1 - m_2 - \cdots - m_{\sigma-1}$ times, where $p_1, p_2, \ldots, p_{\sigma-1}$ are probabilities of the events $E_1{}', E_2{}', \ldots, E'_{\sigma-1}$. [45, p. 38]

Next Stirling's formula is derived: $x! \approx (2\pi)^{1/2} x^{x+1/2} e^{-x}$. Using this formula the expression

$$P_{\mu,m} = \frac{\mu!}{m!(\mu - m)!} p^m (1 - p)^{\mu - m}$$

is transformed to the form

$$P_{\mu,m} = [2\pi p(1 - p)\mu]^{-1/2} \exp\left[-Z^2/2p(1 - p)\mu\right].$$

This expression represents the probability that, in μ cases, an event with probability p will occur m times and hence the complementary event whose probability is $1 - p$ will occur the remaining $\mu - m$ times, where $m = p\mu + Z$. [45, p. 45]

Further, the probability that in μ cases event E will occur at least m times and at most $m + s$ is determined. Denoting this probability by $\prod_{\mu,m}^{m+s}$, Chebyshev obtains

$$\prod_{\mu,m}^{m+s} = \frac{x}{\pi^{1/2}(s + 1)} \sum_{k=0}^{s} \exp\left[-\left(X + \frac{k}{s + 1} x\right)^2\right],$$

---

†Or using modern notation:

$$P(E_1|F) = \frac{P(E_1)\,P(F|E_1)}{P(E_1)\,P(F|E_1) + P(E_2)\,P(F|E_2) + \cdots + P(E_\mu)\,P(F|E_\mu)}$$

(*Translator's remark*).

where

$$X = \frac{Z}{\sqrt{2p(1 - p)\mu}}, \qquad x = \frac{s + 1}{\sqrt{2p(1 - p)\mu}}.$$

This sum can be computed fairly easily, provided the number $s + 1$ is moderate. However, the computations become involved if $s + 1$ is large.

Chebyshev then proceeds to compute $\prod_{\mu,m}^{m+s}$ for large values of $s + 1$. In this case he arrives at the result that

$$\prod_{\mu,m}^{m+s} = T(x) - T(X + x),$$

where

$$T(Z) = \frac{1}{(2\pi)^{1/2}\, 10^{13}} \sum_{k=0}^{10^{15}-1} \exp\left[-\left(Z + \frac{k}{10^{13}} \right)^2 \right].$$

A table of values of $T(Z)$ is presented. We thus easily obtain that

$$\prod_{\mu,\, p\mu - [2p(1-p)\mu]^{\frac{1}{2}}}^{p\mu + [2p(1-p)\mu]^{\frac{1}{2}}} = 1 - 2T(1) = 0.842\ 700\ 8,$$

which is the probability that an event with probability p occurs d times out of μ, where

$$p\mu - [2p(1 - p)\mu]^{1/2} \le d \le p\mu + [2p(1 - p)\mu]^{1/2}.$$

Next we have:

$$\prod_{\mu,\, p\mu - 2[2p(1-p)\mu]^{\frac{1}{2}}}^{p\mu + 2[2p(1-p)\mu]^{\frac{1}{2}}} = 1 - 2T(2) = 0.995\ 322\ 4$$

for

$$p\mu - 2[2p(1 - p)\mu]^{1/2} \le d \le p\mu + 2[2p(1 - p)\mu]^{1/2},$$

and so on, and finally,

$$\prod_{\mu,\, p\mu - 5[2p(1-p)\mu]^{\frac{1}{2}}}^{p\mu + 5[2p(1-p)\mu]^{\frac{1}{2}}} = 1 - 2T(5) = 0.999\ 999\ 999\ 998,$$

where

$$p\mu - 5[2p(1 - p)\mu]^{1/2} \le d \le p\mu + 5[2p(1 - p)\mu]^{1/2}.$$

In other words, these are the probabilities that in μ cases the ratio of the number of repetitions of an event with probability p, to the number of all the cases will differ from $p - (1/2\mu)$ by no more than

$$\frac{[2p(1 - p)]^{1/2}}{\mu^{1/2}} - \frac{1}{2\mu}, \quad \frac{2[2p(1 - p)]^{1/2}}{\mu^{1/2}} - \frac{1}{2\mu}, \dots,$$

$$\frac{5[2p(1 - p)]^{1/2}}{\mu^{1/2}} - \frac{1}{2\mu}, \dots.$$

It is therefore extremely probable that, as the number of cases increases, the ratio of the number of repetitions of an event to the number of all the cases will deviate very little from the probability of the event in these cases. This is the meaning of J. Bernoulli's theorem. [45, p. 53]

We note that Chebyshev not only gave a proof for Bernoulli's theorem, but also derived the bound on the errors committed when applying the theorem.

Next, Chebyshev applies the function $T(Z)$ to various problems, in particular to the estimation of the deviation of the mean from its true value. The example he considers is based on Cavendish's experiments to determine the density of the Earth. Cavendish obtained 29 different measurements having a mean value of 5.48. Chebyshev obtains that one can assert with probability 0.9924794 that this density differs from the actual density of the Earth by no more than 0.1.

Here we clearly detect the basic idea propagated by Chebyshev—to give bounds on approximations suitable for the application of limit theorems.

In his Master's thesis, Chebyshev proves Poisson's theorem for a finite number of different probabilities. But soon thereafter he obtained a general elementary proof of this theorem with the corresponding bounds on the errors. This proof is presented in his paper "Elementary proof of a general proposition in probability theory,"† which was published in *Crelle's Journal* in 1846. The purpose of the paper is stated as follows:

> This note is devoted to a proof of the following proposition: It is always possible to assign a number of trials so large that with probability, arbitrarily close to certainty, the ratio of the number of repetitions of a certain event E to the number of trials will deviate from the arithmetic mean of the probabilities of the event E no more than the given limits, no matter how close these limits are.
>
> This basic proposition of probability theory which includes J. Bernoulli's law as a particular case was derived by M. Poisson. ... However, in spite of the ingenuity of the method of derivation used by this famous geometer, it did not present the bounds on the errors

---

†"Démonstration élémentaire d'une proposition générale de la Théorie des Probabilités."

committed when applying this approximate analysis and since we do not know the magnitude of these errors, his proof does not have the required accuracy. [43, p. 14]

Chebyshev proceeds from the following considerations: The probability P_m that the event E in the course of μ trials will occur at least m times is equal to an expression symmetric with respect to p_k (where p_k is the probability of E in the kth trial, $k = 1, \ldots, \mu$ and is linear relative to each one of the p_ks.† For example, with respect to p_1 and p_2, one can write

$$P_m = U + V(p_1 + p_2) + Wp_1p_2,$$

where U, V, and W do not depend on p_1 and p_2.

If $0 < a = p_1 + p_2 \leqq 1$, then $P_m = U + Va + Wp_1p_2$ attains its maximum at $p_1 = p_2 = a/2$ or at $p_1 = 0$.

Next Chebyshev proves the following theorem: The largest value attained by P_m, subject to the condition that $p_1 + p_2 + \cdots + p_\mu = s$, corresponds to the values of $p_1, p_2, \ldots, p_\mu$ given by the equations

$$p_1 = 0, \quad p_2 = 0, \quad \ldots, \quad p_\rho = 0, \quad p_{\rho+1} = 1, \quad p_{\rho+2} = 1, \quad \ldots, \quad p_{\rho+\sigma} = 1,$$

$$p_{\rho+\sigma+1} = \frac{s - \sigma}{\mu - \rho - \sigma}, \quad p_{\rho+\sigma+2} = \frac{s - \sigma}{\mu - \rho - \sigma}, \quad \ldots, \quad p_\mu = \frac{s - \sigma}{\mu - \rho - \sigma},$$

where ρ and σ are fixed numbers.‡

Using this result, Chebyshev obtains for $m > s + 1$ the inequality:

$$P_m < \frac{1 \cdot 2 \cdots \mu}{1 \cdot 2 \cdots m \cdot 1 \cdot 2 \cdots (\mu - m)} \left(\frac{s}{\mu}\right)^m \left(\frac{\mu - s}{\mu}\right)^{\mu - m + 1} \frac{m}{m - s}$$

$$= \frac{\mu!}{m!(\mu - m)!} \left(\frac{s}{\mu}\right)^m \left(\frac{\mu - s}{\mu}\right)^{\mu - m + 1} \frac{m}{m - s} \tag{IV.1}$$

or $P_m < \rho_m m/(m - s)$, where ρ_m is the probability that the number of occurrences of event E equals m under the condition that $p_1 = p_2 = \cdots = p_\mu = s/\mu$, i.e., in the case when Poisson scheme is reduced to the corresponding Bernoulli scheme.

Utilizing the inequality

$$2.50x^{x + 1/2} e^{-x} < 1 \cdot 2 \cdots (x - 1) x < 2.53x^{x + 1/2} e^{-x + (x/12)},$$

---

†The proof is based on the fact that P_m is obtained as a sum of coefficients at t^m, $t^{m+1}, \ldots, t^\mu$ in the expansion of the product $\prod_{i=1}^{\mu} (p_i t + q_i)$ and $q_i = 1 - p_i$ (*Translator's remark*).

‡This theorem has been widely utilized in recent years in papers dealing with the distribution of the numbers of successes in various trials as well as in problems connected with confidence intervals for Poisson trials. It was independently rediscovered by W. Hoeffding in 1956 (cf. W. Hoeffding [200]; V. A. Sevast'yanov, *Theory of Probability and Its Applications*, Vol. 6, 1961, pp. 220–222) (*Translator's remark*).

Chebyshev transforms the expression of the right-hand side of (IV.1) to the form

$$P_m < \frac{1}{2(m-s)} \left(\frac{m(\mu-m)}{\mu} \right)^{1/2} \left(\frac{s}{m} \right)^m \left(\frac{\mu-s}{\mu-m} \right)^{\mu-m+1}. \qquad \text{(IV.2)}$$

This inequality yields the following theorem:

If the probabilities of the event E in μ consecutive trials are $p_1, p_2, \ldots, p_\mu$ and their sum is equal to s, the magnitude of the expression

$$\frac{1}{2(m-s)} \left(\frac{m(\mu-m)}{\mu} \right)^{1/2} \left(\frac{s}{m} \right)^m \left(\frac{\mu-s}{\mu-m} \right)^{\mu-m+1}$$

for $m > s + 1$ always exceeds the probability that E occurs at least m times in the course of μ trials.

In terms of complementary events, we obtain the following assertion: If the probabilities of the event E in μ successive trials are $p_1, p_2, \ldots, p_\mu$ and their sum equals s, then the magnitude of the expression

$$\frac{1}{2(s-n)} \left(\frac{n(\mu-n)}{\mu} \right)^{1/2} \left(\frac{\mu-s}{\mu-n} \right)^{\mu-n} \left(\frac{s}{n} \right)^{n+1}$$

for $n < s - 1$ always exceeds the probability that E will occur no more than n times in the course of these trials.

Chebyshev then states the third theorem:

> However, repetitions of the event E may yield only one of the following cases: either the event E occurs at least m times, or no more than n times, or, finally, more than n, but less than m times. Therefore, the probability of the latter case is determined by subtracting the probabilities of the first two cases from unity. [43, p. 21]

As a corollary to the last two theorems we obtain: If the probabilities of the event E in μ successive trials are $p_1, p_2, \ldots, p_\mu$ and the sum of probabilities is s, then the probability that the number of repetitions of event E in the course of these μ trials is less than m and greater than n, will exceed for $m > s + 1$ and $n < s - 1$ the magnitude of the expression

$$1 - \frac{1}{2(m-s)} \left(\frac{m(\mu-m)}{\mu} \right)^{1/2} \left(\frac{s}{m} \right)^m \left(\frac{\mu-s}{\mu-m} \right)^{\mu-m+1}$$
$$- \frac{1}{2(s-n)} \left(\frac{n(\mu-n)}{\mu} \right)^{1/2} \left(\frac{\mu-s}{\mu-n} \right)^{\mu-n} \left(\frac{s}{n} \right)^{n+1}.$$

The basic assertion formulated at the beginning of the paper follows as a corollary from this result.

Chebyshev thus established that the value of the lower bound on the number of trials μ required that the probability of the deviation $(m/\mu) - p$ in excess of the given limits will be less than an arbitrarily small value derived for the Bernoulli case is also applicable in the Poisson case.

This paper contains the first general proof of Poisson's theorem for the case of independent events.

The bound given by (IV.2) is sufficient for justification of various practical applications of this theorem.

It should be noted that Chebyshev does not stipulate explicitly in his theorems the condition of independence of the events. It is possible that he was of the opinion that, in general, probability theory is devoted to the study of independent events alone. In none of his papers does he stipulate the condition of independence, although he utilizes this notion effectively and never applies his results to dependent events. (It should be noted that Poisson erroneously applied his theorem to various dependent events.)

Chebyshev's method in this work was the first example of the extremal argument subsequently used repeatedly in his works.

The paper "Elementary proof of a general proposition in probability theory" is subtitled "Extract from a memoir on elementary analysis in probability theory." From this S. N. Bernstein conjectures: "Since the main part of this note does not appear in Chebyshev's dissertation, we may thus conclude that the paper 'Elementary proof of a general proposition in probability theory' was presented in the defence of his Master's thesis which took place in Moscow in 1846'" [41, p. 45]. In our opinion, however, these are insufficient grounds for this assertion, although this paper is indeed a logical continuation of the subject matter of his Master's dissertation.

On 17 December 1866, Chebyshev presented at a session of the Academy of Sciences, his paper "On mean values"† which was published in 1867 in the journal *Matematicheskii Sbornik* [127, pp. 1–9], as well as in Liouville's *Journal de Mathématiques Pures et Appliquées* [88, pp. 177–184]. Here Chebyshev proved an important inequality known nowadays as the Chebyshev inequality. Using this inequality, he obtains a theorem from which Bernoulli's and Poisson's theorems follow as particular cases.

At the beginning of his paper "On mean values" Chebyshev proves the following theorems. ‡

Theorem   If we designate by $a, b, c \ldots$, the mathematical expectations of the quantities $x, y, z \ldots$, and by $a_1, b_1, c_1 \ldots$, the mathematical expectations of their squares $x^2, y^2, z^2 \ldots$, the probability that the sum

---

†"Des valeurs moyennes."

‡The notation used in this book differs from the original Chebyshev's notation (*Translator's remark*).

$x + y + z + \ldots$, is included within the limits

$$a + b + c + \cdots + \alpha[a_1 + b_1 + c_1 + \cdots - a^2 - b^2 - c^2 - \cdots]^{1/2},$$

and

$$a + b + c + \cdots - \alpha[a_1 + b_1 + c_1 + \cdots - a^2 - b^2 - c^2 - \cdots]^{1/2}$$

will always be larger than $1 - 1/\alpha^2$, no matter what the size of α. [43, p. 431]

Next Chebyshev proceeds to the following theorem:

If N be the number of the quantities $x, y, z, \ldots$, and if in the theorem which we have just demonstrated we set

$$\alpha = N^{1/2}/t,$$

and divide by N both the sum $x + y + z + \cdots$ and its limits

$$a + b + c + \cdots + \alpha[a_1 + b_1 + c_1 + \cdots - a^2 - b^2 - c^2 - \cdots]^{1/2}$$

and

$$a + b + c + \cdots - \alpha[a_1 + b_1 + c_1 + \cdots - a^2 - b^2 - c^2 - \cdots]^{1/2}$$

we will obtain the following theorem concerning the mean values:

Theorem   If the mathematical expectations of the quantities $x, y, z, \ldots$ and $x^2, y^2, z^2, \ldots$ be respectively $a, b, c, \ldots, a_1, b_1, c_1, \ldots$, the probability that the difference between the arithmetic mean of the N quantities $x, y, z, \ldots$ and the arithmetic mean of the mathematical expectations of these quantities will not exceed

$$\frac{1}{t}\left[\frac{a_1 + b_1 + c_1 + \cdots}{N} - \frac{a^2 + b^2 + c^2 + \cdots}{N}\right]^{1/2}$$

will always be larger than $1 - t^2/N$ whatever the value of t. [43, p. 435; 216, p. 583]

This is the famous Chebyshev inequality† which can be written in a more modern form as follows:

$$P(|x - \bar{x}| \leqq \varepsilon) \geqq 1 - [D(x)/\varepsilon^2], \qquad \text{(IV.3)}$$

Here the random variable x possesses a finite variance $D(x)$, and ε is an arbitrary positive number.

---

†See also footnote on p. 203.

Indeed, Chebyshev's first theorem can be written in the form

$$P(\bar{x} + \bar{y} + \bar{z} + \cdots - \alpha[\bar{x}^2 + \bar{y}^2 + \bar{z}^2 + \cdots - (\bar{x})^2 - (\bar{y})^2 - (\bar{z})^2 - \cdots]^{1/2}$$

$$\leq x + y + z + \cdots \leq \bar{x} + \bar{y} + \bar{z} + \cdots$$

$$+ \alpha[\bar{x}^2 + \bar{y}^2 + \bar{z}^2 + \cdots - (\bar{x})^2 - (\bar{y})^2 - \cdots)]^{1/2}$$

$$\geq 1 - 1/\alpha^2.$$

where $\bar{x} = E(x)$, $\bar{x}^2 = E(x^2)$, and $(\bar{x})^2 = [E(x)]^2$. Applying this theorem to the random variable x, we obtain

$$P(\bar{x} - \alpha[\bar{x}^2 - (\bar{x})^2]^{1/2} \leq x \leq \bar{x} + \alpha[\bar{x}^2 - (\bar{x})^2]^{1/2}) \geq 1 - 1/\alpha^2.$$

However, $\bar{x}^2 - (\bar{x})^2 = D(x)$, thus

$$P(\bar{x} - \alpha[D(x)]^{1/2} \leq x \leq \bar{x} + \alpha[D(x)]^{1/2}) \geq 1 - 1/\alpha^2,$$

$$P(|x - \bar{x}| \leq \alpha[D(x)]^{1/2}) \geq 1 - 1/\alpha^2.$$

Let $\alpha[D(x)]^{1/2} = \varepsilon$, then $\alpha^2 = \varepsilon^2/D(x)$ and we thus arrive at the customary form (IV.3).

As a corollary to his theorem, Chebyshev obtains the following result:

Theorem If the mathematical expectations of the quantities $U_1, U_2, U_3, \ldots$ and of their squares $U_1{}^2, U_2{}^2, U_3{}^2, \ldots$ do not exceed a given finite limit, the probability that the difference between the arithmetic mean of N of these quantities and the arithmetic mean of their mathematical expectations will be less than a given quantity, becomes unity as N becomes infinite. [43, p.436; 216, p.584]

Indeed, consider the random variable $\tilde{x}$, which is the arithmetic mean of the given random variables:

$$\tilde{x} = \frac{x_1 + x_2 + \cdots + x_n}{n}; \qquad E(\tilde{x}) = \frac{E(x_1) + E(x_2) + \cdots + E(x_n)}{n};$$

$$D(\tilde{x}) = \frac{D(x_1) + D(x_2) + \cdots + D(x_n)}{n^2}.$$

If the mathematical expectations of the random variable and their squares are bounded, so also are their variances, i.e. all $D(x_i) < c$, where c is a fixed number. Then $D(\tilde{x}) < nc/n^2 = c/n$. Applying the Chebyshev inequality to $\tilde{x}$, we obtain

$$P(|\tilde{x} - E(\tilde{x})| \leq \varepsilon) \geq 1 - D(\tilde{x})/\varepsilon^2,$$

or

$$P\left(\left|\frac{x_1 + x_2 + \cdots + x_n}{n} - \frac{\bar{x}_1 + \bar{x}_2 + \cdots + \bar{x}_n}{n}\right| \leq \varepsilon\right) \geq 1 - \frac{c}{n\varepsilon^2}.$$

Approaching the limit with $n \to \infty$, we obtain

$$\lim_{n \to \infty} P\left(\left| \frac{x_1 + x_2 + \cdots + x_n}{n} - \frac{\bar{x}_1 + \bar{x}_2 + \cdots + \bar{x}_n}{n} \right| \le \varepsilon \right) = 1.$$

This result is known as Chebyshev's theorem or as Chebyshev's form of the law of large numbers.

The theorem establishes that for n sufficiently large one can assert with probability (arbitrarily) close to 1 that the arithmetic mean of the random variables varies arbitrarily little from a certain fixed value—the mean of their mathematical expectations.

Poisson's and Bernoulli's theorems are particular cases of Chebyshev's law of large numbers. Indeed, let event A occur in n trials with probabilities $p_1, p_2, \ldots, p_n$ respectively. Define the random variable x_i as the number of occurrences of event A in the ith trial. Then

$$E(x_i) = 1 \cdot p_i + 0 \cdot q_i = p_i, \qquad D(x_i) = p_i q_i \le \tfrac{1}{4} \,;†$$

$$\bar{x} = \frac{x_1 + x_2 + \cdots + x_n}{n} = \frac{m}{n}.$$

The variables x_i satisfy the conditions of Chebyshev's theorem, i.e.

$$\lim_{n \to \infty} P\left(\left| \frac{x_1 + x_2 + \cdots + x_n}{n} - \frac{p_1 + p_2 + \cdots + p_n}{n} \right| \le \varepsilon \right) = 1,$$

or

$$\lim_{n \to \infty} P\big(|(m/n) - \tilde{p}| \le \varepsilon \big) = 1,$$

where $\tilde{p}$ is the arithmetic mean of the probabilities of occurrences of the event in individual trials. And this is the statement of Poisson's theorem. If all the probabilities $p_i = p$, then $\tilde{p} = p$, and we obtain Bernoulli's theorem:

$$\lim_{n \to \infty} P\big(|(m/n) - p| \le \varepsilon \big) = 1.$$

It should be noted that the gist of the proof of the basic inequality (Chebyshev's inequality) was contained in I. J. Bienaymé's paper on the method of least squares. This paper was reprinted in the same issue of Liouville's journal (*J. Math. Pures Appl.* **12**, 1867) in which Chebyshev's paper "On mean values" appeared [83]. (Originally Bienaymé's paper was published in *Comptes Rendus*, **31**, 1853.)‡

---

†Since $p_i = 1 - q_i$, and the function $x(1 - x)$ for $0 \le x \le 1$ attains its maximum at $x = \tfrac{1}{2}$ (*Translator's remark*).

‡Recently published note by Heyde and Seneta (*Biometrika*, 1972, pp. 680–683) discusses some important—largely forgotten—contributions of Bienaymé (1796–1878) in probability and statistics (*Translator's remark*).

This inequality was referred to as the Bienaymé–Chebyshev inequality for a long time. Markov elaborates as follows:

> We associate this remarkable and simple inequality with the two names, Bienaymé and Chebyshev, because Chebyshev was the first to clearly express and prove it, while the basic idea of the proof was pointed out much earlier by Bienaymé in a memoir containing the inequality itself, albeit in a not particularly obvious form.† [121, p. 92]

However, since this inequality was "clearly expressed and proved by Chebyshev first" and since the application of this inequality to a generalization of the law of large numbers is due entirely to Chebyshev, it is referred to in modern literature as "the Chebyshev inequality."

The law of large numbers in its general form was proved by Chebyshev in 1866. The second basic problem that occupied Chebyshev's attention was the central limit theorem. However, only in 1887 in the *Proceedings of the Academy of Sciences* was Chebyshev's paper devoted to this subject published. The paper is entitled "On two theorems concerning probabilities."‡ (In 1891 this paper was reprinted in *Acta Mathematica*.) The first theorem discussed in this paper deals with the law of large numbers, which was proved in Chebyshev's paper "On mean values."

The second theorem, which is proved in this paper, is stated as follows:

> If the mathematical expectations of the variables $u_1, u_2, u_3, \ldots$ are zero and if the mathematical expectations of all their powers are bounded from above by a certain finite limit, then the probability that the sum of n variables
>
> $$u_1 + u_2 + u_3 + \cdots + u_n$$
>
> divided by the square root of the *doubled* sum of the mathematical expectations of the squares, contained between two arbitrary numbers t and t' will approach the limit
>
> $$\pi^{-1/2} \int_t^{t'} \exp(-x^2)\, dx$$
>
> as n increases to infinity. [44, p. 230]

We shall now discuss this statement. Chebyshev considers a sequence of random variables $u_1, u_2, \ldots, u_n$. The mathematical expectations of the μth powers of these variables will be denoted by $a_n^{(\mu)} = M(u_n^\mu)$. Chebyshev's theorem can then be stated in the form: Subject to the conditions:

1. $a_n^{(1)} = 0$;
2. $a_n^{(2)}, a_n^{(3)}, \ldots, a_n^{(\mu)}, \ldots$ are bounded by a fixed constant,

---

†See Heyde and Seneta's note in *Biometrika*, 1972, for a more detailed historical perspective (*Translator's remark*).

‡"**Sur deux** théorèmes relatifs aux probabilités."

$$\lim_{n \to \infty} P\left(t \le \frac{u_1 + u_2 + \cdots + u_n}{[2(a_1^{(2)} + a_2^{(2)} + \cdots + a_n^{(2)})]^{1/2}} \le t' \right) = \frac{1}{\pi^{1/2}} \int_t^{t'} \exp(-x^2)\, dx.$$

The second condition is not stated very clearly. A. N. Kolmogorov writes in this connection: "Judging from the subsequent text, the second condition should be interpreted as follows: The mathematical expectation of $a_n^{(\mu)}$ for each μ does not exceed a certain constant independent of n (which, however, may depend on μ." [44, p. 406]

It should be noted again that Chebyshev does not stipulate here or at any other point the mutual independence of the variables $u_1, u_2, \ldots, u_n$.

Kolmogorov also observes that in Chebyshev's formulation of the theorem a condition is omitted without which the theorem taken literally is incorrect. Chebyshev in his argument neglected the fact that in general the expression

$$1/q_n{}^2 = [a_1^{(2)} + a_2^{(2)} + \cdots + a_n^{(2)}]/n$$

may approach 0 as $n \to \infty$. The simplest way to remedy this omission is to add the following condition: the arithmetic means

$$1/q_n{}^2 = [a_1^{(2)} + a_2^{(2)} + \cdots + a_n^{(2)}]/n$$

approach a finite positive limit $1/q^2$ as $n \to \infty$ [44, p. 406].

With this minor modification and additional condition Chebyshev's theorem can be proved in a completely rigorous manner.

In this paper Chebyshev widely utilizes the theory of moments. The moment problem arises in various branches of mathematics, such as in problems connected with approximation theory, approximate calculations of definite integrals, theory of continued fractions, and others. The problem can be stated as follows: Given the numbers $C_i (i = 1, 2, \ldots, k)$. It is required to find a function $\varphi(x)$ such that $C_k = \int_a^b x^k\, d\varphi(x)$. The problem may have different aspects depending on whether the numbers a and b are finite or infinite. The quantities $C_k = \int_a^b x^k\, d\varphi(x)$ are called the moments of order k. If $\varphi(x)$ possesses a derivative which is integrable in the Riemann sense, then these moments can be rewritten in the form

$$C_k = \int_a^b x^k f(x)\, dx.$$

Many scientists have been engaged in investigations of this problem. Chebyshev considered it his paper "Sur les valeurs limitées des integrales" (*J. Math. Pures Appl. Ser. 2* **19**, 157–160 (1874)).† (See [130, sec. II].)

---

†In this paper the *one-sided* Chebyshev inequality is proved [cf. W. Hoeffding, *J. Amer. Statist. Assoc.* **58**, 1–21 (1963)] (*Translator's remark*).

In the paper "On two theorems concerning probabilities"† [*Acta Math.* **14**, 305–315 (1890–1891)], Chebyshev essentially constructed the method of moments in probability theory.

At the beginning of this paper, Chebyshev refers to his article "On integral residua which yield approximate values of the integrals"‡ presented on 18 November 1866, and published in *Acta Mathematica*, 1887, and states the obtained result.

If the function $f(x)$ being positive, yields

$$\int_{-\infty}^{+\infty} f(x)\,dx = 1, \qquad \int_{-\infty}^{+\infty} xf(x)\,dx = 0,$$

$$\int_{-\infty}^{+\infty} x^2 f(x)\,dx = \frac{1}{q^2}, \qquad \int_{-\infty}^{+\infty} x^3 f(x)\,dx = 0, \ \dots,$$

$$\int_{-\infty}^{+\infty} x^{2m-2} f(x)\,dx = \frac{1\cdot 3\cdot 5 \cdots (2m-3)}{q^{2m-2}}, \qquad \int_{-\infty}^{+\infty} x^{2m-1} f(x)\,dx = 0,$$

then the value of the integral $\int_{-\infty}^{v} f(x)\,dx$ is contained within the limits

$$\frac{1}{\pi^{1/2}} \int_{-\infty}^{qv/2^{\frac{1}{2}}} e^{-x^2}\,dx - \frac{3^{3/2}(m^2 - 2m + 3)^{3/2}(q^2 v^2 + 1)^3}{2(m-3)^3(m-1)^{1/2}},$$

and

$$\frac{1}{\pi^{1/2}} \int_{-\infty}^{qv/2^{\frac{1}{2}}} e^{-x^2}\,dx + \frac{3^{3/2}(m^2 - 2m + 3)^{3/2}(q^2 v^2 + 1)^3}{2(m-3)^3(m-1)^{1/2}}$$

for all values of v. [44, pp. 229–230]

This assertion is valid under the condition that all the moments of the function $f(x)$ up to $(2m-1)$th coincide with the moments of the function $q(2\pi)^{-1/2} \exp(-q^2 x^2/2)$, i.e.

$$\int_{-\infty}^{+\infty} f(x)\,dx = q(2\pi)^{-1/2} \int_{-\infty}^{+\infty} \exp(-q^2 x^2/2)\,dx;$$

$$\int_{-\infty}^{+\infty} xf(x)\,dx = q(2\pi)^{-1/2} \int_{-\infty}^{+\infty} x \exp(-q^2 x^2/2)\,dx;$$

$$\vdots$$

$$\int_{-\infty}^{+\infty} x^{2m-2} f(x)\,dx = q(2\pi)^{-1/2} \int_{-\infty}^{+\infty} x^{2m-2} \exp(-q^2 x^2/2)\,dx;$$

$$\int_{-\infty}^{+\infty} x^{2m-1} f(x)\,dx = q(2\pi)^{-1/2} \int_{-\infty}^{+\infty} x^{2m-1} \exp(-q^2 x^2/2)\,dx.$$

†"Sur deux théorèmes relatifs aux probabilités."
‡"Sur les résidus intégraux qui donnent des valeurs approchées des intégrales."

It is then proved (after stating the basic limit theorem) that under conditions of the theorem the moments of the random variable

$$x_n = (u_1 + u_2 + \cdots + u_n)/n^{1/2}$$

approach the moments of the Laplace–de Moivre distribution.

The problem is thus reduced to the proof of the following relationships:

$$\lim_{n \to \infty} A_n^{(\mu)} = \frac{q}{\sqrt{2\pi}} \int_{-\infty}^{+\infty} x^\mu \exp(-q^2 x^2/2)\, dx$$

$$= \begin{cases} \dfrac{1 \cdot 3 \cdot 5 \cdots (\mu - 1)}{q^\mu} & \text{for} \quad \mu \text{ even} \\[2mm] 0 & \text{for} \quad \mu \text{ odd}, \end{cases}$$

where $A_n^{(\mu)}$ are the moments of x_n^μ, i.e., $A_n^{(\mu)} = E(x_n^\mu)$, and, as before,

$$x_n = (u_1 + u_2 + \cdots + u_n)/n^{1/2}.$$

It follows from the inequalities presented at the beginning of the paper that the continuous function $f(x)$, all of whose moments $\int_{-\infty}^{+\infty} x^\mu f(x)\, dx$ equal $A^{(\mu)}$, coincides with the function $q(2\pi)^{-1/2} \exp(-q^2 x^2/2)$.

However, actually it is necessary to prove that from the fact that, given a sequence of nonnegative functions $f_n(x)$, the moments $A_n^{(\mu)} = \int_{-\infty}^{+\infty} x^\mu f_n(x)\, dx$ converge as $n \to \infty$ to the moments

$$A^{(\mu)} = q(2\pi)^{-1/2} \int_{-\infty}^{+\infty} x^\mu \exp(-q^2 x^2/2)\, dx$$

of the function $\tilde{f}(x) = q(2\pi)^{-1/2} \exp(-q^2 x^2/2)$, it follows that the integral $\int_{-\infty}^{t} f_n(x)\, dx$ converges to the integral $\int_{-\infty}^{t} \tilde{f}(x)\, dx$ as $n \to \infty$ for each and every value of t.

Chebyshev did not prove this assertion. He regarded as self-evident the fact that, given a sequence of probability distributions $P_n(x)$, the convergence $P_n(x) \to P(x)$ follows from the convergence of $\lim_{n \to \infty} \int_{-\infty}^{+\infty} x^k\, dP_n(x)$ to C_k, where C_k are moments which uniquely determine the distribution $P(x)$

$$\left(C_k = \int_{-\infty}^{+\infty} x^k\, dP(x) \right).$$

(This assertion was proved shortly thereafter by Markov, utilizing Chebyshev's inequality.)

After proving the convergence of the moments of

$$x_n = (u_1 + u_2 + \cdots + u_n)/n^{1/2},$$

Chebyshev asserts that from this convergence and the inequalities presented in the beginning of the paper, the assertion of the basic theorem, namely, the fact that

$$\lim_{n \to \infty} P(t < x_n < t') = q(2\pi)^{-1/2} \int_t^{t'} \exp(-q^2 x^2/2)\, dx$$

immediately follows.

As was indicated above, this assertion requires certain revisions and some additional proofs.

This paper is included in the third volume of Chebyshev's "Complete Collected Works" supplemented by Kolmogorov's commentaries. These commentaries provide a deep analysis into this memoir. A general appraisal of this work was given by Kolmogorov as follows:

> In spite of certain deficiencies, this memoir highlights one of Chebyshev's greatest achievements and the realization of a project which occupied him for a number of years. The results of Chebyshev's investigations on the problem of moments are applied here to the determination of the form of the probability distribution law of a sum of a large number of independent random variables and it is established that, under certain very general conditions, this distribution law, with the increase in the number of summands, approaches in the limit the normal distribution law of de Moivre–Laplace (the so-called basic limit theorem of probability theory); moreover, the possibility of a further refinement of this result is pointed out in this paper, although without a rigorous proof. [44, p. 404]

Gnedenko gives the following evaluation of the work:

> Chebyshev did not present a rigorous proof of his theorem; moreover, he did not introduce the restrictions required for the validity of the theorem. However, in spite of these logical gaps, he deserves great praise for focusing attention on this important problem and constructing a method for its proof (the method of moments). He generated interest in this problem among his students who not only completed the proof of this assertion, but also extended the conditions for its applicability almost to its natural limitations. [158, p. 399]

The two basic problems of probability theory—the law of large numbers and the limit theorem for the sum of independent random variables—were the subject of Chebyshev's investigations in the field of probability theory. The task that confronted him was to prove these theorems for a broader class of random variables. These were the central problems of probability theory. The future path of the development of probability theory was dependent upon their solution.

The problem of the law of large numbers was completely solved by Chebyshev in 1866 in his paper "On mean values." The solution of the central limit theorem problem was achieved only after 20 years.

Random phenomena take place, as a rule, due to the effect of a large number of causes, where each single cause individually has only an insignificant effect on the phenomenon.

Since the number of causes is large, and the distribution of probabilities of each one of these qualities is in general unknown, it is usually very difficult and often impossible to determine the distribution function of the sum. For this reason, the limiting distribution function is usually determined, i.e., the distribution function obtained when the number of causes $n \to \infty$. This limiting function replaces the one that actually describes the behavior of the phenomenon. It was Laplace who, as early as the beginning of the past century, conjectured that the existence of a normal law governing random errors in observations points out that measurement errors are subject to the action of a large number of independent causes. However, he did not develop this idea in his works.

The feasibility of replacing the exact distribution by its limit follows from the so-called central limit theorem. The essence of this theorem is in establishing the conditions under which the distribution function of the sum of independent random variables approaches the normal distribution as the number of summands increases. There are in nature a vast number of phenomena subject to the action of a large number of causes where each individual cause acts independently and exerts only a very small influence on the course of the phenomenon. That is why this theorem is of such importance for the science.

In his last paper on probability theory "On two theorems concerning probabilities" (1887), Chebyshev actually summarizes all his research in this field. First he states the first of his theorems—the law of large numbers.

In the second theorem, one of his most important results, he establishes that under certain very general conditions the distribution law of probabilities of the sum of a large number of independent random variables approaches the normal distribution in the limit as the number of summands increases.

The second basic problem of probability theory was thus also solved by Chebyshev.

Kolmogorov writes that:

> P. L. Chebyshev impelled Russian probability theory into first place in the world. From the methodological point of view the basic change, due to Chebyshev, is not the fact that he was the first who strongly insisted on complete rigor in proving theorems, but mainly that he always strove to obtain exact estimates on deviations from the limiting laws (which arise when the number of trials is large, but *finite*) in the form of inequalities applicable for any number of trials. [90, p. 56]

By establishing the restrictions on applicability of the basic theorems of probability theory, Chebyshev ascribed a clearly defined mathematical character with real meaning to random variables. In each particular case this permitted the determination of whether the limit theorems were applicable to the given random variables.

Probability theory is not an abstract structure in Chebyshev's works; his works found swift immediate and widespread applications.

Chebyshev was a materialist by nature; he arrived at materialism through the natural sciences, mainly through mathematics and mechanics. Some of his views were very close to dialectic tenets. In particular, his opinion of the role of practice in the development of a theory belongs in this category.

Chebyshev created the materialistic Russian school of probability theory. The further development of probability theory proceeded according to the path projected by him.

We conclude this investigation of Chebyshev's works on probability theory by recalling A. Ya. Khinchin's statement which points out that, starting from the second half of the nineteenth century, Russia was "the only country in which the mathematical foundations of probability theory were cultivated with the seriousness it deserved, in view of its prominent role in the natural sciences and engineering. It is entirely due to the works of Chebyshev that the Russian school of probability theory attained this exceptional position" [61, p. 36].

2   Prominent representatives of the St. Petersburg school

Andrei Andreevich Markov (1856–1922) was Chebyshev's closest disciple and the best spokesman for his ideas in probability theory.

Bernstein characterizes him as follows:

> Undoubtedly the most colorful spokesman for the ideas and direction of Chebyshev in probability theory was A. A. Markov—the closest to Chebyshev by nature and the sharpness of his mathematical talent.... His original memoirs are models of rigor and clarity of exposition, and have contributed to a very great extent in transforming probability theory into one of the most perfect fields in mathematics and to the widespread popularity of Chebyshev's methods and ideas. [41, pp. 59–60]

The principal works of Markov in probability theory are related to the limit theorem for the sum of independent random variables, as well as to dependent variables, in particular those connected in a chain.

The limit theorem problem for the sum of independent random variables consists in establishing the conditions under which the limiting relation

$$\lim_{n \to \infty} P\big(S_n \leq M(S_n) + t[D(S_n)]^{1/2}\big) = (2\pi)^{-1/2} \int_{-\infty}^{t} \exp(-x^2/2)\, dx$$

is valid. Here $D(S_n) = M[S_n - M(S_n)]^2$ is the variance of the sum S_n.†

Chebyshev proved this assertion for a certain class of random variables. In his proof he utilized the method of moments; however, as has been pointed out, there are certain gaps in his argument.

Dealing with the problems connected with limit theorems, Markov, starting from 1898 and during the course of a period of years, applied Chebyshev's method of moments in his work.

Markov presented his first proof of a limit theorem in 1898 in a series of letters to A. V. Vasiliev, a professor at Kazan University. Excerpts from this letter were published in the same year [119].

In a letter dated 23 October 1898, Markov writes: "Chebyshev's memoir "On theorems concerning probabilities" is of great importance. Unfortunately, its value is diminished due to the following two features: (1) complexity of proof, (2) insufficient rigor of the arguments" [122, p. 233].

Next, he relates that for a long time he had contemplated a simplification of Chebyshev's proof, at the same time making it completely rigorous.

After this introduction, Markov proceeds to the proof of the limit theorem:

> First we take the theorem on mathematical expectations, which constitutes the main content of Chebyshev's memoir "On two theorems concerning probabilities."
>
> This theorem can be stated as follows:
>
> If the mathematical expectations of independent variables $x_1, x_2, x_3, \ldots, x_n$ are zero and the mathematical expectation of the variable $(x^n)^k$ remains finite for each integer-valued k as n approaches infinity, then each one of the differences
>
> $$M\left(\frac{x_1 + x_2 + \cdots + x_n}{n^{1/2}}\right)^m - A_m\left\{M\left(\frac{x_1 + x_2 + \cdots + x_n}{n^{1/2}}\right)^2\right\}^{m/2}$$
>
> approaches the limit zero with an indefinite increase of n. Here m is a given integer and
>
> $$A_m = 2^{m/2}\pi^{-1/2}\int_{-\infty}^{+\infty} t^m \exp(-t^2)\,dt. \quad [122, \text{p. } 234]$$

In other words, Markov investigates a sum of independent random variables x_i ($i = 1, 2, \ldots, n$) with $M(x_i) = 0$, such that for every integer k, the kth moment $M(x_n^k)$ exists and is bounded in absolute value by a constant C_k independent of n. Let, moreover,

$$Y_n = \frac{x_1 + x_2 + \cdots + x_n}{n^{1/2}}.$$

† M is the expectation operator.

Under these conditions, Markov proves that

$$\lim_{n \to \infty} \{M(Y_n^m) - A_m[M(Y_n^2)]^{m/2}\} = 0,$$

where

$$A_m = 2^{m/2}\pi^{-1/2} \int_{-\infty}^{+\infty} t^m \exp(-t^2)\,dt.$$

This statement of the limit theorem differs somewhat from Chebyshev's formulation and is free of the inaccuracies contained in the latter.

Markov's proof is based on the application of the properties of mathematical expectation and the generalized binomial formula; the proof is completely rigorous, without the logical gaps of Chebyshev's proof.

In the paper "On the roots of the equation $e^{x^2}\,\partial^m e^{-x^2}/\partial x^m = 0$" published in 1898 [120], Markov proves the limiting theorem in the following formulation:

The probability that the sum $u_1 + u_2 + \cdots + u_n$ of the independent variables $u_1, u_2, \ldots, u_n$ is contained between the limits

$$\alpha[2(a_1 + a_2 + \cdots + a_n)]^{1/2} \quad \text{and} \quad \beta[2(a_1 + a_2 + \cdots + a_n)]^{1/2},$$

where $a_1, a_2, \ldots, a_n$ are the mathematical expectations of the variables $u_1^2, u_2^2, \ldots, u_n^2$ and α and β are two arbitrarily chosen quantities, approaches as n tends to infinity to the limit

$$\pi^{-1/2} \int_{\alpha}^{\beta} \exp(-x^2)\,dx,$$

provided the infinite sequence of independent variables $u_1, u_2, \ldots, u_n$ satisfy the following conditions:

1. the mathematical expectations of $u_1, u_2, \ldots, u_n$ are zero;
2. the mathematical expectations of $u_n^2, u_n^3, u_n^4, \ldots$ remain finite for finite values of k as k increases indefinitely;
3. the mathematical expectation of u_k^2 does not become arbitrarily small as k increases indefinitely. [122, pp. 267–268]

In the modern formulation Markov's theorem is usually stated as follows: If a sequence of mutually independent random variables $\xi_1, \xi_2, \ldots, \xi_n$ is such that for all integer valued $r \geq 3$ the condition

$$\lim_{n \to \infty} C_n(r)/B_n^2 = 0$$

is satisfied, where

$$B_n^2 = \sum_{k=1}^{n} D\xi_k; \qquad C_n(r) = \sum_{k=1}^{n} M|\xi_k - M\xi_k|^r,\dagger$$

†This condition is weaker than condition 3 of the original formulation (*Translator's remark*).

then

$$\lim_{n \to \infty} P\left\{ B_n^{-1} \sum_{k=1}^{n} (\xi_k - M\xi_k) < x \right\} = (2\pi)^{-1/2} \int_{-\infty}^{x} \exp(-z^2/2)\, dz.$$

Thus Markov proved the limit theorem under the same assumptions as those stipulated by Chebyshev. The main assumption is the existence of finite moments of all orders.

In 1900 and 1901 two papers of A. M. Lyapunov appeared in the journal of the St. Petersburg Academy of Sciences: "On a theorem in the calculus of probabilities" [103] and "A new form of a theorem on the limit of probabilities" [104]. In these papers, Lyapunov proves the limit theorem with substantially weaker restrictions than those required by Markov. First, he waives the requirement of the existence of the moments of all orders. Also he proves his theorem utilizing the method of characteristic functions which he developed,† rather than by the method of moments used by Markov. Lyapunov's formulation of the limit theorem is known nowadays as the central limit theorem of probability theory.

It would appear that, using the method of moments, one cannot obtain Lyapunov's result, since the characteristic functions exist for any random variable, while the moments, i.e. the mathematical expectation of the powers of the random variables, may not exist in all cases.

Commenting on this state of affairs, Markov remarks:

> In the important memoir [referring to Chebyshev's "On two theorems concerning probabilities"] which clearly pointed up the value of the method of mathematical expectations, certain gaps remained in connection with the statement, as well as the proof of the theorem; these defects were straightened out in my papers "The law of large numbers and the method of least squares" and "On the roots of the equation $e^{x^2} \partial^m e^{-x^2}/\partial x^m = 0$."
>
> Thus, conditions have been established under which the theorem on the limit of probabilities holds; they are also necessary, provided we utilize for its proof certain well-known simple arguments.
>
> Later, academician A. M. Lyapunov endeavored to arrive at this theorem using another method, appropriately supplementing the usual derivation of the approximate formula and, at the same time, establishing this theorem for a possibly wider class of variables. He accomplished this in his memoirs "Sur une proposition de la théorie des probabilités" and "Nouvelle forme du théorème sur la limite de probabilité."

†This method originated in the works of Laplace and Lagrange (cf. p. 146). Attempts to utilize this method were made in 1892 by I. Sleshinskiĭ of the University of Odessa (This fact is mentioned in Lyapunov's 1900 paper [103] cf. pp. 220–221.) (*Translator's remark*).

In the latter work, the generality of results obtained by A. M. Lyapunov exceeded the generality achieved using the method of mathematical expectations. It seemed that these results could not be attained using the method of mathematical expectations due to the fact that this method requires the consideration of infinitely many mathematical expectations whose existence is not assumed in the cases studied by Lyapunov.

To restore the importance of the method of mathematical expectations which was shattered by these developments, it was necessary to show that this method was not completely exhausted in the above mentioned works. I contemplated this problem for quite a while and was able to solve it. [122, pp. 321–322]

Markov proved Lyapunov's theorem in the following formulation *by the method of moments*:

Let $z_1, z_2, \ldots, z_k, \ldots, z_n, \ldots$ be an infinite sequence of independent random variables; let, moreover, there exist for each k the quantities

$$a_k = M(z_k) \qquad \text{and} \qquad b_k = M(z_k - a_k)^2$$

and

$$b_k^{(2+\delta)} = M|z_k - a_k|^{2+\delta},$$

where δ is a positive number and the symbol $|V|$ denotes the absolute value of the quantity V.

Finally, let us assume that the ratio

$$\frac{b_1^{(2+\delta)} + b_2^{(2+\delta)} + \cdots + b_n^{(2+\delta)}}{(b_1 + b_2 + \cdots + b_n)^{1+\delta/2}}$$

approaches the limit zero as n increases to infinity.

These are A. M. Lyapunov's conditions. We must show that under these conditions the limit probability theorem is valid. Namely, for any given t_1 and t_2, with $t_2 > t_1$, the probability of the inequality

$$t_1 < \frac{z_1 + z_2 + \cdots + z_n - a_1 - a_2 - \cdots - a_n}{[2(b_1 + b_2 + \cdots + b_n)]^{1/2}} < t_2$$

tends to the limit

$$\pi^{-1/2} \int_{t_1}^{t_2} \exp(-t^2)\, dt$$

as n approaches infinity. [122, pp. 322–323]

This theorem was proved by Markov in his paper "The theorem on the limit of probability for the cases of A. M. Lyapunov" (1913). This paper is closely related to his paper "Chebyshev's inequalities and the basic theorem." Both papers were published in the supplement to the third edition of his "Calculus of Probabilities." In the second of these papers the expansion of the integral

$$\int_{-\infty}^{\infty} \frac{\exp(-t^2)}{z - t}\, dt$$

into a continued fraction and its convergent is discussed. Also the Chebyshev relation that connects the value of the integral

$$\pi^{-1/2} \int_{-\infty}^{\alpha} \exp(-t^2)\, dt$$

for any α with the coefficients in the expansion of the convergent *of a continued fraction* into partial fractions is investigated.

In the first paper, the method of moments is applied in such a manner that the proof of Lyapunov's version of the central limit theorem goes through.

The basic idea for application of the method of moments is the introduction of so-called truncated random variables. These variables are defined as follows:

$$x_k = \begin{cases} z_k - a_k & \text{for } |z_k - a_k| < N, \\ 0 & \text{for } |z_k - a_k| \geq N, \end{cases}$$

where the z_k are the given random variables and $a_k = M(z_k)$.

The truncated variables x_k possess moments of all orders. For a suitable choice of N, the distribution of the sum of x_k will deviate little from the distribution of the original sum under Lyapunov's conditions and both sums have the same limiting distribution.

The method of truncated variables enabled Markov to prove Lyapunov's theorem. This method was often utilized in many subsequent investigations, where truncated variables are introduced in the following manner:

$$x_k' = \begin{cases} x_k & \text{if } x_k < N, \\ 0 & \text{if } x_k \geq N. \end{cases}$$

For N sufficiently large, the equality $x_k' = x_k$ is almost certain,† while the x_k' possess moments of all orders.

---

†More precisely, the probability that one or more $x_k' \neq x_k$ tends to zero as $N \to \infty$ (*Translator's remark*).

At the conclusion of his paper, Markov presents an example where the central limit theorem is not valid. Evidently in this case Lyapunov's conditions are violated.

Markov was also the originator of a very important branch of probability —the study of dependent random variables. He was interested in the following two problems: the applicability of the law of large numbers and of the central limit theorem to sums of dependent variables.

Although in the above we have used the term "law of large numbers" on numerous occasions, the meaning of this term has not been given precisely. J. Bernoulli did not give this name to his theorem; Poisson coined this term. It is of interest that Chebyshev does not refer to his theorem as "the law of large numbers," although Poisson's theorem is obtained from Chebyshev's as a particular case. Markov interprets this law as a law

> in view of which it can be asserted with probability as close to certainty as desired that the arithmetic mean of several variables, provided the number of these variables is large, will deviate in an arbitrarily small amount from the arithmetic mean of their mathematical expectations. [122, p. 341]

If we accept this interpretation of the law of large numbers, all three theorems—Bernoulli's, Poisson's, and Chebyshev's—will be different forms of this law. The last theorem is the most general form of it. This interpretation is the one generally accepted nowadays.

Describing Chebyshev's form of the law of large numbers, Markov observes: "Clearly, Chebyshev's conditions do not cover all the cases to which the above mentioned law may be applied" [122, p. 342].

As we know, Chebyshev extended the law of large numbers for the case of independent random variables, with uniformly bounded variances, $D(x) \leq c$.

Markov extended these conditions in his paper "Extension of the law of large numbers to dependent variables".† Markov showed that if the sequence of mutually independent random variables is such that

$$\lim_{n \to \infty} \frac{D(x_1) + D(x_2) + \cdots + D(x_n)}{n^2} = \lim_{n \to \infty} \frac{1}{n^2} \sum_{k=1}^{n} D(x_k) = 0,$$

then

$$\lim_{n \to \infty} P\left(\left| \frac{x_1 + x_2 + \cdots + x_n}{n} - \frac{M(x_1) + M(x_2) + \cdots + M(x_n)}{n} \right| < \varepsilon \right) = 1.$$

---

†This paper was first published in the *Notices (Izvestiya) of the Physical-Mathematical Society at Kazan University Ser. 2* **15** (No. 4), 155–156. The date indicated on the cover of the volume is 1906; however, the date at the end of Markov's paper is March 25, 1907.

Later, Kolmogorov† proved that this condition is close to the necessary condition that he obtained in 1928 and in a different form by Khinchin in 1929‡.

In this paper, Markov also proves that the law of large numbers is applicable to $S_n = \sum_{i=1}^{n} x_i$, provided $D(x_i) < C$, and the dependence between the variables is such that the increase of any one of them yields a decrease in the mathematical expectation of the others.

Next Markov remarks that "the same conclusion concerning the applicability of the law of large numbers can be obtained also in the case when the mathematical expectation of x_k for each k decreases with the increase of the sum

$$x_1 + x_2 + \cdots + x_{k-1},\text{"} \quad [122, \text{p. } 344].$$

Further, Markov investigates a sequence of random variables which form a "chain." These chains of dependent random variables are now referred to as Markov chains. The study of Markov chains has become a large branch of probability theory with an enormous literature.

A sequence of random variables, each one of which admits any number of values (outcomes), is called a simple Markov chain if the probabilities of the outcomes in the $(n + 1)$th trial [$(n + 1)$th variable] depend only on the outcome of the nth trial. If these probabilities depend on the outcomes of the previous k trials, then the chain is called a compound Markov chain of the kth order.

In his 1907 paper Markov considers a simple chain with a finite number of states (i.e., each x_i admits only a finite number of values—actually the case when each x_i admits only two values 0 or 1 is considered here). He shows that these random variables obey the law of large numbers. It should be mentioned that in his paper Markov requires that all the "transition probabilities" $P_{\alpha\beta}$ be strictly positive. However, Markov's result remains valid also under the much weaker condition when for each α only one $P_{\alpha\beta}$ is strictly positive.

This paper concludes with the remark that the independence of the variables is not a necessary condition for the validity of the law of large numbers.

In a series of papers "Extension of limit theorems of probability theory to a sum of variables connected in a chain" (1908), "On connected variables not forming a genuine chain" (1911),§ "On a case of trials connected into a

†A. N. Kolmogorov, "Ueber die Summen durch die Zufall bestimmter unabhaengiger, *Math. Ann.* **99**, 309 (1928) (*Translator's remark*).

‡A. Ya. Khinchin. Sur la loi des grands nombres, *Compt. Rend. Acad. Sci.* **189**, 477–479 (1929). Khinchin's conditions pertain to the case of *identically* distributed independent random variables. (*Translator's remark*).

§*Mém. Acad. Sci. St. Petersburg Ser. 8* **22** (1908).

compound chain" (1911),† "On trials connected in a chain from unobservable events" (1912) Markov proves the applicability of the central limit theorem to the sums of random variables forming a simple homogeneous chain or a compound homogeneous chain, as well as to two-dimensional vectors forming a chain, and to the case of sums forming the so-called Markov–Bruns chain.

In his later works Markov introduced several types of chains.

Concerning Bruns' contributions, Markov writes:

> In H. Bruns' book (H. Bruns. Wahrscheinlichkeitsrechnung und Kollektivmasslehre) and in his paper "Das Gruppenschema für zufällige Ereignisse" (*Abhandlungen der mathematisch-physischen Klasse der Königlich Sächsischen Gesellschaft der Wissenschaften*, 1906, B. XXIX) remarkable cases of dependent trials are investigated, which do not fall into the category of chain of trials introduced by us. [122, p. 401]

Here he has in mind the following situation: Let a sequence $Y_1, Y_2, \ldots, Y_n, \ldots,$ of independent random variables Y_n, admitting two values 1 or 0 with the corresponding probabilities α and β ($\alpha + \beta = 1$) be given. Consider the sequence of dependent variables $x_n = Y_n Y_{n+1}$, $n = 1, 2, \ldots$; each term in this sequence is a product of two adjacent terms of the original sequence. The sequence $x_1, x_2, \ldots, x_n$ is the simplest case of a Markov–Bruns chain.

Markov proves that the central limit theorem is applicable to the sum $S_n = \sum_{i=1}^{n} x_i$, provided $B_n = \text{Var}(S_n) > a_n$, where $a_n > 0$.

In his paper "Investigation of a general case of trials connected in a chain" (1910), Markov investigates the limiting distribution of the sum $S_n = \sum_{i=1}^{n} x_i$, where x_i form a nonhomogeneous chain, admitting only two values with the transition probabilities

$$p_i' = P(x_i = 1 | x_{i-1} = 1), \qquad p_i'' = P(x_i = 1 | x_{i-1} = 0).$$

He proves the central limit theorem for the sum S_n in the case when $p_0 < p_i' < 1 - p_0$, $p_0 < p_i'' < 1 - p_0$, where $p_0 > 0$ is a constant independent of n. The proof utilizes the method of moments.

Markov does not discuss the limiting case when $p_0 \to 0$.

In 1926, S. N. Bernstein obtained the same result. Moreover, he also showed that the central limit theorem is applicable to S_n, provided $p_0 = 1/n^a$, where $a < \frac{1}{5}$, however the theorem may not apply if $a = \frac{1}{3}$.

Investigations of Markov chains and later on of Markov's processes are of great significance in the application of probability theory to various branches of the natural sciences and engineering. Markov chains are utilized for systems that move from one state to another only at definite

---

† *Izv. Akad. Nauk.*

time points $t_1, t_2, \ldots, t_k, \ldots$ and when the probabilities p_{ij} are given. These are the (conditional) probabilities that the system will be at the time t_{k+1} in the state ω_j given that at time t_k it was in the state ω_i.†

The model of an atom proposed by Bohr is an example of such a system. In this model the electron of the hydrogen atom can be located in one of the admissible orbits. Let the electron be located in the ith orbit. Assume that the change in state of the atom may occur only during the time points $t_1, t_2, t_3, \ldots$. The probability of the transition from the ith orbit to the jth depends only on i and j. The difference $j - i$ depends on the change in the amount of energy charge. This transition probability does not depend on the previous location of the electron.

Markov processes are ones in which the state of the system at the initial time determines the probability distribution of the possible states of this system at subsequent time points. Markov chains are special cases of such processes corresponding to the situation in which the system changes stepwise (rather than continuously) at particular time points and when the number of possible states in the system is finite. The theory of Markov processes is applicable in the theory of Brownian motion, in diffusion theory, quantum mechanics, and many other branches of science.

Markov himself did not investigate any applications of the chains in the natural sciences or engineering. In his paper "An example of statistical investigation of the poem 'Eugene Onegin.'"‡ illustrating the connection of events in a chain which was included in the appendix of the fourth edition of "The Calculus of Probabilities" (1924) [121], Markov studies the interchange of vowels and consonants in the Russian language. He chooses a sequence of 20,000 letters from Pushkin's "Eugene Onegin"§ and asserts that this sequence may be approximately considered as a simple chain. He also investigates the sequence of 100,000 letters from a novel by the Russian author S. T. Aksakov.

Markov sought the probability that a letter randomly chosen from a Russian text would turn out to be a vowel. This probability depends on whether the preceding letter is a vowel or a consonant. For the text of "Eugene Onegin," the probability of the occurrence of a vowel after a vowel is equal to $\alpha = 0.128$ and a vowel after a consonant is $\beta = 0.663$.††

A simple Markov chain possesses the ergodic property, i.e., the property that the limiting distribution does not depend on the initial state

---

†For homogeneous Markov chains these probabilities do not depend on the time points (*Translator's remark*).

‡*Izv. Akad. Nauk.* **7** (6) (1913).

§The first chapter and the first stanzas of the second (*Translator's remark*).

††This type of investigation has been carried out in various languages in recent years, as a result of development of the information-theoretic concept of *entropy* by C. E. Shannon in 1948 [210] (*Translator's remark*).

$\left(\lim_{t \to \infty} F(x, t | \zeta, \tau) = F(x)\right)$. In this case

$$\lim_{s \to \infty} \alpha_s = \lim_{s \to \infty} \beta_s = \frac{\beta}{(1 - \gamma)},$$

where $\gamma = \alpha - \beta$ (α_s and β_s are the s-step transition probabilities).

Applying this formula to our case, we obtain that the limiting (stationary) probability of the occurrence of a vowel is

$$\lim_{n \to \infty} \alpha_n = \frac{\beta}{1 - \alpha + \beta} = \frac{0.663}{1 - 0.128 + 0.633} = 0.432.$$

Markov's computations show that this quantity coincides with the frequency of occurrences of a vowel in the above mentioned text of "Eugene Onegin."

Markov began his teaching career at the University of St. Petersburg in 1880 as an assistant professor. He read various mathematical courses. When Chebyshev retired from the University in 1883, Markov replaced him in his course on probability theory. Even after his retirement in 1905 with the title of distinguished professor, he continued to lecture on probability theory.

The main characteristic of his pedagogical activity was his desire to present the material in a rigorous manner without piling up a vast amount of information, but rather aiming to lay a solid foundation based on which the students could develop a critical approach to the subject. His active and sharp mind was in evidence in all his lectures and he was always thoroughly prepared for his classes.

The high quality of his lectures is reflected in his book "The Calculus of Probabilities" (first edition in 1913). In this volume many new ideas and results were included. This book, although a highly scholarly treatise, was intended for beginners as well. All the proofs are presented with impeccable rigor and thoroughness. Significant attention is devoted to applied problems. In the introduction to the fourth (posthumous) edition of the book, it is stated: "The most valuable feature of the book is that it does not present the standard dry, overworked smooth scientific material, but is teeming with original research contributions. These features render the book as a classical work with no equal in the theory of probability" [122, p. XIV].

The book starts with the basic notions and theorems. The definition of probability is classical. However, after presenting the definition of equipossibility (two events are equipossible if there is no reason to prefer one over the other; several events are called equipossible if each pair of them is equipossible), Markov remarks:

> In my opinion, the various notions can be defined not so much by words, which may require, in turn, additional explanation, as by our attitude towards them which is gradually elucidated. [121, p. 2]

Markov next discusses the basic theorems, which are the multiplication and addition rules for probabilities.

The second chapter is devoted to sequences of trials. Here Bernoulli's formula is derived ($P_{m,n} = C_n^m p^m q^{n-m}$) and the most probable number of occurrences of an event is obtained.

Markov attaches great importance to Bernoulli's theorem and presents three different proofs of this result, the first proof being Bernoulli's original proof. He then remarks:

> It is usually concluded from Bernoulli's theorem that, as the number of trials increases indefinitely, the ratio of the number of occurrences of an event to the number of trials approaches to the probability of the event in a single trial. This conclusion is, however, not entirely correct in the cases when the conditions of the Bernoulli theorem are not satisfied, as well as in the cases when this theorem is applicable.
>
> The conditions under which Bernoulli's theorem is valid are the independence of the trials and the constancy of the probability of the event.
>
> In this case, Bernoulli's theorem reveals that the occurrence of significant deviations of the ratio m/n from p for large values of n is very unlikely. But it does not completely eliminate the possibility of such deviations; these unlikely deviations may indeed take place. [121, p. 67]

This is a very interesting remark. Markov here protests against the definition of probability as the $\lim_{n\to\infty} m/n = p$, which later on became the basis of the unsubstantiated definition of probability proposed by von Mises.

In this chapter he also derives Stirling's formula and the de Moivre–Laplace theorem.

The third chapter is devoted to the law of large numbers. In the beginning of this chapter some properties of mathematical expectations are considered; next Markov proves the Chebyshev inequality and Chebyshev's theorem. He writes: "In my opinion, the term 'law of large numbers' should be used for the totality of all generalizations of Bernoulli's theorem" [121, p. 98]. At the end of the chapter certain fair and biased games are considered.

In the fourth chapter are solved various problems, including those dealing with computations of lottery outcomes in various countries. Problems of division of stakes, gambler's ruin, and others are also considered.

The fifth chapter is called "Limits, irrational numbers, and continuous variables in the calculus of probabilities." In this chapter such problems as the determination of the probability of irreducible rational fractions with randomly chosen numerators and denominators, Buffon's problem, and others are discussed. Also in this chapter the notion of mathematical expectation for continuous variables is introduced.

In the sixth chapter probabilities of hypothesis and future events are considered.

The last two chapters are devoted to the problems of least squares and insurance applications. Moreover, a number of Markov's papers, mentioned above, are included in this volume. All chapters are supplemented with references and all examples and problems are solved in detail.

This remarkable book combines a simplicity and clarity of exposition along with prominent new contributions to probability theory.

Lyapunov describes the state of probability theory at the beginning of the twentieth century as follows:

> Chebyshev has shown in one of his memoirs that the results of his investigations on limiting values of integrals led to the proof of Laplace's and Poisson's well-known theorem concerning the probability that the sum of a large number of independent random variables is contained within given limits.
>
> This theorem was the subject of a large number of investigations. However, the attempts to prove it rigorously under somewhat more general conditions were unsuccessful for a long period of time, and it was, to the best of my knowledge, Chebyshev who first succeeded in this undertaking.
>
> However, this famous scientist presented only a sketch of the proof.... For this reason, certain supplementary arguments were required for the completion of his memoir. Markov performed this task impeccably in a recent paper.
>
> Nevertheless, it must be admitted that this proof, based on a special theory, is somewhat complicated and cumbersome. It will, therefore, be necessary to continue these investigations and, in any case, a direct proof would seem desirable.
>
> I therefore found it worthwhile to reexamine previous methods in use in this connection....
>
> I was able to obtain a very general result which proved not only Markov's version of Chebyshev's theorem but even a substantially more general theorem. Moreover, this general result was obtained using an argument independent of any special theory and based only on the most elementary considerations. [112, pp. 181–183]

Lyapunov's more general results were obtained using a new method— the so-called method of characteristic functions.†

The characteristic function $\varphi_x(t)$ of a random variable x is the mathe-

---

†Cf. B. V. Gnedenko and A. N. Kolmogorov, "Limit Distributions for Sums of Independent Random Variables" (1954, English translation), p. 5 and the footnote on p. 211 (*Translator's remark*).

matical expectation of the variable e^{itx}, i.e.,

$$\varphi_x(t) = M(e^{itx}) = \int_{-\infty}^{+\infty} e^{itx} \, dF(x).$$

The integral $\int_{-\infty}^{\infty} e^{itx} \, dF(x)$ is convergent for all real values of t and for every distribution function $F(x)$.

The method of characteristic functions is more general than the method of moments. Characteristic functions exist for any random variable and determine completely the moments of the distribution (provided the latter exist). Moreover, the characteristic function determines uniquely the distribution function $F(x)$, independently of whether the moments exist or not. The method of characteristic functions evolved as a very powerful tool. It soon became the basic method for solution of problems on sums of random variables, mainly due to the following property: the characteristic function of sums of independent random variables equals the product of their characteristic functions, i.e., $\varphi_{x+y}(t) = \varphi_x(t) \, \varphi_y(t)$.

In his first paper on probability theory, published in 1900 [103], Lyapunov proves the central limit theorem under a substantially more general set-up than Chebyshev's or Markov's.

He considers an infinite sequence of independent random variables $x_1, x_2, x_3, \dots$, which admit real values. Without these values, it is possible to calculate for each one of the variables the probability that they will be contained within the given limits and, moreover, this probability is independent of the values taken by other variables. This is a consequence of the fact that the variables are independent. Under these assumptions Lyapunov proves the following theorem:

Assuming the existence of the mathematical expectation of the variables

$$x_i, \qquad x_i^2, \qquad |x_i|^3, \qquad i = 1, 2, 3, \dots$$

and denoting them by

$$\alpha_i, \qquad a_i, \qquad l_i, \qquad i = 1, 2, 3, \dots,$$

correspondingly, we set

$$a_1 - \alpha_1^2 + a_2 - \alpha_2^2 + \cdots + a_n - \alpha_n^2 = A$$

and denote by L^3 the maximum of the n quantities $l_1, l_2, \dots, l_n$. Then, if the expression $(L^2/A) \, n^{2/3}$ approaches zero, as n increases indefinitely, the probability of the inequalities

$$z_1(2A)^{1/2} < x_1 - \alpha_1 + x_2 - \alpha_2 + \cdots + x_n - \alpha_n < z_2(2A)^{1/2}$$

for arbitrary values of z_1 and $z_2 > z_1$, approaches, as $n \to \infty$, to the limit

$$\pi^{-1/2} \int_{z_1}^{z_2} \exp(-z^2)\, dz$$

uniformly in z_1 and z_2. [112, pp. 184–185]

It is not required in the above statement that the moments of all orders be uniformly bounded (as is the case in Markov's theorem); the only requirement is the existence of moments of the first three orders.

In two short notes "On a theorem in the calculus of probabilities" [105] and "A general proposition in the calculus of probabilities" [106] published in 1901, Lyapunov points out the possibility of further generalizations of this theorem.

These investigations resulted in an article "Nouvelle forme du théorème sur la limite de probabilité [*Notes Acad. Sci. St. Petersburg Ser. 8* **12** (No. 5), 1–24 (1901)] [104]. In this paper the assumptions on the random variables were further weakened. The final form of Lyapunov's theorem is as follows:

If δ denotes a positive number and d_i is the expectation of the variable $|x_i - \alpha_i|^{2+\delta}$, then, provided there exists a value of δ such that the ratio

$$\frac{(d_1 + d_2 + \cdots + d_n)}{(a_1 + a_2 + \cdots + a_n)^{1+\delta/2}} = \frac{\sum\limits_{i=1}^{n} M(x_i - \alpha_i)^{2+\delta}}{\left[\sum\limits_{i=1}^{n} D(x_i)\right]^{1+\delta/2}}$$

tends to zero as n increases to infinity, the probability of the inequalities

$$z_1 < \frac{x_1 - \alpha_1 + x_2 - \alpha_2 + \cdots + x_n - \alpha_n}{[2(a_1 + a_2 + \cdots + a_n)]^{1/2}} < z_2$$

approaches, with $n \to \infty$, to the limit

$$\pi^{-1/2} \int_{z_1}^{z_2} \exp(-z^2)\, dz$$

uniformly in z_1 and $z_2 > z_1$. [112, pp. 223–224]

Here $\{x_1, x_2, x_3, \ldots\}$ is an infinite sequence of independent random variables; $\alpha_1, \alpha_2, \alpha_3, \ldots$, are their mathematical expectations; $a_1, a_2, a_3, \ldots$ are the mathematical expectations of the variables $(x_1 - \alpha_1)^2$, $(x_2 - \alpha_2)^2$, $(x_3 - \alpha_3)^2$, $\ldots$, and z_1 and z_2 are given numbers.

The main difference between Markov's and Lyapunov's conditions is as follows: Markov requires that for all $p > 2$,

$$\lim_{n \to \infty} \frac{\sum\limits_{i=1}^{n} M(x_i^p)}{\left[\sum\limits_{i=1}^{n} D(x_i) \right]^{p/2}} = \lim_{n \to \infty} M_n^{(p)}$$

approaches 0, while Lyapunov's condition is the existence of only one value of $\delta > 0$ such that

$$\lim_{n \to \infty} \frac{\sum\limits_{i=1}^{n} M(x_i^{2+\delta})}{\left[\sum\limits_{i=1}^{n} D(x_i) \right]^{1+\delta/2}}$$

approaches 0.

Lyapunov's achievement is, in S. N. Bernstein's view, "a classical result which constitutes a culmination point of Lyapunov's investigations in probability theory" [112, p. 480].

In his paper "On a theorem in the calculus of probabilities,"† Lyapunov considers the case $\delta = 1$ ($p = 3$). Here he was unable to replace Markov's conditions by the single condition $M_n^{(3)} \to 0$. In the paper [104] mentioned above the following assertion is proved: "The condition of this theorem is such that, if it is satisfied for some given value of δ, it will also be satisfied for all smaller values" [112, p. 226], i.e. the condition $M_n^{(p)} \to 0$, where $p = 2 + \delta > 3$, yields the condition $M_n^{(p_1)} \to 0$ for all $p_1 < p$.

Moreover, Lyapunov obtained an upper bound on the error committed in replacing the exact distribution of the sum by its limiting distribution.

Lyapunov's theorem is of great importance for probability theory and its applications. For this reason, it is called the central limit theorem of probability theory. It explains, in particular, why so many random variables obey the normal law. It follows from Lyapunov's theorem that, if the random variable X is a sum of a large number of independent random variables, each one of which has only an insignificant contribution to the sum, then the distribution of X will be close to normal. Moreover, the distribution laws of the summands may be unknown and arbitrary.

This type of random variable occurs very often in practice. For example, errors in measurements, indices of quality, the sizes and weights of production items, and many physical quantities which are subject to random changes, etc.

Lyapunov's contributions triggered a number of investigations firstly toward further relaxation of his conditions for the validity of the central limit theorem. However, it took 20 years to improve his results. Y. W.

---

†*Izv. Akad. Nauk. Ser. 5* **13,** No. 4, 359–386 (1900) [103]; the short note and the longer paper bear the same title (*Translator's remark*).

Lindeberg in 1922† obtained a new sufficient condition, and in 1935 W. Feller‡ showed the necessity of this condition. This necessary and sufficient condition of Lindeberg and Feller is stated as follows: If $F_n(x)$ is the distribution function of X_n, and h is a fixed positive number, then, as $n \to \infty$

$$(1/n) \int_{|x| > h} x^2 \, d \sum_{i=1}^{n} F_i(x) \to 0.$$

This condition is necessary and sufficient for the convergence of the distribution of $(1/n) \sum_{i=1}^{n} X_i$ to the normal distribution.

Attempts were made to extend Lyapunov's theorem to the case of dependent random variables. These were culminated by S. N. Bernstein's paper "Sur l'extension du théorème limité du calcul des probabilités aux sommes de quantités dépendantes" [21] where Lyapunov's results are extended to weakly correlated variables and to the case of random vectors.§

Kolmogorov, in his address "The role of Russian science in the development of probability theory," presented at the conference at Moscow University, states:

> The significance of Chebyshev's, Markov's, and Lyapunov's contributions was, after a great delay, properly appraised in the West only in the twenties or even thirties of the present century.... This may be partially attributed to the fact that the ideas of the St. Petersburg's school were very remote from the science of statistics.
>
> It should not, however, be concluded from the above remark that the works of the St. Petersburg's school were not in touch with the needs of the mathematical aspects of the natural sciences. But, due to the backwardness of Russian physics in the second half of the nineteenth century, the interests of mathematicians in the St. Petersburg's school were not directed toward the most interesting perspectives in the applications of probability theory. (Boltzmann's investigations in physics relate to the years 1866–1898.)
>
> Chebyshev, in particular, was gifted with an innate feeling for reality in posing mathematical problems. Starting with relatively elementary and sometimes even old-fashioned applied problems, he extracted the general mathematical concepts which potentially encompassed a very wide range of technical and scientific problems. ([158, p. 404]; see also [90, p. 59].)

---

†*Math. Z.* **15**, 211–225 (1922).

‡*Math. Z.* **40**, 521–559 (1935).

§Additional details and some more recent results are presented in M. Loève's "Probability Theory" [203] (*Translator's remark*).

3   Probability in physics

As early as 1827 the English botanist Robert Brown, while observing through a microscope plant spores floating in water, detected the movement of minute suspended particles. Later it was discovered that every sufficiently small grain suspended in fluid constantly moves in a most unpredictable manner. This type of movement was called Brownian motion (or movement). It is due to random impacts or bombardment of chaotically moving molecules upon the particle in suspension. Only by means of probabilistic arguments was it possible to develop a sound theory of Brownian motion.† In general, molecular theory is based on probabilistic considerations. The presence of a large number of molecules and the chaotic nature of its thermal movement lend themselves to this type of argument.

One can assert with a high degree of accuracy that gas molecules are uniformly distributed in their volume. On the average, the deviations from the uniform distribution are negligible. There are, however, some noticeable deviations, but these occur very seldom and the larger the deviations the less frequently they take place. All these results, as well as some other results in molecular physics, may be obtained basically using statistical methods. The mean velocity of molecules, their mean kinetic energy, and many other quantities are derived in this manner. The basic problem of statistical physics is the determination of various average values for physical quantities and the establishment of laws connecting these values. Statistical considerations constitute one of the most widely utilized methods of physics.

If, before the second half of the nineteenth century, the basic areas for application of the methods of probability theory were the processing of observations, statistics (especially demography), and some other related problems, the situation completely changed in the second half of that century. This was mainly due to the appearance of the works of an Austrian physicist, Ludwig Boltzmann (1844–1906), and the American scientist Josiah Willard Gibbs (1839–1903).

Boltzmann was one of the most prominent theoretical physicists of the second half of the nineteenth century and one of the founders of modern physics. His name is primarily connected with the initiation and development of statistical physics. His main contribution was the molecular-kinetic interpretation of the second law of thermodynamics and the derivation of the statistical interpretation of entropy.

The state of physical bodies is characterized by temperature, pressure,

---

†This theory was developed in 1905 by A. Einstein and M. von Smoluchowski. The analytical results derived by Einstein were later experimentally verified and extended.

density, etc. To each state there correspond many various cases of distributions of molecules and atoms (the so-called "atomic pictures" of the body. The more the atomic pictures correspond to a particular state, the more often is this state encountered, i.e. the greater is the probability of encountering this particular state. We denote this probability by *W*.

These states, whose probability is maximal, occur most often. The smaller its probability, the less frequently does a state occur. However, even very unlikely states do occur. In particular, the state of the gas at the present moment is very unlikely. Boltzmann writes in this connection:

> The event that all molecules in a gas have exactly the same velocity in the same direction is not a hair less probable than the event that each molecule has exactly the velocity and direction of motion that it actually has at a particular instance in the gas. [31, p. 56]†

And later in the same treatise:

> Each distribution of states, even if very unlikely, has a probability different from zero, though very small. Similarly, when one has ... the case that one molecule has the velocity that it actually has at that time, likewise for the second, third, and other molecules, is not in the least more probable than the case that all molecules have the same velocity. [31, p. 59]†

A state in which all the molecules of a gas are gathered in one half of the volume is unlikely, but possible. Boltzmann calculated that for a volume of 1 cm³ the probability that all the molecules will be located in one-half of this volume is equal to $(\frac{1}{10})^{10^{10}}$.

If there are no forces from outside, the system moves from less probable states to more probable ones. In this manner it achieves its most probable state, around which it would undergo certain variations, i.e. fluctuations. The most probable state is the state of equilibrium.

Thus, the probability constantly tends toward its maximal value. The same is true of the behavior of the entropy, which always increases as a result of any processes that occur in the system.

Boltzmann introduces into consideration the function H which is an analogue of the entropy and is of significance in statistical physics. Let

$$f(\xi, \eta, \varepsilon, t)\, d\xi\, d\eta\, d\varepsilon = f\, d\omega$$

denote the number of molecules m at the instant of time t for which the components of the velocity in the direction of the three coordinate axes are situated within the limits

$$\xi \quad \text{and} \quad \xi + d\xi, \qquad \eta \quad \text{and} \quad \eta + d\eta, \qquad \varepsilon \quad \text{and} \quad \varepsilon + d\varepsilon.$$

†Originally published by the University of California Press; reprinted by permission of The Regents of the University of California.

A parallelepiped whose vertex has the coordinates ξ, η, and ε, and the edges parallel to the coordinate axis are equal to $d\xi$, $d\eta$, and $d\varepsilon$ will be called parallelepiped $d\omega$.

If the function f is known for some value of t, then the distribution of velocities of the molecules m is determined at the instant of time t.

Analogously, let

$$F(\xi_1, \eta_1, \varepsilon_1, t)\, d\xi_1\, d\eta_1\, d\varepsilon_1 = F_1\, d\omega_1$$

denote the number of molecules m_1 whose velocity components lie within the limits:

$$\xi_1 \quad \text{and} \quad \xi_1 + d\xi_1, \qquad \eta_1 \quad \text{and} \quad \eta_1 + d\eta_1, \qquad \varepsilon_1 \quad \text{and} \quad \varepsilon_1 + d\varepsilon_1.$$

Then

$$H = \int f \ln f \, d\omega + \int F_1 \ln F_1 \, d\omega_1.$$

Next it is proved under the assumption: "that the velocity distribution is molecular-disordered at the beginning and remains so ..., the quantity H can only decrease" [31, p. 55].†

A molecular-disordered distribution (state), according to Boltzmann, is a state in which the location and velocity before the collision of either one of two colliding molecules does not depend on the location and velocity of the other.

> Each molecule flies from one collision to another so far away that one can consider the occurrence of another molecule, at the place where it collides the second time, with a definite state of motion, as being an event completely independent (for statistical calculations) of the place from which the first molecule came (and, similarly, for the state of motion of the first molecule). [31, p. 41]†

After proving that the function H cannot increase as a function of time, Boltzmann interprets it (actually this quantity with a minus sign) as an analogue of the entropy. In 1877 Boltzmann pointed out the connection between the function H and the probability of a given distribution.

Heated discussions followed Boltzmann's H-theorem. As a result, Boltzmann was required to consider more carefully the connection between this theorem and other problems. This led him to the realization of the statistical nature of the second law of thermodynamics.

The total amount of entropy increases in the course of any physical processes (through any occurrence in nature). The law of the inevitable increase in entropy is called the second principle of thermodynamics. What is then the relation between the entropy S and the probability W?

†Originally published by the University of California Press; reprinted by permission of The Regents of the University of California.

The entropy of two bodies is equal to the sum of entropies of the individuals, while their joint probability is equal to the product of probabilities of the individual states ($W = W_1 \cdot W_2$). It thus follows that the relation between S and W should be logarithmic: $S = k \ln W$. To determine the coefficient of proportionality k, it is necessary to compute S and W in a certain particular case. This computation was carried out by Boltzmann for gases. He obtained that $k = 1.38 \times 10^{-16}$ erg/deg—this the famous Boltzmann constant—which is one of the basic universal constants of physics.

Thus, Boltzmann showed that entropy is a measure of probability of a sojourn of the system in a given state.

Concerning the second law of thermodynamics, Boltzmann writes:

> The fact that in nature the entropy tends to a maximum shows that for all interactions (diffusion, heat conduction, etc.) of actual gases the individual molecules behave according to the laws of probability in their interactions, or at least that the actual gas behaves like the molecular-disordered gas which we have in mind.
>
> The second law is thus found to be a probability law. [31, p. 85]†

These ideas on the connection between entropy and probability, which were intensively and deeply developed by Boltzmann, were not immediately accepted by physicists.

Boltzmann had to explain the most elementary points and principles of probability theory (see, e.g. [31, pp. 55, 91]). In his arguments he often uses the notion of a number of molecules possessing a given property rather than the notion of probability. In his opinion, this approach is more convenient since "an actual number of objects is a more perspicuous concept than a mere probability" [31, p. 61].†

According to Boltzmann, the increase is merely the most probable change of entropy. Let a system be given whose state at the instant t_0 is of low probability and which in the course of time with a high probability moves toward more probable states, which leads with a very high probability to an increase in the entropy, although fluctuations are possible when the entropy is decreased.

From the reversibility principle of mechanics, it then follows that before the instant t_0 the entropy should have decreased with the same high probability. The observed invertability of statistical processes is connected with the fact that, at the present time, *fading* rather than growth of cosmic fluctuations is taking place.

Boltzmann felt that probability theory is applicable to such cosmic processes. "Since, however, the probability calculus has been verified in

---

†Originally published by the University of California Press; reprinted by permission of The Regents of the University of California.

so many special cases, I see no reason why it should not also be applied to natural processes of a more general kind" [31, p. 448].†

He also presents detailed discussion and clarification concerning the problem of the return of a system to its previous state:

> The transition from an ordered to a disordered state is only extremely improbable. Also, the reverse transition has a definite calculable (though inconceivably small) probability, which approaches zero only in the limiting case when the number of molecules is infinite. The fact that a closed system of a finite number of molecules, when it is initially in an ordered state and then goes over to a disordered state, finally after an inconceivably long time must again return to the ordered state is therefore not a refutation but rather indeed a confirmation of our theory.
>
> One should not, however, imagine that two gases in a $\frac{1}{10}$ liter container, initially unmixed, will mix, then again after a few days separate, then mix again, and so forth. On the contrary, one finds by the same principles which I used for a similar calculation that not until after a time enormously long compared to $10^{10^{10}}$ years will there be any noticeable unmixing of the gases. [31, pp. 443–444]†

Boltzmann believed that statistical methods are applicable also to the study of electromagnetic waves. He wrote: As in the theory of gases, one can determine the most probable state for the radiation... at this state the waves are not ordered, but interact among themselves in a most different manner [30, p. 617]. Max Planck shared this opinion. In his "Wege zur physikalische Erkenntnis" he mentions that "the attempt to comprehend the meaning of the experimental radiation laws naturally led me to a consideration of the connection between entropy and probability, i.e., to Boltzmann's way of thinking" [149, p. 102].

Boltzmann's ideas greatly influenced the development of physics. In particular, statistical notions developed by him paved the way for quantum theory, since statistical notions are among the initial concepts of this theory.

Boltzmann was a proponent of the philosophical grounds of materialism and was opposed to the idealistic philosophy of Mach, Ostwald, and Schopenhauer; in general, he attacked idealism. In one of his essays he wrote:

> The highly respected English philosopher Berkeley was responsible for the invention of the greatest stupidity ever contrived by the human mind, i.e., philosophical idealism which rejects the existence of the material world. [35, p. 51]

He also polemicized very sharply with Schopenhauer in his lecture

---

†Originally published by the University of California Press; reprinted by permission of The Regents of the University of California.

concerning a thesis of Schopenhauer [29] which contained numerous personal attacks showing his low opinion of this idealistic philosopher.

Timiryazev[†] praised Boltzmann highly as a physicist–antimetaphysicist, as a physicist–philosopher, and as one of the most brilliant representatives of this science.

Even before Boltzmann, a number of physicists visualized solids as consisting of a large number of particles which can, therefore, be examined only using statistical methods.

D. Bernoulli explained the pressure of the gas on the sides of a vessel by movements of its molecules. M. V. Lomonosov,[‡] in his investigations, arrived at the same conclusion. The immediate predecessor of Boltzmann was James Clerk Maxwell who thought of molecules as elastic solids.

Starting from this premise, Maxwell constructed a theory of gases that was related to the works of Clausius. In a survey paper read in 1875 before the London Chemical Society, Maxwell states that the special contribution of Clausius is in the development of methods for investigating systems consisting of infinitely many molecules in motion and thus "opened up a new field of mathematical physics."

Clausius distributed the molecules in groups according to their velocities, and it thus became feasible to substitute for the impossible task of following every individual molecule through all its encounters, that of registering the increase or decrease of the number of molecules in the different groups.

Maxwell then continues in his lecture:

> By following this method, which is the only one available either experimentally or mathematically, we pass from the methods of strict dynamics to those of statistics and probability.
>
> When an encounter takes place between two molecules, they are transferred from one pair of groups to another, but by the time that a great many encounters have taken place, the number which enter each group is, on an average, neither more nor less than the number which leave it during the same time. When the system has reached this state, the numbers in each group must be distributed according to some definite law. [93, pp. 148–149]

This distribution law of the velocities of molecules was derived by Maxwell.[§] For this purpose, he proceeds from the following consideration: Let

[†]K. A. Timiryazev (1843–1920), a well-known Russian scientist-botanist–physiologist; his main research was in the field of photosynthesis and its relation to physiological processes in plants; as a proponent of Darwinism and a well-known historian of the natural sciences, he propagated materialistic philosophy (*Translator's remark*).

[‡]M. V. Lomonosov (1711–1765), Russian scientist and writer, in his experiments anticipated the mechanical nature of heat and kinetic theory of gases (*Translator's remark*).

[§]More on the history of the Maxwell distribution can be found in Sheynin's article [212a] (*Translator's remark*).

$\varphi(x)\,dx$ be the probability that the projection of the velocity of a molecule on the x axis is contained between x and $x + dx$, and let the corresponding definitions be given for $\varphi(y)\,dy$ and $\varphi(z)\,dz$. The probability that the vector from the origin representing the velocity will be contained between

$$x, \quad y, \quad z \quad \text{and} \quad x + dx, \quad y + dy, \quad z + dz,$$

is equal to

$$P = \varphi(x)\,\varphi(y)\,\varphi(z)\,dx\,dy\,dz. \tag{IV.4}$$

This probability, on the other hand, should be a function of the distance from the origin, i.e.,

$$\varphi(x)\,\varphi(y)\,\varphi(z) = f(x^2 + y^2 + z^2).$$

Taking logarithms on both sides, we obtain:

$$\ln \varphi(x) + \ln \varphi(y) + \ln \varphi(z) = \ln f(x^2 + y^2 + z^2).$$

Differentiation with respect to x yields:

$$\frac{\varphi'(x)}{\varphi(x)} = \frac{2xf'(x^2 + y^2 + z^2)}{f(x^2 + y^2 + z^2)} \quad \text{or} \quad \frac{\varphi'(x)}{x\varphi(x)} = \frac{2f'(x^2 + y^2 + z^2)}{f(x^2 + y^2 + z^2)}.$$

Analogously,

$$\frac{\varphi'(y)}{y\varphi(y)} = \frac{\varphi'(z)}{z\varphi(z)} = \frac{\varphi'(x)}{x\varphi(x)} = \frac{2f'(x^2 + y^2 + z^2)}{f(x^2 + y^2 + z^2)}.$$

Taking into account certain additional physical considerations, we easily determine a function that satisfies this relation:

$$\varphi(x) = \exp(-\kappa^2 x^2); \qquad \varphi(x)\,\varphi(y)\,\varphi(z) = f(x^2 + y^2 + z^2) = e^{-\kappa^2(x^2 + y^2 + z^2)}.$$

This formula represents the Maxwell law of velocities.

Bertrand initially objected to this type of argument. His point was that it is uncertain that the coordinates y and $y + dy$ are independent of the coordinates x and $x + dx$. Therefore, Eq. (IV.4) may not be valid. Later it was proved that these events are indeed independent, so that Bertrand's objection was invalidated.

Boltzmann proved the Maxwell distribution law by appraising collisions of molecules. Even if it is assumed that at a certain initial time all molecules have the same velocity, nevertheless, after a certain period, as a result of collisions, the distribution of the velocities of molecules will be random, following the Maxwell distribution.

Clausius, Maxwell, and others did obtain certain results in the field of statistical physics, however, the probabilistic arguments and the statistical approach were only incidental in their works.

Boltzmann and Gibbs were the first to introduce probabilistic arguments and statistics into physics consistently.

Problems of statistical mechanics occupied Gibbs' attention as early as the eighties of the previous century, but only after ten years did they become the focal point of his scientific activity.

In 1892 Gibbs observed in a letter to Lord Rayleigh:

> Just now I am trying to get ready for publication something on thermodynamics from the a priori point of view, or rather on "Statistical Mechanics" of which the principal interest would be in its application to thermodynamics—in the line therefore of the work of Maxwell and Boltzmann. [60, p. 38]

At that time Gibbs started lecturing on statistical physics at Yale University.

In 1902, Gibbs published the book "Basic Principles of Statistical Mechanics." This monograph was of great importance to the development of theoretical physics. Classical statistical physics originated primarily in the works of Maxwell and Boltzmann and was pursued further in Gibbs' book where it achieved its logical conclusion.

Gibbs regards properties of solids as those of an ensemble consisting of a large number of minute particles subject to the laws of mechanics. He studies the properties of this ensemble using the methods of probability theory and investigates the basic role of the notion of probability as applied to the physical sciences. This notion makes it possible to carry out a penetrating analysis of macroscopic properties of substances. Gibbs connects these properties with the average statistical properties of ensembles. He widely utilizes Boltzmann's idea, interpreting entropy as the probabilistic state of the system. He properly poses the problem, which originated with Boltzmann, concerning the contradiction between thermodynamical non-invertability resulting from the law of the increase of entropy and the reversibility of the laws of motion, i.e., the reversibility in time of all purely mechanical processes. This problem remains a topic for investigation up to the present time.

As has been emphasized in this book on various occasions, all physical observations are subject to error. The behavior of the action of a certain system is not the consequence of precisely defined initial positions, but rather positions specified only by a distribution law. Even if solely for this reason, probabilistic considerations and the study of random variables should be introduced into the physical sciences.

Gibbs was of the opinion that physics does not deal with results that will definitely occur, but only with those whose occurrence is highly probable.

Although Gibbs' ideas penetrated deeply into theoretical physics, some of his works were not completed. This was due, in part, to the absence or inadequate development of the corresponding branches of probability theory. As Norbert Wiener points out: "Gibbs' introduction of probability into physics occurred well before there was an adequate theory of probability he needed" [181, p. 10].

In his book "Basic principles of statistical mechanics" Gibbs describes the aims and object of statistical methods in physics: "The usual point of view in the study of mechanics is that where the attention is mainly directed to the changes which take place in the course of time in a given system" [66, p. VII, Preface]. The state of the system is taken here "at any required time." In certain cases, states are studied which differ "infinitesimally" from the actual state of the system.

> For some purposes, however, it is desirable to take a broader view of the subject. We may imagine a great number of systems of the same nature, but differing in the configurations and velocities which they have at a given instant, and differing not merely infinitesimally, but it may be so as to embrace every conceivable combination of configuration and velocities. And here we may set the problem, not to follow a particular system through its succession of configurations, but to determine how the whole number of systems will be distributed among the various conceivable configurations and velocities at any required time, when the distribution has been given for some one time. The fundamental equation for this inquiry is that which gives the rate of change of the number of systems which fall within any infinitesimal limits of configuration and velocity. Such inquiries have been called by Maxwell *statistical*. [66, pp. VII–VIII]

Among his predecessors Gibbs names Clausius, Maxwell, and Boltzmann. Next Gibbs considers the probability that an arbitrary system in the ensemble will be contained within the given limits and establishes the principle of density-in-phase (or probability of phase):

> In the general case, the fundamental equation admits an integration, which gives a principle which may be variously expressed, according to the point of view from which it is regarded, as the conservation of density-in-phase, or of extension-in-phase, or of probability of phase. . . . In other words, we combine the principle of conservation of probability of phase, which is exact, with those approximate relations, which it is customary to assume in the "theory of errors." [66, pp. X–XI]

Maxwell and initially Boltzmann believed that a gaseous mass consists of a large number of molecules and that one should study the statistical characteristics of their motion. Later Boltzmann and after him Gibbs were of the opinion that one should consider a very large number of identical gaseous masses. However, in these identical masses the movement of molecules was not the same.

By investigating sufficiently large numbers of these masses and studying their most often encountered properties, Boltzmann and Gibbs arrived at a statistical mechanics interpretation.

Gibbs' ideas were received enthusiastically by many scientists. "Gibbs lives," says R. A. Millikan, "because profound scholar, matchless analyst that he was, he did for statistical mechanics and for thermodynamics what Laplace did for celestial mechanics and Maxwell did for electrodynamics, namely, made his field a well-nigh finished theoretical structure" [60, p. 436].

During the first decade of the twentieth century, due to various articles by Planck, Lorentz, Ehrenfest, and others, Gibbs' statistical approach became accessible to a wide circle of scientists.

Thus, Boltzmann and Gibbs widely utilized probabilistic notions in their investigations, in particular, in the notion of probability. But this notion and the probabilistic arguments were not quite the same as those used by mathematicians. For example, Gibbs introduces such notions as the coefficient of the probability of the phase $p = D/N$, where D is the density of the phase and N the total number of systems, etc. Subsequent advances in problems of statistical physics required additional development of the mathematical techniques and tools of probability theory.

4   Bertrand's paradoxes

Mathematicians repeatedly pointed out the necessity of rendering a precise meaning to the basic notions of probability theory. In this connection Joseph Bertrand's position is of interest.

Bertrand constructed a number of paradoxes related to the basic notions of probability theory [23]. One of these is as follows: A chord of a circle with radius r is chosen at random. What is the probability that its length exceeds the length of a side of the inscribed equilateral triangle?

Interpreting differently the words "chosen at random," we obtain differing solutions for this problem. For example, by considering chords which are parallel to some given direction, we obtained that the required probability is $\frac{1}{2}$. Indeed, the chords whose length exceeds the side of the equilateral triangle are less than $r/2$ units distant from the center of the circle (Fig. 11).

If we consider that the chords chosen at random emanate from a fixed point on the circle, then the required probability turns out to be $\frac{1}{3}$ (Fig. 12). If we interpret the words "chosen at random" in the sense that the probability that the midpoint of a chord lies within a certain part of the circle is proportional to the area of this part, we obtain the required probability to be equal to $\frac{1}{4}$. Indeed, since any point of the circle may serve as the midpoint of the chord, and the midpoints of chords larger than the side

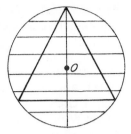

Figure 11

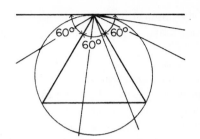

Figure 12

of the triangle lie within the concentric circle with radius one-half that of the given circle (Fig. 13), the required probability is $\pi(r/2)^2/\pi r^2 = \frac{1}{4}$. Additional interpretations of the notion "randomly chosen" may result in still different values of probabilities.

This ambiguity in the final answer may be explained in the following manner: The word "probability" presupposes certain definite experiments. In many problems this experiment is not explicitly described, but is usually evident from the content of the problem. However, in the case under consideration the words "a chord chosen at random" in the statement of the problem do not give any indication of a specific experiment, unless additional explanations are included.

Bertrand prefers the first of the three solutions presented above. The following general problems also lead to this answer: If a circular disk is tossed on a plane ruled with parallel lines, the probability that one of the intersecting chords exceeds the sides of the inscribed equilateral triangle is $\frac{1}{2}$. The same result would be obtained if one looked upon the chord as the intersection of the moon and the trajectory of a star, or if one considered the chords described in the circular field of a telescope by stars to which one has not aimed and which, thus, occupy an arbitrary position.

We describe yet another of Bertrand's paradoxes. Two points M and M' are selected at random on the surface of a sphere. What is the probability that the smaller of the great circles MM' is less than α?

This probability is the same regardless of the position of M and, hence, if M is fixed, M' must fall on the curved surface of the segment that corresponds to a semicentral angle $MOA = \alpha$ (see Fig. 14). Designating the radius of the sphere by R, one thus has $MP = OM - OP = R(1 - \cos \alpha)$, and the ratio of the area of the curved surface of the segment to the area of the sphere is

$$MP/2R = (1 - \cos \alpha)/2 = \sin^2(\alpha/2).$$

This is the desired probability. If α is very small, one can replace $\sin(\alpha/2)$ by $\alpha/2$ and approximate the probability by $\alpha^2/4$.

In connection with the method just presented, Bertrand indicated an alternative kind of reasoning which leads to an entirely different result [23, Ch. 1].

Given two points M and M', the great circle joining them is determined, and since all great circles on the surface of a sphere are equivalent, the probability is not affected if one fixes the great circle in advance. As we have already seen, the probability that two points on a circle are such that MM' is less than α equals α/π, and this differs from the result $(\alpha^2/4)$ obtained before, especially if α is very small. If α is $1°$ (or $\pi/180$ radians) one has

$$\alpha^2/4 = \pi^2/360^2 \quad \text{and} \quad \alpha/\pi = 1/180.$$

The ratio of the second value to the first is $720/\pi^2$, which is greater than 72.

Bertrand thus concludes that the given problem is insoluble and therefore both solutions are incorrect.

However, this assertion of Bertrand is erroneous. The first solution is correct, while the second is not. It is presupposed in this problem that all parts of the surface of the sphere equal in size are equivalent as far as the location of the points M and M' on these parts is concerned. The error in the second solution is that it is assumed that the probability of the point M' lying on a given arc of the great circle is proportional to the length of the arc. However, since the arc of the great circle is of thickness zero, the probability that the points will lie on this circle is also zero.

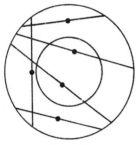

Figure 13

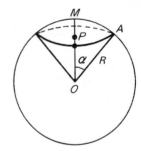

Figure 14

To avoid this last conclusion, it is necessary to consider instead of lines (and arcs) thin layers with the sides emanating from the point M (Fig. 15).

It follows from this figure that the probability for a point to lie on this layer in the neighborhood of point M is less than the corresponding probability at distances of $90°$ from this point.

Bertrand commented critically on various problems in probability theory. For example, the following question was discussed: If certain stars are situated very closely together in the firmament, does it follow from this fact that they are actually close in space?

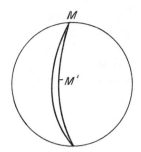

Figure 15

If one studies the number of stars of a specific size, it is possible to calculate the probability of disposition of a set number of stars in the firmament within a given small circle. If the probability obtained is very small, one may assume that this grouping of stars is due to a cause, i.e., the stars are indeed close together in the space. Bertrand objects to this argument:

> The Pleiades seem to be closer to one another than would be natural. This assertion is worthy of interest, but if one wants to translate the consequence into figures, one runs into difficulties. In order to make this "drawing closer" precise, must one look for the smallest circle containing the group? The greatest angular distance? The sum of the squares of all the distances? The area of the spherical polygon whose vertices are some of these stars while the others are in its interior? So far as the Pleiades group is concerned, these quantities are all much smaller than should be expected, so which one gives the appropriate measure of unlikelihood? If three of the stars happened to form an equilateral triangle, must one include this (which surely has a very small a priori probability) among the circumstances revealing a cause? [23, p. 170]

Concerning Bertrand's remark about the equilateral triangle we merely note that any other triangle has the same "small" probability.

Bertrand (and later Poincaré) discussed the following problem. Given three identical boxes each one of which has two compartments. The first box contains a gold medal in each of its compartments, the second a silver one and the third a gold medal in one compartment and a silver in the second. A box is chosen at random. What is the probability that its compartments will contain different medals? Clearly the answer is $\frac{1}{3}$. We now pose another question: What is the probability that in the second compartment of this box there will be a medal made of a different material than that in the first opened compartment?

Bertrand solves this problem in the following manner: Let, for example, a gold medal be contained in the first opened compartment, then the other compartment may contain either a gold or a silver medal; hence the required probability is $\frac{1}{2}$. This solution is, however, incorrect since Bertrand does not establish the equipossibility of the cases under consideration.

The correct solution of the problem will be evident if we represent the boxes by the following diagram:

If a gold medal is in the opened compartment, we then have one of three equiprobable cases (1, 2, 3) of which only the third one is favorable (for the event where probability is to be determined). The required probability is thus equal to $\frac{1}{3}$. The same argument is valid if a silver medal was in the first opened compartment.

However, if it is known which compartment is opened, the left or the right, then the required probability is $\frac{1}{2}$.†

Bertrand does not accept the theory of the average man proposed by Quetelet. He argues as follows: Can there exist a man whose height is equal to the average height, his weight to the average weight and so on? Let us take for example two balls, one with radius $r_1 = 1$ and the other with $r_2 = 3$. If they are made of the same material and the first weighs 1 gm, the weight of the second is then 27 gm. The average ball then has a radius of $r = 2$ and weight $(27 + 1)/2 = 14$ gm. However, if the average ball is made of the same material, then, having a radius of $r = 2$, it should weigh 8 and not 14 gm.

There cannot, therefore, exist a ball with an average radius, as well as an average weight. Thus we cannot be at all certain that there exists a man with various average characteristics.

In his text [23] Bertrand discusses the argument which arose originally in connection with smallpox inoculation. Originally one person in 200 died as a result of inoculation. Various opinions were expressed concerning the advisability of this procedure. D. Bernoulli came up with the calculation that the average lifespan was increased by three years and concluded from this that inoculation was worthwhile. Bertrand cites d'Alembert's objection to Bernoulli's argument and supports d'Alembert's view. He writes:

---

†The fallacies in these arguments arise from a not sufficiently careful definition of the *sample space* related to the event under consideration. "One should never speak of probabilities, except in relation to a given sample space" (W. Feller, Vol. I, p. 13 [195]). In the first case the sample space consists of three points (*Translator's remark*).

Assume that it is possible surgically to increase the mean duration of life not only by four years but by 40 years, with the proviso, however, that one quarter of the patients face possible immediate death. To sacrifice one quarter of the lives in order to double the remaining three-quarters' lives would seem to be a great gain. But who would want to take advantage of this opportunity? What physician would agree to operate? Who would undertake to invite 4000 healthy, strong people to undergo an operation and order 1000 coffins the next day? What director of an educational institution would dare to inform 50 mothers that by increasing the mean duration of life of his 200 students, he played a "favorable" game, but that their sons were among the losers? Most reasonable parents would accept one chance against 200, but no one would expose himself to the risk of one chance against four. [23, p. XII]

Poincaré took note of Bertrand's paradoxes and generalized them. He suggested that one consider, in place of a continuous variable in Bertrand's paradoxes, an arbitrary continuous function $f(x)$ of these variables. In this case all the problems concerning x could be replaced by the corresponding problems with $f(x)$. This remark is also applicable to functions of several variables. It would follow that the definition of the probability of an event is quite arbitrary, since it depends on an arbitrary continuous function.

For example, let the position of a point depend on two variables; introduce an arbitrary positive function $\varphi(x, y)$ satisfying the condition

$$\int_{-\infty}^{\infty} \int_{-\infty}^{\infty} \varphi(x, y)\, dx\, dy = 1.$$

Then the probability that a point (x, y) lies within the area S will be $\int\int_S \varphi(x, y)\, dx\, dy$. With such a definition the probability will depend on $\varphi(x, y)$. However, Poincaré observes that in many known cases the final result of the calculations hardly depends on the choice of this arbitrary function, provided it satisfies certain nonrestrictive conditions.

Neither Bertrand nor Poincaré aimed at undermining the authority of probability theory. Using their paradoxes, they attempted to emphasize the inaccuracy and vagueness in some of the basic notions of probability theory and in this manner provided a motivation for their clarification and precision.

V

The Axiomatic Foundations of Probability Theory

1 The need for axiomatization

Toward the beginning of the twentieth century, probability theory developed enormously as a result of the contributions of the Russian school and was applied successfully to physics. On the other hand, during the very same period various completely unfounded applications of this theory appeared.

Certain mathematicians began using probability theory for political purposes. Especially active in this connection was P. A. Nekrasov (1853–1924). From 1893 on Nekrasov was rector of Moscow University; he was appointed superintendent of the Moscow Educational District shortly thereafter and later became a member of the Scientific Council of the Ministry of Public Education.

In his numerous works, Nekrasov was a proponent of the idealistic philosophy. The following examples from his work illustrate his point of view.

> A slave feels a stationary dependence on his master, a criminal on the court and police, and so on, and the measure of this dependence may be given by means of probability theory. [136, p. 19]

> Probability theory gives a numerical measure of stationary, as well as nonstationary effects of dependences. [136, p. 21]

In 1896 Nekrasov published "Probability Theory" containing his lectures at Moscow University and the Land-Surveying Institute. The second substantially revised and supplemented edition of this book was published in 1912 [138].

When reading this book, one is amazed by the abundance of disconnected pseudo-scientific sentences, which often seem to pertain to basic and fundamental notions of probability theory.

Discussing social problems, Nekrasov sharply opposes political changes in which the masses participate. He considers private property a prime principle, which it is the czarist regime's province to protect.

Nekrasov writes that a man of science will necessarily arrive at a belief in God, a very close union between God and man being predestined. He often refers to commentaries on statements in the Scriptures. All this is included in his book "Probability Theory." Moreover, Nekrasov "corroborates" and "deduces" his assertions using the laws of probability theory, and often resorts to complicated mathematical calculations to make his arguments convincing.

In 1915 Nekrasov in his capacity as a member of the Council of the Ministry of Public Education organized a commission for the introduction of elementary theory into the high school curriculum. One of the program's aims was to utilize probability theory to nurture obedience to the Czar, to praise the czarist autocracy, and so on.

When Markov found out about the creation of this commission, he initiated another commission in the Academy of Sciences which included D. K. Bobylev, A. N. Krylov, A. M. Lyapunov, A. A. Markov, and V. A. Steklov. The new commission drafted and passed a resolution which contained, among others, the following passage:

> Nekrasov's views have long been known to mathematicians. However, as long as they were confined to specialized mathematical journals, they could be considered harmless. It is quite another matter when an official body sponsors these ideas. Therefore, the Academy of Sciences feels duty-bound to express its views concerning the basic errors and false and therefore harmful ideas which are being propagated by P. A. Nekrasov with the aim of putting them into practice in the high schools. . . . It is this Commission's opinion that the above mentioned blunders and . . . abuse of mathematics with the preconceived aim of transforming this science into a tool for religious and political persuasion . . . will cause irreparable harm to education. [76, p. 178]

After suffering this rebuff, Nekrasov decided to accuse Markov of advocating materialism. He quotes a passage from Markov's "The Calculus of Probabilities"† in which Markov disagrees with Bunyakovskiĭ, who claimed that it is reprehensible to doubt the validity of certain kinds of stories even if they seem to be improbable. He then asserts:

> By invalidating the above-quoted basic proposition of academician Bunyakovskiĭ, Markov is facilitating the propagation of historical materialism ... One does not need a better manual for systematic dissemination of the most groundless materialism than Markov's book. [139, p. 16]

In this controversy we are not so much interested in the reactionary views of Nekrasov, who masked his pseudo-scientific deductions with references to probability theory, as in the progressive materialistic position taken by Markov. In this struggle the noble image of the scientist–materialist came to light.‡

The necessity of reevaluating the logical foundations of probability theory in order to secure its position as a genuine mathematical discipline became more and more evident.

In view of the existing ambiguity concerning the subject matter of probability theory, even such a prominent mathematician as Emile Borel (1871–1956) was carried away with unjustified applications of this theory, extending it to areas of no relevance to its subject.

In 1914 his highly interesting book entitled "Le Hasard" appeared; it was translated into Russian in 1923 [32]. Many important problems are discussed in this book. After a detailed discussion of the basic laws of probability theory which includes profound historical and philosophical observations, Borel describes the penetration of probabilistic methods into physics, biology, and other sciences; he discusses the relationship between probability theory and other branches of mathematics. In spite of the definite positive qualities of this book, one should mention that certain fundamental problems of probability theory are erroneously interpreted in this volume.

For example:

> Imagine one thousand Parisians passing by a seven story immovable property; they all agree to call it a house; however, they refuse to call a stone structure serving as a shelter to two rabbits and three hens a house. Let us consider an average structure (of these two); here opinion may be divided; if 748 out of 1000 voters call this structure a house, it would therefore be correct to assert that the probability that this structure is a house is 0.748 and the opposite probability is 0.252. [32, p. 88]§

†See p. 218.
‡More details on the progressive role of Markov are given in [88, 123, 145].
§Translated here from the French Edition: Borel, E., "Le Hasard." Presses Universitaires de France, Paris, 1948.

Such an arbitrary interpretation of probability could have arisen only as a result of the ambiguity and vagueness of this notion.†

Borel applies probability theory to social, moral, and other similar problems.

> Can one go one step further and construct on the basis of probability theory a real individualistic and social morality? The most exalted moral dictum ever suggested to mankind would seem to be contained in the Gospel's commandment: "Love thy neighbor as thyself." [32, pp. 168–169] ‡

He then discusses this commandment and concludes:

> The only reasonable interpretation which can be given to this commandment is as follows: consider your neighbor as a quantity equivalent not to yourself, in any case, but to a certain part of yourself, contained between zero and one, never reaching the lower limit zero, but sometimes reaching the upper limit one. I don't think that such a statement can be called egotistic. As a matter of fact, after careful deliberation, it becomes clear that this interpretation expresses the approach of reasonable altruism. Different degrees of altruism and egotism are determined by the value of the coefficients: to some people the coefficient 1 may be attributed; others will be assigned 0.9, or 0.5 ... or the values of the coefficient 0.00001, 0.000001. I shall not discuss these values; it is a matter of practical morality, rather than probability theory. The important point is that these coefficients should not be equal to zero. This assertion can be taken as the basis of theoretical morality. In the establishment of this fact, as well as in the investigation of its consequences, we constantly come across the ideas of probability theory. [32, pp. 169–170] ‡

Nowadays no student of probability theory even on the most elementary level will seriously accept this type of assertion. These quotations are brought up not to create the impression that the whole book consists of similar propositions. We repeat that this is a very interesting and useful book. But even a mathematician of Borel's stature had a rather vague idea concerning the subject matter and the methods of probability theory. He was aware, however, that probability theory should be improved. Borel writes in the introduction to this monograph that "applications to exact sciences will refine probability theory" [32, p. VI]. ‡

It is worthy of note that in his later book on probability, entitled "Probability and Certainty," written in 1950 [33], he also touches upon a number

---

†We note, however, that this type of argument is basic for a subjective definition of probability. See the end of Section 4 of this chapter (*Translator's remark*).

‡Translated here from the French Edition: Borel, E., "Le Hasard." Presses Universitaires de France, Paris, 1948.

of basic and important problems of probability theory, but does not give any unjustified applications. By that time probability theory had become a full-fledged mathematical discipline.†

However, ambiguity, lack of comprehension, and confusion concerning probabilistic and statistical methods remained for a long time. We cite yet another example. In the introduction to the Russian translation of Borel's book "Le Hasard" the editor of the translation V. A. Kostitzin writes:

> Achievements in physics and astronomy put statistical methods in the forefront of the modern natural sciences [32, p. VIII].

He then continues a few pages later:

> The scope of hypotheses concerning the molecular world is constantly shrinking and in the not too distant future, the structure of the atom and interatomic forces will be well known. Then the era of sound scientific determinism will return and statistical regularities in the natural sciences will turn out to be just a temporary stage in our knowledge. [32, p. X]‡

As we have seen, this prediction has not materialized and laws of statistics have penetrated into constantly widening domains of modern natural sciences. Moreover, this process is determined not by the level of our knowledge, but by the actual structure of phenomena investigated in these sciences.

A direct predecessor of the founders of axiomatization in probability theory was Henri Poincaré (1854–1912), a prominent mathematician and well-known physicist. He made remarkable contributions to the fields of differential equations, integral equations, algebra, theory of numbers, geometry, theory of electricity, thermostatics, theory of Hertz's waves, the kinetic theory of gases, and so on. Within these various problems, probability theory occupies a comparatively modest place.

Poincaré wrote a number of books and articles of a philosophical nature in which he sometimes discusses philosophical and methodological problems of probability theory as well. He is also the author of the book "Calcul des Probabilités" ("Calculus of Probabilities") published in 1912 [151].

Poincaré's philosophical views were based on idealism and machism.‡ Lenin in his treatise "Materialism and Empiriocriticism" justly criticized this philosophy of Poincaré, calling him the French machist and terming his gnosiological conclusions as idealistic (see [151]). In Soviet literature he was also subject to criticism. However, one should not diminish his

---

†See, however, the first footnote on p. 243 (*Translator's remark*).

‡Ernst Mach (1838–1916) was an Austrian physicist and philosopher and one of the leaders of modern positivism. He felt that science should confine itself to a description of phenomena that could be perceived by the senses (*Translator's remark*).

his specialized contributions to physics and mathematics. He was indeed a prominent scientist.

Poincaré's "Calcul des Probabilités" is one of the most rigorous and interesting books on probability theory written at the beginning of the twentieth century. We shall now discuss some of the general theses given therein.

In his definition of random events he uses a deterministic approach:

> If a minute cause which escapes our notice determines a considerable effect that we cannot miss, then we say that this effect is due to chance.
>
> If we had an exact knowledge of the laws of nature and the position of the universe at the initial moment, we could predict exactly the position of that same universe in a succeeding moment.

And further,

> It may happen that small differences in the initial conditions produce very great ones in the final phenomena. A small error in the former will produce an enormous error in the latter. Prediction becomes impossible, and we have a fortuitous phenomenon. [151, pp. 1–5; 150]

Poincaré presents several examples of random phenomena: (1) The unsteady equilibrium of a cone standing on its vertex. It is unknown on which side this cone will fall. (2) Meteorological phenomena. "One cannot predict accurately where a cyclone originated. Here again we find the same contrast between a very trifling cause that is inappreciable to the observer and considerable effects which are sometimes terrible disasters" [151, p. 6; also 150]. (3) Allocation of small planets among zodiacal constellations: "Very small initial differences in their distances from the sun or, what amounts to the same thing, in their mean motions, have resulted in enormous differences in their actual longitudes. Here again, we have a small cause and a great effect or, more precisely, small differences in the cause and great differences in the effect" [151, p. 6]. (4) The game of roulette.

In all these examples the random events were interpreted as events in which very small differences in the initial conditions led to appreciable differences in the results. In addition to this type of random event, Poincaré also investigates events whose random outcome is due to a complexity and multitude of causes. This type of random event is encountered in the kinetic theory of gases; also belonging to this category are the random distribution of rain drops on a certain surface, the distribution of suspended particles in a vessel full of liquid, the distributions of cards in a deck after a thorough shuffle, random errors in observations, and so on. "Here again we have only small causes, but each of them would produce only a small effect; it is by their union and their number that their effects become formidable" [151, p. 10; also 150, p. 75].

Poincaré also considers a third type of random events, which however, as he himself observes, is reduced to the first two types. He presents the following example: A roofer (slater) drops a tile which kills a passerby.

But the man has no thought for the slater, nor the slater for him; they seem to belong to two worlds completely foreign to one another. Nevertheless the slater drops a tile which kills the man, and we should have no hesitation in saying that this was chance.

Our frailty does not permit us to take in the whole universe, but forces us to cut it up in slices. We attempt to make this as little artificial as possible, and yet it happens, from time to time, that two of these slices react upon each other, and then the effects of this mutual action appear to us to be due to chance. . . . Each time that two worlds, generally foreign to one another, thus come to act upon each other, the laws of this reaction cannot fail to be very complex, and moreover a very small change in the initial conditions of the two worlds would have been enough to prevent the reaction from taking place. How very little it would have taken to make the man pass a moment later, or the slater drop his tile a moment earlier! [151, p. 11; also 150]

Next Poincaré poses the question: But why should chance be subject to a law? He then returns to an example using roulette. The pointer of the roulette wheel is whirled with force and, after having made a great number of revolutions, stops in a certain subdivision. The probability that the initial push is within a and $a + \varepsilon$ is equal to the probability that it will be comprised between $a + \varepsilon$ and $a_2 + 2\varepsilon$, provided that ε is very small. "This is a property common to all analytic functions. Minute variations of the function are proportional to minute variations of the variable" [151, p. 12; also 150, p. 77]. Poincaré believes that "chance obeys laws," since the distribution of chances is continuous in nature. But on the other hand, explaining the existence of continuity in nature, he asserts that the continuity itself is determined by the interaction of a multitude of chances.

In the course of this history, complex causes have been at work and they have been at work for a long time. They have contributed to bringing about the mixture of the elements, and they have tended to make everything uniform, at least in a small space. They have rounded off the corners, leveled the mountains, and filled up the valleys. However capricious and irregular the original curve may have been given, they have worked so much to regularize it that they will finally give us a continuous curve, and that is why we can quite confidently admit its continuity. [151, p. 67; also 150, p. 83]†

†The previous quotations which appeared in Poincaré's "Calcul des Probabilités" may also be found in his "Science and Method" as indicated by the double references (*Translator's remark*).

It is Poincaré's opinion that "science is deterministic a priori; it postulates determinism" [152, p. 127]. Poincaré assigns to probability only those areas which have not been investigated as yet, in which our knowledge is incomplete: "Thus the problems of probability may be classed according to the greater or lesser depth of our ignorance" [154, p. 189].

"Chance is only the measure of our ignorance. Fortuitous phenomena are, by definition, those laws we are ignorant of" [151, p. 51; also 150, p. 65].

Starting from these propositions, Poincaré arrives in his book on probability theory at some rather strange conclusions: "We do not to what are due accidental errors, and precisely because we do not know, we are aware they obey the law of Gauss. Such is the paradox" [151, p. 16; 153, p. 406]. In the spirit of the deterministic philosophy of the end of the eighteenth century positive conclusions are derived from ignorance.

Poincaré extends his probabilistic concept to the whole history of humanity.

What does the phrase "very slight" mean? . . . A difference is very slight, an interval is very small, when within the limits of this interval the probability remains sensibly constant. And why may this probability be regarded as constant within a small interval? It is because we assume that the law of probability is represented by a continuous curve.

And what gives us the right to make this hypothesis? . . . It is because, since the beginning of the ages, there have always been complex causes ceaselessiy acting in the same way and making the world tend toward uniformity without ever being able to turn back. These are the causes which little by little have flattened the salients and filled up the reentrants, and this is why our probability curves now show only gentle undulations. In milliards of milliards of ages another step will have been made toward uniformity, and these undulations will be ten times as gentle; the radius of mean curvature of our curve will have become ten times as great. And then such a length as seems to us today not very small, since on our curve an arc of this length cannot be regarded as rectilineal, should on the contrary at that epoch be called very little, since the curvature will have become ten times less and an arc of this length may be sensibly identified with a sect.

Thus the phrase "very slight" remains relative; but it is not relative to such or such a man, it is relative to the actual state of the world. It will change its meaning when the world shall have become more uniform, when all things shall have blended still more. But then doubtless men can no longer live and must give place to other beings [151, 16–17; 153, p. 409].

Proceeding to the definition of probability and discussing its classical definition, Poincaré writes:

How can we determine that all the cases are equally probable? Mathematical determination is not possible in this case; in each application we must put conditions and stipulate that we shall consider these particular cases as equiprobable. These assumptions are not completely arbitrary, but they may escape the mathematician, if he does not analyze them after they have been made [151, p. 28].

We thus observe that Poincaré in his support of Bertrand's paradoxes and by his other critical remarks on the foundations of probability theory demanded a more rigorous approach to these concepts. His "Calcul des Probabilités" was one of the best textbooks of its period. However, his idealistic philosophical views affected his interpretation of many basic problems of probability theory.†

2 Prerequisites for axiomatization

The development of probability theory in the beginning of the twentieth century necessitated a reevaluation and refinement of its logical foundations. This was required by the progress in statistical physics, as well as the development of probability theory itself. The Laplace-type classical foundations were obviously insufficient and inadequate. The expansion of other sciences in which probabilistic notions were used demanded further clarification and justification of these notions.

In order to establish a logical order and consistency for any kind of inference, it is necessary to single out the initial concepts, establish the rules for inference, and show the absence of contradiction in all the results obtained.

This can be achieved using an axiomatic method that makes it possible to encompass the totality of objects studied by a given mathematical theory. The essence of this method is that certain propositions—to be called axioms—are postulated as the basis of the theory, and all the subsequent propositions are deduced from these axioms; moreover, the rules of deduction should be distinctly formulated. The use of axiomatic methods was stepped up after N.I. Lobachevskiĭ pointed out the possibility of constructing geometry on axioms different from Euclid's. A variety of mathematical theories then appeared which were constructed using axiomatization. Toward the beginning of the twentieth century, the axiomatic method penetrated various branches of mathematics. A thorough analysis of systems of axioms for geometry was carried out by David Hilbert, Giuseppe

---

†See the recent book by J. L. Boursin and P. Coussat "Autopsie du Hasard," Bords. Coll. Études Supérieures, 1970, where many of Poincaré's ideas are elaborated (*Translator's remark*).

Peano, and V. F. Kagan, and systematic investigations of axiomatization for arithmetic were initiated by Peano and Hilbert.

The axiomatic method is not confined to various branches of mathematics; it is a method for discovering new facts in general. The axiomatization, along with a summary of obtained results, serves as a stimulus to the further development of a theory.

In the beginning of the twentieth century all texts on probability theory presented the classical foundations of probability theory stemming from Laplace. However, the inadequacy of this system became more and more evident for the level of the theory at that period. The classical definition of probability using equipossible events is actually a tautology,† since equipossibility is in essence equiprobability.

It should be noted that, for a limited class of events for which "symmetry" is clearly evident, this definition can be justified, but one cannot extend it to other types of events. From here yet another essential deficiency of the classical concept follows—the highly restrictive nature of its applicability. The classical definition is not applicable to a majority of problems in physics, statistics, biology, and the technical sciences. In all these problems it was not feasible to determine the equiprobable cases without which one cannot discuss probability (in the classical sense).

Moreover, based on classical foundations, one cannot predict the behavior of processes in real life. Here additional special assumptions which do not follow logically from the basic notions are necessary.

Thus new logical foundations for probability theory based on the axiomatic method were required.

3   Bernstein's contributions

Reevaluation of the logical foundations of probability theory served as the beginning of a new and most fruitful stage in its development. The first works in this direction are due to S. N. Bernstein (1880–1968). In 1917 he published in the *Proceedings of the Kharkov Mathematical Association* (Vol. 15) a paper entitled "An essay on the axiomatic foundations of probability theory." In later years he also devoted some of his works to the axiomatization of probability theory. ‡

In the compendium "Thirty Years of Mathematics in the USSR" [73], Bernstein's works are evaluated as the beginning of a new stage in the development of probability theory in Russia.

---

†Saying the thing twice over in different words.

‡See Bernstein's obituary in *Uspehi Mat. Nauk* 24, 211–219 (1969) [185] concerning up-till-now unknown facts of his work on applications of probability theory to biology (also *Ann. Math. Statist.* 13, 53–61 (1942)) (*Translator's remark*).

In 1927 the first edition of Bernstein's "Probability Theory" was published. The last (fourth) edition appeared in 1946.† This text is one of the best books on probability theory and for many years it served as a textbook not only for mathematicians and physicists, but for multitudes of students of various other disciplines.

In his book, Bernstein presents a detailed axiomatization of probability theory. He assumes the basic scheme which our inference in the natural sciences operates on, "that based on previous experience, we assert that the occurrence of an event belonging to a known class A is certain, whenever a given set of conditions α is realized, independently of any other factors" [22, p. 7]. However, in general, the occurrence of an event is not an absolute certainty. One cannot predict the behavior of real phenomena with unshakable certainty. The laws that connect α with A have a practical meaning only in those cases when the set of conditions α is not too large and can be subject to observations. If this condition is not fulfilled, event A is called a random event. We then attempt to introduce in place of α a simpler set of conditions β in whose presence the occurrence of A acquires a certain definite probability.

> The basic assumption of probability theory (the postulate of the existence of mathematical probability) is that there exist such sets of conditions β which (at least theoretically) may be repeatedly realized infinitely many times in whose presence the occurrence of event A in a given experiment has a definite probability which is numerically expressed. [22, p. 8]

If the probability is also defined for event B, then one of the three relations is valid:

$$P(A) = P(B); \qquad P(A) > P(B); \qquad P(A) < P(B).$$

> It is the experience which determines whether under the realization of a given set of conditions β and complete arbitrariness of all other circumstances it is possible to ascribe a definite probability to event A. [22, p. 8]

Bernstein introduces three axioms: (1) the axiom of comparability of probabilities; (2) the axiom of incompatible (disjoint) events; (3) the axiom of combination of events. The first two axioms are concerned with the case when the set of conditions β is fixed. The third axiom associates the probability of A under conditions α with the probability of the same event under a different set of conditions β.

Before proceeding to a statement of the axioms, we introduce certain preliminary notions.

---

†In 1948 the fifth edition was planned, but did not materialize due to Bernstein's refusal to omit sections devoted to biological applications. See [185] (*Translator's remark*).

If the occurrence of a yields also to the occurrence of A, a is then called a particular case of event A. If event A can occur without its particular case A_1, A_1 is then called a particular case of A in the strict sense. Otherwise we consider A_1 to be a particular case of A in the wide sense.

We now state the first axiom:

The axiom of comparability of probabilities. If a is a particular case of A in the strict sense, then $P(a) < P(A)$; conversely, if for events a_1 and A the the inequality $P(a_1) < P(A)$ holds, then $P(a_1) = P(a)$, where a is a certain particular case of A in the strict sense.

Two obvious corollaries follow from this axiom:

1. The probability of a certain event is larger than the probability of a possible event.

2. The probability of a possible event is larger than that of an impossible one.

This means that all certain events have the same maximal probability and that all impossible events have the same minimal probability.

As far as the second part of the axiom is concerned, some difficulties may arise in establishing event a. However, it is assumed that, in principle, it will always be possible to determine such an event.

Second axiom:

The axiom of incompatible (disjoint) events. If it is known that events A and A_1 are incompatible, and, moreover, that events B and B_1 are also incompatible, while $P(A) = P(B)$ and $P(A_1) = P(B_1)$, then the probability of event C, which consists in the occurrence of event A or event A_1, is equal to the probability of event C_1 consisting in the occurrence of B or B_1, i.e. $P(A \text{ or } A_1) = P(B \text{ or } B_1)$.

The meaning of the second axiom is that the probability of the occurrence of one out of two disjoint events is determined by the probabilities of each one of them separately, i.e. is a function of these probabilities and does not depend on the particular nature of these very events.

This axiom is easily extended to an arbitrary number of incompatible (disjoint) events.

As a corollary of these two axioms the following conclusion follows:

> If out of the n possible, disjoint, and equiprobable cases, m cases are favorable to event X, then the probability of event X depends only on the numbers m and n (and not on the nature of the experiment under consideration), i.e., prob $X = F(m, n)$ where $F(m, n)$ is a certain function. [22, p. 13]

Can any arbitrary function $F(m, n)$ satisfy the first two axioms? It turns out that only functions of the form $F(m/n)$ satisfy these axioms, and moreover $F(m/n)$ should be an increasing function of the ratio m/n. Any such function can be chosen to represent the probability of X. The common procedure is to set $F(m/n) = m/n$. This is the probability of event X under above stated conditions.

The axiom of combination of events: If α is a particular case of event A, then the probability of α under given conditions depends only on the probability of event A under the same conditions and on the probability acquired by α in the case when event A occurs.

This means that if α_1 is a particular case of event A_1, then $P(\alpha) = P(\alpha_1)$, if $P(A) = P(A_1)$ under the given conditions, provided the probability acquired by α after event A occurs is equal to the probability acquired by α_1 after event A_1 occurred.

If α is a combination (join) of events A and B, then under the assumption of the occurrence of event A, the occurrence of α is equivalent to the occurrence of B.

The axiom of combination of events can be formulated also as follows: The probability of combination of A and B (under given conditions) depends only on the probability of A (under the same conditions) and on the probability acquired by B after the occurrence of A.

For independent events, these axioms assert: If events A and B are independent, then the probability of the combination of A and B depends only on the initial probabilities of these events.

The axiom of combination of events can be expressed as

$$(A,B) = \Phi\left[(A),(B)_A\right] = \Phi\left[(B),(A)_B\right],$$

where (A) is the probability of A; $(A)_B$ is the probability of A after the occurrence of B; (A, B) is the probability of the combination of A and B; Φ is a function fixed in advance (the form of this function is determined by the multiplication rule: $(A, B) = (A)(B)_A$ and depends on the form of function F).

On the basis of these axioms Bernstein constructs the whole edifice of probability theory. As Kolmogorov points out:

> The first systematically developed axiomatization of probability theory, based on the notion of qualitative comparison of (random) events according to their (larger or smaller) probability is due to S. N. Bernstein. The numerical value of the probability appears in this conception as a derived rather than as a primary notion. [90, p. 60]

This idea of Bernstein's was further developed by V.I. Glivenko† and the American mathematician B.O. Koopman.

In his numerous works on the application of probability theory to problems of the natural sciences, Bernstein adhered to the opinions he

---

†Glivenko showed in 1939 the equivalence of Bernstein's axiomatization with Kolmogorov's set-theoretical axioms to be described in Section 5 (*Translator's remark*).

expressed at the First All Russian Conference of Mathematicians in 1927 in Moscow.

> Purely mathematical probability theory cannot be concerned whether the coefficient called mathematical probability has any practical value, either subjective or objective. The only condition to be fulfilled is the absence of contradictions, namely: various methods of calculating this coefficient under given conditions and provided the axioms are not violated, should lead to the same value of this coefficient.
>
> Moreover, if we want the conclusions of probability theory not to degenerate into a simple mental exercise, but allow for empirical verification, it is necessary to consider only those propositions or assertions which can actually be established as false or true. The cognitive process, which is irreversible in its nature, actually means that certain propositions became veritable, i.e. are realized and simultaneously in this case their negation becomes false or impossible.
>
> Therefore, in order to construct probability theory as a unified cognitive method, it is required that the truth of a proposition be uniquely and without exception characterized by a definite maximal value of probability, which is set to equal unity, and the falsity of the proposition should be equal to the smallest probability which is taken to be zero. [157]

We note that the requirement of consistency is a materialistic requirement. A system of axioms is considered consistent if there exist mathematical objects such that the relations between them are expressed by this system of axioms. One cannot prove consistency in a purely logical manner; each proof of consistency is a relative proof; namely, one can prove merely that one system is consistent provided the other is. In the final analysis the consistency of any system can be reduced to the consistency of the arithmetic. To prove the consistency of the arithmetic, one must resort to an experiment. Arithmetic is consistent because all its laws are manifestations of quantitative relations between the objects of the real world and these laws have been verified billions of times by practical experiences of humanity. Thus, the requirement of consistency of axioms in the final analysis reduces to the requirement of the conformity of the axioms to reality.

Bernstein's ideas on axiomatization and the application of probability theory to problems in the natural sciences served as the basis of his "Probability Theory," which is one of the best texts on probability theory in world literature.†

---

†Bernstein's contributions to probability theory also include works on limit theorems for dependent variables as well as new results in the theory of Markov chains (see p. 216) (*Translator's remark*).

4 The frequency approach of R. von Mises

The definition of the notion of probability serves as the basis for any axiomatization of this theory. The shortcomings of the classical definition of probability had been acknowledged for a long time. Also the subjective interpretation of probability was found to have serious faults. Criticism of these defects had been generally welcomed. The most wellknown and widespread critique, especially among the natural scientists, was in the works of the German scientist Richard E. von Mises (1883–1953). Von Mises fled Hitler's Germany to the United States where he became head of an Institute of Applied Mathematics. He is the founder of the so-called frequency approach in probability theory. Von Mises consistently and persistently emphasized the fundamental defects in the classical concepts. He and his school were the first to underline clearly the idea that the notion of probability makes sense only in the presence of mass phenomena. One of the main controversies between the frequency school and the primary direction of the development of probability theory consists in the answer to the question, is probability theory a mathematical discipline or rather merely a branch of science utilizing methods extensively?

Von Mises believed that probability is not a mathematical discipline. To substantiate this assertion he argues that some real process is always associated with each probabilistic problem, therefore probability theory is a science investigating phenomena of the real world, while mathematics, according to von Mises, does not deal with these problems. However, the whole modern development of probability theory irrefutably establishes that it does indeed belong to the realm of the mathematical sciences.

A number of Soviet investigators categorically rejected von Mises' philosophical postures but at one stage did accept the thesis that probability theory is not a branch of mathematics. For example, Ernest Kolman considers it related to mathematics but nevertheless different, because, in his view, the subject matter of probability theory is the study of possibility (or chance).

> This category is not studied in mathematics, although mathematical methods are utilized for its investigation. Before using mathematical methods in probability theory, we must first establish the equipossibility of individual events, and this cannot be achieved mathematically. [89, p. 229–230]

We shall not make a detailed comment on this assertion. We only point out that probability theory studies not the "category of possibility (or chance)," but rather mass random phenomena, and that equipossibility does not play as fundamental a role as was felt previously.

The basic notion in von Mises' frequency theory of probability theory is the concept of a collective. A collective is an infinite sequence K of similar observations, each of which determines a certain point belonging to a given finite-dimensional space R. According to von Mises, one can speak of probability if and only if there exists a definite population of events—the collective.

> The word "collective" ... denotes a series of similar events or processes which differ by certain observable attributes, i.e. colors, numbers, or anything else. ... The principle which underlines the whole of our treatment of the probability problem is that a collective must exist before we begin to speak of probability. [131, p. 16]

Each collective must satisfy the following two conditions: (1) There must exist limits of the relative frequencies of events with particular attributes within the collective. (2) These limits are invariant with respect to the choice of any subsequence of the collective which is arbitrary (except that it must not be based on distinguishing the elements of the collective in their relation to the attribute under consideration).

These axioms of von Mises may also be stated in a different manner: The *first* axiom of von Mises postulates the existence of the limit

$$\lim_{n \to \infty} m/n = P(S),$$

where m is the number of cases among the first n observations, for which the point determined by the observations belongs to the subset S. This limit exists for every proper subset $S \subset R$.

The *second* axiom is equivalent to the following statement: It is required that an analogous limit should exist and have the same value $P(S)$ for every subsequence K' that can be formed from K according to a rule such that it can always be decided whether or not the nth observation of K should belong to K' *without knowing the result of this particular observation*. This latter part of the second axiom has no precise mathematical meaning.

> Attempts to express the second axiom in a more rigorous way do not, so far, seem to have reached satisfactory and easily applicable results. ... I think that these difficulties must be considered sufficiently grave to justify, at least for the time being, the choice of a fundamentally different system (of axiomatization). [46, p. 4]

Proceeding from the premise that probability theory is not a mathematical discipline, von Mises regarded his axioms as properties of a collective and did not interpret them as axioms of a mathematical theory.

As Khinchin points out in his article on von Mises' theory: "von Mises nowhere and never proceeds toward a complete formalization of his theory, i.e. he does not put it in a purely axiomatic form" [86, p. 86].

Having formulated these two axioms, von Mises completes the construction of the foundations of probability theory and presumes that we can now carry on solving specific problems and determining general regularities. But, as we have already pointed out, one cannot build the axiomatic foundations of probability theory on these axioms.

Von Mises accepts the basic premise that probability and frequency are interrelated quantities, and defines probability as a limiting value of the frequency.

> We will say that a collective means a mass phenomenon or a repetitive event, or, briefly, a long series of observations for which there are sufficient reasons to believe the hypothesis that the relative frequency of an attribute would tend to a fixed limit if it were indefinitely continued. This limit will be called the probability of the attribute considered within the given collective. ([131, p. 20]; also [206])

Actually there are no "sufficient reasons to believe this hypothesis." We may never know whether a given frequency possesses a limit because, for one thing, to determine this would require an infinite number of experiments. This definition is mathematically inconsistent, since it is impossible to determine the functional relationship between the number of trials n and the frequency m/n of the occurrence of the events, where m is the number of occurrences of the event, while we cannot calculate the limit $\lim_{n \to \infty} m/n$, which is "defined" as the probability without stipulating such a dependence.

According to von Mises the events do not possess probabilities prior to the experiment; the probability is not an objective property of the phenomenon. Probabilities of events arise only as a result of an experiment. Thus, in von Mises' view, we do not determine the existing objective properties by means of an experiment, but rather attribute them to the phenomena. Von Mises believes that no additional validity for the concept of probability is necessary. Probability thus is deprived of its meaning as an objective numerical characteristic of real-world phenomena.

Concerning von Mises' definition, the well-known Swedish mathematician H. Cramér observes:

> The probability definition thus proposed would involve a mixture of empirical and theoretical elements, which is usually avoided in modern axiomatic theories. It would, for instance, be comparable to defining a geometrical point as the limit of a chalk spot of infinitely decreasing dimensions, which is usually not done in modern axiomatic geometry. [47, pp. 150–151]†

Kolmogorov comments concerning von Mises' thesis as follows:

---

†"Mathematical Methods of Statistics," by Harald Cramér (Princeton University Press, © 1946); #9 Princeton Mathematical Series. Reprinted by permission of Princeton University Press.

The assumption concerning the probable nature of trials, i.e., concerning the tendency of frequencies to group around a fixed value may be valid on its own (as the assumption concerning "randomness" of a certain phenomenon is) only if certain conditions are presented which cannot be retained for an indefinitely long time and with indefinite precision.

Therefore, the limiting transition $\mu/n \to p$ cannot have real meaning. Moreover, the formulation of the stability of frequencies principle using this limiting process requires the availability of admissible methods for determining infinite sequences of trials which can be a mere mathematical fiction. [128, pp. 274–275]

Next Kolmogorov points out that the foundations of mathematical probability theory can be worked out in a logically simpler and more rigorous manner.

Khinchin remarks concerning the two axioms of von Mises:

The possibility of total formalization of the frequency theory, while retaining both of these requirements, is at best questionable, since the requirement of the existence of a limit is meaningless with respect to the concepts modern mathematics associates with the notion of irregular sequence. [86, No. 2, p. 86]

Von Mises' ideas were widely publicized but his concepts were never popular among mathematicians in view of the defects mentioned above. As Khinchin points out: "If von Mises' theory has no followers, to the best of our knowledge, among our mathematicians (mainly for its purely mathematical defects), it continues to be quite successful among physicists" [85]. This was written as late as 1952. However, at present the situation has changed and the frequency postulates of von Mises are also unsatisfactory for the majority of physicists.

A detailed critique of von Mises' ideas can be found in the works of B.V. Gnedenko and A. Ya. Khinchin (see, e.g. [85, 86, 70]).

A number of attempts were made to formalize frequency theory completely in the course of which von Mises' assumptions were modified somewhat. For example, Kamke suggested replacing the infinite collectives by finite ones leaving out the requirement of irregularity. K. Dörge, E. Tornier, H. Copeland, and others partially relaxed the requirement of irregularity, i.e. they require that the same value of the limit be preserved not for any choice of a subsequence, but only for a certain limited class. Tornier refused the use of schemes which do not fit into the frequency interpretation. For this purpose, he constructed a cumbersome formal calculus and was forced to abandon the possibility of formulating and solving a number of elementary problems of probability theory, within the framework of his theory.

Von Mises' attitude to all these modifications of his frequency concepts was negative. In his opinion the requirement of irregularity is the basic cornerstone of the theory. It should be noted that these attempts may possibly evolve into a formalization of probability theory, but any kind of frequency formalization is of necessity very cumbersome. This is due to the fact that the frequency interpretation is not sufficiently abstract and carries certain concrete meaning.†

The more abstract an axiomatic system, the simpler it is; the more substantial, the more complex it is, and it is more difficult to make inferences within such a theory. This is the main defect of all types of frequency theories.

The most prominent probabilists never adhered to the frequency school, while the followers of this school did not contribute significantly to probability theory.

There were many attempts to lay the foundations of probability theory using various approaches. For example, an Italian mathematician Bruno de Finetti proposed a subjective interpretation of probability theory. By means of this approach to probability he attempted to overcome contradictions which arose in the classical theory as well as in von Mises' frequency school.

According to de Finetti, probability is a purely subjective quantity. Each person estimates in his own way the probability of a certain event. In this theory, not only the notion of probability but other basic notions, such as dependence, independence, equiprobability, and others are all defined subjectively. De Finetti asserts that the relationship between frequency and probability is also subjective. "No relationship between probabilities and frequencies is of an empirical nature" [58, p. 26].

The definition of probability in terms of frequencies is rejected by de Finetti since in the case of such a definition one must assume the existence of the objective probability which is unacceptable from a subjective point of view. Naturally, being a mathematician, de Finetti requires that the consistency condition be fulfilled in the definition of subjective probabilities.‡

Somewhat later, H. Jeffreys (see [82]) developed the notion of probability as the degree of likelihood. This concept was first proposed by John Maynard Keynes in 1921.§ According to this theory every proposition possesses a certain probability. These types of probabilities do not lend themselves to a frequency interpretation. The development of the theory of degrees of likelihood is being continued by some mathematicians up to the present time.††

However, Jeffreys' attempts were not widely accepted.

†Some recent investigations and a review of the voluminous literature on this subject are presented in [205] (*Translator's remark*).

‡The most recent exposition of these ideas may be found in de Finetti's article [196] and his book to appear in 1973 (Wiley, New York).

§"A Treatise on Probability," London, 1921 (*Translator's remark*).

††The third edition of Jeffreys' "Theory of Probability" appeared in 1961 (*Translator's remark*).

5   The beginning of a new stage in the development of probability theory

Simultaneous with attempts to lay the foundations of probability theory were rapid new developments in the science. By analyzing the trends in this development, Kolmogorov was able to construct an axiomatization of probability theory which was a decisive stage in its further development.

In 1923, A. Ya. Khinchin derived the law of the iterated logarithm, which is a distinctive generalization and refinement of the law of large numbers. We shall now briefly discuss his results (as well as previous results leading to them):

According to Bernoulli's theorem as $n \to \infty$ for any $\varepsilon > 0$, we have, in an obvious notation,

$$P(|m/n - p| < \varepsilon) \to 1.$$

In 1909, E. Borel showed that for $p = \frac{1}{2}$

$$P\left(\lim_{n \to \infty} \frac{m - np}{n} = 0 \right) = 1,$$

i.e. the difference $m - np$ for large n with almost complete certainty is small as compared with n [namely, $m - np = o(n)$].

In 1917, F. P. Cantelli (*Atti. Accad. naz. Lincei Rc.* **26,** 39–45) generalized Borel's result for the case of an arbitrary $0 < p < 1$.

In 1913, F. Hausdorff ("Grundzüge der Mengenlehre," Leipzig) obtained the following estimate for the Bernoulli case: With probability 1, $m - np = 0(n^{\varepsilon + 1/2})$, where $\varepsilon > 0$ is an arbitrary number.

In 1914, G. H. Hardy and J. E. Littlewood (*Acta Mathematica* **37,** 155–239) showed that with probability 1, $m - np = 0((n \ln n)^{1/2})$.

Finally, in 1924,† Khinchin published the following theorem: If the probability of the occurrence of event A in each of n independent trials is equal to p, then the number m of occurrences of event A in n trials satisfies the following relation as $n \to \infty$:

$$P\left(\lim_{n \to \infty} \sup \frac{|m - np|}{(2npq \ln \ln n)^{1/2}} = 1 \right) = 1.$$

The function $(2npq \ln \ln n)^{1/2}$ is the exact upper bound on the random variable $|m - np|$ in the sense indicated in the formula.

We shall present this result geometrically: The values of n are marked

---

†A. Khinchin, "Über einen Satz der Wahrscheinlichkeitsrechnung," *Fundamenta Mathematicae* **6,** 9–20 (1924).

on the x-axis and the values of $m - np$ on the y-axis. We then draw two straight lines $y = \varepsilon n$ and $y = -\varepsilon n$. Borel and Cantelli's theorem asserts that for n sufficiently large with almost complete certainty the value $m - np$ will be confined between the lines $y = \varepsilon n$ and $y = -\varepsilon n$. However, these limits turn out to be too wide and Khinchin obtained more precise bounds on the possible values of $m - np$. If we draw the curves

$$y = (1 + \varepsilon)(2npq \ln \ln n)^{1/2} \tag{l}$$

and

$$y = -(1 + \varepsilon)(2npq \ln \ln n)^{1/2}, \tag{l'}$$

then, according to Khinchin's theorem, for any $\varepsilon > 0$ and n sufficiently large, the difference $m - np$ will almost certainly be confined within these curves. If, however, we plot the curves

$$y = (1 - \varepsilon)(2npq \ln \ln n)^{1/2} \tag{ll}$$

and

$$y = -(1 - \varepsilon)(2npq \ln \ln n)^{1/2}, \tag{ll'}$$

then the difference $m - np$ will almost certainly fall outside these limits infinitely often (see Fig. 16).

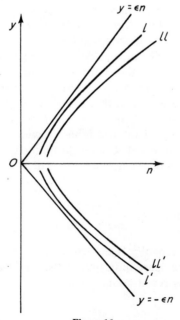

Figure 16

In later years generalizations of this law were investigated by Khinchin and Kolmogorov† and others.

These works of Khinchin and Kolmogorov marked the beginnings of the Moscow school of probability theory.

Starting from the twenties of the present century, the nature of the investigations in probability theory was determined to a great extent by the concepts of set theory and the theory of functions. A careful analysis of basic notions of probability theory has revealed that a far-reaching analogy can be established between these notions and the basic concepts of set theory, as well as the metric theory of functions. As is pointed out in the first edition of Gnedenko's "Course in Probability Theory,"

> These analogies between such seemingly different branches of science made it possible to present the logical foundations of probability theory in a different light and enrich its contents with new problems and new methods of investigation, as well as to complete the solution of certain classical problems. [69, p. 363]

Although Markov extended the limits of applicability of the law of large numbers, he did not, however, obtain a definite solution of this problem. Only by means of the methods and notions of the theory of functions were the necessary and sufficient conditions for applicability of the law of large numbers established.

In 1926, Kolmogorov‡ derived these conditions. He proved the following theorem:

A sequence of mutually independent random variables ξ_1, ξ_2, ..., ξ_n obeys the (weak) law of large numbers if, and only if, the following relations are satisfied as $n \to \infty$:

$$(1) \quad \sum_{k=1}^{n} \int_{|x| \geq n} dF_k(x) \to 0; \qquad (2) \quad (1/n) \sum_{k=1}^{n} \int_{|x| < n} x\, dF_k(x) \to 0;$$

$$(3) \quad (1/n^2) \sum_{k=1}^{n} \int_{|x| < n} x^2 \, dF_k(x) \to 0.$$

Here $F_k(x)$ denotes $P(\xi_k - M\xi_k < x)$.

This theorem presents a complete solution of one of the central problems of probability theory—the problem of the (weak) law of large numbers.

In 1928 Khinchin showed§ that if the random variables $\xi_1, \xi_2, \ldots$ were

---

†A.N. Kolmogorov, Über das Gesetz des iterierten Logarithmus, *Math. Ann.* **101**, 126–135 (1929) (*Translator's remark*).

‡The paper was published in *Math. Ann.* **99**, 309–319 (1928) and was supplemented by a note in the same journal, **102**, 484–488 (1930) (*Translator's remark*).

§The paper was published in *Compt. Rend. Acad. Sci.* **189**, 477–479 (1929).

not only independent but also identically distributed, then the existence of the expectation $M\xi_n$ was a necessary and sufficient condition for the (weak) law of large numbers to apply. In 1909, Borel posed the problem concerning the conditions that should be imposed on a sequence of random variables $\xi_1, \xi_2, \ldots$ in order that the *strong* law of large numbers will hold, i.e., in order that

$$P\left(\lim_{n\to\infty} \frac{1}{n} \sum_{k=1}^{n} (\xi_k - M\xi_k) = 0\right) = 1.$$

Borel solved this problem for the Bernoulli scheme in the case of $p = \frac{1}{2}$. The most general results for this problem were obtained by Kolmogorov (for independent random variables) and by Khinchin (for the dependent case).

Kolmogorov discovered in particular that for independent identically distributed random variables the necessary and sufficient conditions for the strong law of large numbers are identical to those required for the weak law, i.e., the existence of the mathematical expectation.

In all these investigations the analogy with the metric theory of functions played a significant role. In particular, the analogue of the weak law of large numbers is the notion of convergence in measure and of the strong law of large numbers is the convergence everywhere.

Thus the ideas of the metric theory of functions began to penetrate deeper and deeper into probability theory. Starting from the middle twenties, Kolmogorov was engaged in the logical formulation of these new ideas. This work resulted in the publication of the book "Grundbegriffe der Wahrscheinlichkeitsrechnung" in 1933.† Here the analogies between the notions of the measure of a set and the probability of an event, between the integral and the mathematical expectation, orthogonality of functions and the independence of random variables, and others were established. The necessity of axiomatization of probability theory based on measure-theoretic notions became evident. This was carried out by Kolmogorov in his book. As a result of this axiomatization, probability theory acquired an equitable position among other mathematical disciplines.

We now briefly consider Kolmogorov's axiomatization.

Let observations or trials be made, which at least theoretically admit the possibility of being repeated indefinitely. Each trial may result in a certain outcome which depends on chance. The totality of all the possible outcomes forms the set (space) E which is the first basic notion of Kolmogorov's axiomatization.

---

†English translation: "Foundations of the Theory of Probability" was published by Chelsea (New York) in 1950.

This set E is called the *set of elementary events*. What the elements of this set are is immaterial for the logical development of probability theory, much as the meaning of the words "point," "line," etc. are immaterial for the axiomatic development of geometry. (Only after such an axiomatization is available, can probability theory have various interpretations, including those which are not associated with the notion of random events.) Any subset of the set E, i.e., an arbitrary set of possible outcomes, is called a *random event*. In other words, random events are elements of a collection F of subsets of the set E. The notion of a random variable is defined here as a function of elementary events; prior to Kolmogorov this notion was considered as the initial one. Furthermore, not all the random events are considered, but only a certain *field* of events.

As we pointed out in this book on numerous occasions, probability theory deals only with those events whose frequency of occurrence is stable in a certain sense. This proportion is formalized in the axiomatic theory of Kolmogorov by the fact that a certain positive number, which is called the probability of the event, corresponds to each event in the field. We note that this definition is abstract and is not related in any way to the notion of frequency or any other notion which may serve as a real-world motivation for this definition. This allows us to interpret axiomatization not solely from a probabilistic point of view. The possibilities for application of probabilities have thus been extended.

In conclusion we state Kolmogorov's axioms:

1. If two random events A and B belong to the collection F, then events A or B [denoted $(A \cup B)$ or $(A + B)$], A and B [$(A \cap B)$], not A and not B [$(\bar{A} \cap \bar{B})$] are also contained in F.†

2. F contains the set E and all its singletons.

3. To each set A in F is assigned a non negative real number $P(A)$. This number is called the *probability* of the event A.

4. $P(E)$ equals 1.

5. If A and B have no elements in common, then

$$P(A + B) = P(A) + P(B).‡$$

In case of an infinite field F an additional axiom is supplemented, which for the case of finite fields follows as a corollary from the above stated five axioms.

6. For a decreasing sequence of events

$$A_1 \supset A_2 \supset ... \supset A_n \supset ...$$

---

†In other words, F is a field (in the algebraic sense) (*Translator's remark*).

‡The system of sets F together with a definite assignment of numbers $P(A)$, satisfying axioms 1 − 5, is called a *probability field* or *probability space* (*Translator's remark*).

of F, the product (intersection) of which is empty, the following equation holds:

$$\lim_{n \to \infty} P(A_n) = 0. \dagger$$

Since each axiomatically constructed discipline may have various interpretations, axiomatic probability theory can be construed in different terminologies. The axiomatization resulted in abstracting the notion of probability from its frequency interpretation, but at the same time made it possible always to pass over from a formal system to real-world processes. Naturally enough, every inference from this theory can be interpreted in frequency terminology.

In recent years attempts have been made to interpret probability from a more general points of view including the information-theoretical approach.‡

---

†This axiom known as the *axiom of continuity* is equivalent to the *extended axiom of addition* (extension of axiom 5 for a denumerable case), see, e.g., Gnedenko's "A Course in Probability Theory" [69] for details (*Translator's remark*).

‡See the paper by Martin-Löf [204].

Bibliography

1. Adrian, R., Research concerning the probabilities of the errors which happen in making observations, *The Analyst, or Mathematical Museum* 1 (No. 4), 83–87, 93–109, Philadelphia, 1808.
2. d'Alembert, J., Croix ou pile, *in* "Encyclopédie," Vol. 4, Paris, 1754.
3. d'Alembert, J., Reflexions sur le calcul probabilités, *in* "Opuscules Mathématiques," Vol. 2, pp. 1–25. Paris, 1761.
4. d'Alembert, J., Sur l'application du calcul des probabilités a l'inoculation de la petite vérole, *in* "Opuscules Mathématiques," Vol. 2, pp. 26–95. Paris, 1761.
5. d'Alembert, J., "Opuscules Mathématiques," Vol. 4, Paris, 1770.
6. d'Alembert, J., Sur le calcul des probabilités, *in* "Opuscules Mathématiques," Vol. 7, pp. 39–60. Paris, 1780.
7. d'Alembert, J., "Oeuvres Complètes," Vol. 1. Paris, 1821.
8. "Algebra and Mensuration from the Sanscrit of Brahmegupta and Bhascara" (H. Th. Colebrooke, transl.). Murray, London, 1817.
9. Andronov, A., and Andronov, E., «Лаплас» ("Laplas"; "Laplace"). Moscow, 1930.
10. Arago, F., «Биографии знаменитых астрономов, физиков и геометров» ("Biografii znamenityh astronomov, fizikov i geometrov"; "Biographies of Famous Astronomers, Physicists and Geometers"). St. Petersburg, 1860.
11. *Archives of the Moscow State University* (1815–1917), No. 173.
12. Bagratuni, G.V., Введение (Vvedeniye; Introduction to the Russian translation of G.F. Gauss, "Selected Works," Vol. 1). Moscow, 1957.
13. Bayes, T., Thomas Bayes essay towards solving a problem in the doctrine of chances. (Studies in the history of probability and statistics 9). Reproduced from *Philos. Trans. Roy.*

Soc. London Ser. A **53** (1763), with a biographical note by G. A. Barnard, in *Biometrika* **45**, 293–315 (1958).

14. Bazikovich, A. S., Биографический очерк (Biograficheskiĭ ocherk; Biographical sketch), *in* Markov's "The Calculus of Probabilities," 4th ed. Moscow, 1924.

15. Bernoulli, D., Specimen theoriae novae de mensura sortis, *Commentarii Academiae Scientiae Imp. Petropolitanae* **V**, 175–192 (1738) (*Papers of the Imperial Academy of Sciences of St. Petersburg* **5**, 175–192 (1738)). English translation in *Econometrica* **22**, 23–36 (1954).

16. Bernoulli, D., De usu algorithmi infinitesimalis in arte conjectandi specimen, *Novi Commentarii Academiae Scientiae Imp. Petropolitanae* **XII** (1768).

17. Bernoulli, D., Гидродинамика (Gidrodinamika; Hydrodynamics), *Akad. Nauk SSSR Trudy Jakutsk. Filial. Ser. Fiz.* (1959) ("Hydrodynamics, with 'Hydraulics' by J. Bernoulli" (T. Carmody and H. Kobus, transls.). Dover, New York, 1968.)

18. Bernoulli, D., Dijudicatio maxime probabilis plurium observationum discrepantium atque verisimillima inductio inde formanda, *Biometrika* **48**, 3–18 (1961) (English translation including Commentary by L. Euler).

19. Bernoulli, J., "Ars Conjectandi." Impensis Thurnisiorum, Fratrun, Basileae, 1713.

20. Bernoulli, J., "Pars Quarta, tradens usum et applicationem proecedentis Doctrinae in Civilibus, Moralibus, et Oeconomicus" ("Application of the Previous Study to Civil, Moral and Economic Problems") (Ya. V. Upsenkiĭ, transl.; A. A. Markov, ed.). St. Petersburg, 1913.

21. Bernstein, S. N., Sur l'extension du théorème limité du calcul des probabilités aux sommes des quantités dépendantes, *Math. Ann.* **97**, 1–59 (1927).

22. Bernstein, S. N., «Теория Вероятностей» ("Teoriya Veroyatnosteĭ"; "Probability Theory"), 4th ed. Moscow-Leningrad, 1946.

23. Bertrand, J., "Calcul des Probabilités." Gauthier-Villars, Paris, 1899.

24. Bespamyatnkh, N. D., Математика в Вильнюсском Университете (1803–1832) (Matematika v Vil'nyusskom Universitete (1803–1832); Mathematics in Vilnus University (1803–1832)), *Uchen. Zap. Karelsk. In-ta* **14** (1963).

25. Biermann, K., Aus der Geschichte der Wahrscheinlichkeitsrechnung, *Wiss. Ann.* **5**, No. 6 (1956).

26. Biermann, K., Spezielle Untersuchungen zur Kombinatorik durch G. W. Leibniz, *Forsch. Fortschr.* (No. 12, 1954; No. 6, 1956).

27. Biermann, K., The problem of the Genoise lottery in the works of classical writers in probability theory, *Istor-Mat. Issled.* (*Historical-Mathematical Investigations*) **10**, 649–670 (1957).

28. Boev, G. P.,«Теория Вероятностей»("Teoriya Veroyatnostei"; "Probability Theory"). Moscow-Leningrad, 1950.

29. Boltzmann, L., Concerning a thesis of Schopenhauer, *in* «Насущные задачи современного естествознания» ("Nasuschnye Zadachi sovremennogo estestvoznaniya; "Basic Problems of the Contemporary Natural Sciences") (K. Timiryazev, ed.). Moscow, 1908.

30. Boltzmann, L., *Wiss. Abhandl.* **3** (1909).

31. Boltzmann, L., "Lectures on Gas Theory" (S. G. Brush, transl.). University of California Press, Berkeley, 1964.

32. Borel, E., «Случай» ("Sluchaĭ"; "Le Hasard"). Moscow-Petrograd, 1923. ("Le Hasard," Presses Universitaires de France, Paris, 1948.)

33. Borel, E., «Вероятность и достоверность» ("Veroyatnost' i dostovernost'"; "Probability and Certainty"). Moscow, 1964.

34. Borel, E., "Probabilité et Certitude." Presses Universitaires de France, Paris, 1950.

35. Broda, E., Ludwig Boltzmann, *Voprosy Istor, Estestvoznan. Tehn.* **4** (1957).

36. Buffon, G. L. L., "Natural History, General and Particular" (W. Smellie, transl.), 2nd

ed., Vol. 1. W. Strahan and T. Cadell, London, 1785. (Russian edition, St. Petersburg, 1789.)

37. Buffon, G. L. L., "Natural History, General and Particular" (W. Smellie, transl.), 2nd ed., Vol. 4. W. Strahan and T. Cadell, London, 1785. (Russian edition, St. Petersburg, 1972.)

38. Bunyakovskiĭ, V. Ya., «Основания математической теории вероятностей» ("Osnovaniya Matematicheskoĭ teorii veroyatnosteĭ"; "Foundations of the Mathematical Theory of Probabilities"). St. Petersburg, 1846.

39. Bunyakovskiĭ, V. Ya., On totaling numerical tables using approximations, *Notes of the Academy of Sciences* 12 (Suppl.), No. 4. St. Petersburg, 1867.

40. Cardano, H., "Opera Omnia," Vol. 1, 1663. (Cf. Ore [141].)

41. Chebyshev, P. L., *Scientific Legacy,* Issue 1. Moscow-Leningrad, 1945.

42. Chebyshev, P. L., «Избранные математические труды» ("Izbrannye matematicheskie trudy"; "Selected Mathematical Works"). Moscow-Leningrad, 1946.

43. Chebyshev, P. L., «Полное собрание сочинений» ("Polnoe sobranie sochineniĭ"; "Complete Collected Works"), Vol. 2. Moscow-Leningrad, 1947.

44. Chebyshev, P. L., «Полное собрание сочинений» ("Polnoe sobranie sochineniĭ"; "Complete Collected Works"), Vol. 3. Moscow-Leningrad, 1948.

45. Chebyshev, P. L., «Полное собрание сочинений» ("Polnoe sobranie sochineniĭ"; "Complete Collected Works"), Vol. 5. Moscow-Leningrad, 1951.

46. Cramér, H., "Random Variables and Probability Distributions" (Cambridge Tracts in Mathematics No. 36.). Cambridge Univ. Press, London and New York, 1937.

47. Cramér, H., "Mathematical Methods of Statistics." Princeton Univ. Press, Princeton, New Jersey, 1946.

48. Dante Alighieri "The Divine Comedy" (G. L. Bickersteth transl.). Harvard Univ. Press, Cambridge, Massachusetts, 1965.

49. David, F. N., Studies in the history of probability and statistics I, Dicing and Gaming, *Biometrika* 42, 1–15 (1955).

50. Davidov, A. Yu., An application of probability theory to medicine. Московский врачебный журнал (Moscovskiĭ vrachebnyĭ jurnal; *Moscow Med. J.*) 1 (1854).

51. Davidov, A. Yu., An application of probability theory to statistics, in «Учебно-литературные статьи к 100-летнему юбилею МУ ("Uchebno-literaturnye stat'i k 100–letnemu yubileyu MU"; "The Scientific Volume Dedicated to the Centennial of Moscow University"). Moscow, 1855.

52. Davidov, A. Yu., The theory of mean values with application to the construction of mortality tables. Speech and report presented at the meetings of Moscow University. Moscow, 1857.

53. D'živelegov, A. K., «Данте Алигьери. Жизнь и творчество.» ("Dante Alig'eri. Žizn' i tvorchestvo"; "Dante Alighieri, His Life and Creative Work"). Moscow, 1946.

54. Dutka, J., Spinoza and the theory of probability, *Scripta Math.* 19, 24–32 (1953).

55. Engels, F., «Диалектика Природы» ("Dialektika prirody"; Introduction to "The Dialectics of Nature," in "K. Marx and F. Engels: Selected Works," Vol. 3.). Progress Publishers, Moscow, 1955.

56. Euler, L., Opera Omnia, series prima, 7. Lipsiae et Berolini, 1923.

57. Euler, L., "Introduction to Infinitesimal Analysis," Vol. 1. Moscow, 1961.

58. de Finetti, B., La prévision, ses lois logiques, ses sources subjectifs, *Ann. Inst. H. Poincaré* 7, 1–68 (1936). (English translation in "Studies in Subjective Probability" (O. Smokler and O. Kyburg, eds.). Wiley, New York, 1964.)

59. Fischer, H. W., "Katalog des ethnographischen Reichsmuseums," Vol. 8. Leiden, 1914.

60. Frankfurt, U. I., and Frank, A. M., «Джозайя Виллард Гиббс» ("Džozaja Villard Gibbs"; "J. W. Gibbs"). Moscow, 1964.

61. Фронт Науки и Техники (Front nauki i tehniki; *Frontiers of Science and Engineering*) 7 (1937).

62. Galilei, G., "Opera," Vol. XIV. Firenze, 1855.
63. Galilei, G., "Dialogue Concerning the Two Chief World Systems—Ptolemaic and Copernican" (S. Drake, transl.). Univ. of California Press, Berkeley, 1953; 2nd rev. ed., 1962.
64. Gauss, C. F., «Сборник статей к 100-летию со дня смерти» ("Sbornik stateĭ k stoletiyu so dnya smerti"; "A Collection of Articles Commemorating the Hundredth Anniversary of his Death"). Moscow, 1956.
65. Gauss, C. F., «Избранные геодезические сочинения» ("Izbrannye geodezicheskie sochineniya"; "Selected Geodesic Works"), Vol. 1. Moscow, 1957.
66. Gibbs, J. W., «Основные принципы статистической механики» ("Osnovnye prinčipy statističeskoĭ mehaniki"; "Basic Principles of Statistical Mechanics"). Moscow-Leningrad, 1946.
67. Gmurman, V. E., «Введение в теорию вероятностей» ("Vvedenie v teoriyu veroyatnostei"; "Introduction to Probability Theory"). Moscow, 1959. (English translation by Scripta Technica, American Elsevier, New York, 1968.)
68. Gnedenko, B. V., О работах Н.И. Лобачевского по теории вероятностей ("O rabotah N. I. Lobachevskogo po teorii veroyatnosteĭ"; "On Lobachevskiĭ's works in probability theory"), *Istor.-Mat. Issled.* **2** (1949).
69. Gnedenko, B. V., «Курс теории вероятностей» ("Kurs Teorii veroyatnosteĭ"; "A Course in Probability Theory"), 1st ed. Moscow-Leningrad, 1950.
70. Gnedenko, B. V., Probability theory and recognition of the real world, *Uspehi Mat. Nauk* **5**, 1–23 (1950).
71. Gnedenko, B. V., «Михаил Васильевич Остроградский», ("Michail Vasil'ievich Ostrogradskiĭ"). Moscow, 1952.
72. Gnedenko, B. V., On L. Euler's contributions to probability theory, the theory of processing observations, demography and insurance, *in* «Леонард Эйлер, в честь 250-летия со дня рождения» ("L. Euler"; "In honor of Euler's 250th birthday"), pp. 184–208. Acad. Sci. USSR, Moscow, 1958.
73. Gnedenko, B. V., and Kolmogorov, A. N., Probability theory, *in* «Математика в СССР за 30 лет» ("Matematika v SSSR za 30 let"; "Thirty Years of Mathematics in the USSR"), pp. 701–727. Moscow-Leningrad, 1948.
74. Gnedenko, B. V., and Pogrebysskiĭ, I. B., «Михаил Васильевич Остроградский» ("Michail Vasil'ievich Ostrogradskiĭ"). Moscow, 1963.
75. Gravé, D., «Математика социального страхования» ("Matematika sočial'nogo strahovaniya"; "The Mathematics of Social Insurance"). Leningrad, 1924.
76. Markov, A. A., Bobylev, D. K., Krylov, A. N., and Steklov, V. A., *Istor.-Mat. Issled.* **1,** 178 (1948).
77. Gnedenko, B. V., *Istor.-Mat. Issled.* **4** (1951).
78. "The History of the Natural Sciences in Russia," Vol. 2. Moscow, 1960.
79. Hobbes, T., "The English Works of Thomas Hobbes," Vol. 4, 2nd reprint. Scientia Verlag Aalen, Germany, 1966.
80. Hotimskiĭ, V., Historical roots of probability theory, *PZM*, Nos. 1 and 6 (1936).
81. Huygens, C. I., "Oeuvres Complètes," Vol. 14. 1920. La Haye, Martinus Nijhoff (for the Société Hollandaise des Sciences).
82. Jeffreys, H., "Theory of Probability." Oxford Univ. Press, London and New York, 1939 (3rd ed., 1969).
83. Chebyshev, P. L., Des valeurs moyennes (On mean values), *J. Math. Pures Appl.* **12,** 177–184 (1867).
84. Kendall, M. G., Studies in the history of probability and statistics II, The beginnings of a probability calculus, *Biometrika* **43**, 1–14 (1956).
85. Khinchin, A. Ya., The method of arbitrary functions and the struggle against idealism in probability theory, *in* "Philosophical Problems in Contemporary Physics," 1952.
86. Khinchin, A. Ya., The frequency theory of R. von Mises and contemporary ideas in probability theory, *Voprosy Filosofii* **12** (1961).

87. Khinchin, A. Ya., and Yaglom, A. M., The science of chance, *in* «Детская энциклопедия»("Detskaya Enčiklopediya" ; "Children's Encyclopedia"), Vol. 3. Moscow, 1959.

88. Kol'cov, A. V., Some materials concerning the biography of the academician A. A. Markov, *Voprosy Istor. Estestvoznan. Tehn.* **1** (1956).

89. Kol'man, E., «Предмет и метод современной математики» ("Predmet i metod sovremënnoĭ matematiki"; "The Subject and Methods of Modern Mathematics"). Moscow, 1936.

90. Kolmogorov, A. N., The role of Russian science in the development of probability theory, *Ucen. Zap. Moskov. Gos. Inst.* **91** (1947).

91. Kudryavčev, P. S., «История физики» ("Istoriya fiziki"; "The History of Physics"), Vol. 1. Moscow, 1948.

92. Kutlumuratov, D., «О развитии комбинаторных методов» ("O razvitii kombinatornyh metodov"; "On the development of Combinatorial Methods in Mathematics"). Nukus, 1964.

93. Kuznecov, B. G., Maxwell's electrodynamics, its origins, development and historical value, *Trudy Inst. Istor. Estestvoznan. Tehn. (Proc. Inst. History Natural Sci. Eng.)* **5** (1955).

94. Lagrange, J. L., Mémoire sur l'utilité de la méthode de prendre le milieu entre les résultats de plusieurs observations, *Misc. Taurinesia* **5**, 167–232 (1770–1773). (Also in "Oeuvres," Vol. 2. Paris, 1868.)

95. Laplace, P. S., «Изложение системы мира» ("Izloženie sistemy mira"; "Exposition du Système du Monde"), Vols. 1 and 2. St. Petersburg, 1861.

96. Laplace, P. S., "Exposition du Système du Monde," *in* "Oeuvres," Vol. 6. Imprimerie Royale, Paris, 1846. ("The System of the World." Longmans, Green, Dublin, 1830.)

97. Laplace, P. S., "Théorie Analytique des Probabilités." Courcier, Paris; Ist ed., 1812, 2nd ed., 1814; 3rd ed., 1820.

98. Laplace, P.S., "A Philosophical Essay on Probabilities" ("Essai Philosophique sur les Probabilités [*English Translation* (F. W. Truscott and F. L. Emory, transls.). Dover, New York, 1951.]

99. Leibniz, G. W., "Die philosophische Schriften," Vol. 4. Berlin, 1880.

100. Leibniz, G. W., «Новые опыты о человеческом разуме» ("Novye opyty o chelovecheskom razume"; "Nouveaux Essais sur l'Entendement Humain"). Moscow-Leningrad, 1936.

101. Leibniz, G. W., "New Essays Concerning Human Understanding" (A.G. Langley, transl.). Open Court, Chicago, Illinois, 1916.

102. Lenin, V. I., «Сочинения» ("Sochineniya" ; "Collected Works"), 4th ed., Vols. 14 and 25. Moscow, 1949.

103. Lyapunov, A. M., Sur une proposition de la théorie des probabilités, *Izv. Akad. Nauk. Ser. 5* **13**, 359–386 (1900).

104. Lyapunov, A. M., Nouvelle forme du théorème sur la limite de probabilité, *Notes Acad. Sci. Phys.-Math. Sect. Ser. 8* **12**, 1–24 (1901).

105. Lyapunov, A. M., Sur un théorème du calcul des probabilités. *C. R. Acad. Sci.* **132** (1901).

106. Lyapunov, A. M., Une proposition générale du calcul des probabilités, *C. R. Acad. Sci.* **132** (1901).

107. Lobachevskiĭ, N. I., «Полное собрание сочинений» ("Polnoe sobranie sochineniĭ"; "Complete Collected Works"), Vol. 1. Moscow-Leningrad, 1946.

108. Lobachevskiĭ, N. I., «Полное собрание сочинений» ("Polnoe sobranie sochinenii"; "Complete Collected Works"), Vol. 2. Moscow-Leningrad, 1949.

109. Lobachevskiĭ, N. I., «Полное собрание сочинений» ("Polnoe sobranie sochineniĭ"; "Complete Collected Works"), Vol. 5. Moscow-Leningrad, 1951.

110. Lurie, S. Ya., Approximate calculations in ancient Greece, *Arkh. Istor. Nauk. i Tehn. Ser. 1* **4** (1934).

111. Lyapunov, A. M., The life and works of P. L. Chebyshev, *in* «П.Л.Чебышев. Избран-

ные математические труды» ("P. L. Chebyshev Izbrannye matematicheskie trudy"; "The Selected Mathematical Works of P. L. Chebyshev"). Moscow-Leningrad, 1946.

112. Lyapunov, A. M., «Избранные труды» ("Izbrannye trudy"; "Selected Works"). Acad. Sci. USSR, Moscow, 1948.

113. Maïstrov, L. E., On mathematical signs and terms appearing in archeological monuments in ancient Russia, *Istor.-Mat. Issled.* **10**, 595–616 (1957).

114. Maïstrov, L. E., The struggle between idealism and materialism in probability theory, *in* «Философские вопросы естествознания» ("Filosofskie voprosy estestvoznaniya"; "Philosophical Problems of the Natural Sciences"), Vol. 2. Moscow State Univ., Moscow, 1959.

115. Maïstrov, L. E., The role of gambling in the origination of probability theory, *Acta Univ. Debrecen* **7** (1961).

116. Maïstrov, L. E., Elements of probability theory in Galileo's works, *Voprosy Istor. Estestvoznan. Tehn.* **16**, 94–98 (1964).

117. Mahalanobis, P. C., The foundations of statistics, *Sankhya (Indian J. Statist.) Ser A* **18**, 183–194 (1957).

118. Makovel'skiĭ, A. O., «Досократики» ("Dosokratiki"; "Presocrates"), Part. I. Kazan, 1914.

119. Markov, A. A., The law of large numbers and the method of least squares, *Izd. Fiz.-Mat. Ob-va Pri Kazan. Ser. 2* **8**, 110–128 (1898).

120. Markov, A. A., Sur les racines de l'équation $e^{x^2} \partial^m e^{-x^2} / \partial x^m = 0$, *Izv. Akad. Nauk. Ser. 5* **9**, 435–446 (1898).

121. Markov, A. A., «Исчисление вероятностей» ("Ischislenie veroyatnosteĭ"; "The Calculus of Probabilities"), 4th ed. Moscow, 1924.

122. Markov, A. A., «Избранные труды» ("Izbrannye trudy"; "Selected Works"). Acad. Sci. USSR, Moscow, 1951.

123. Markov, A. A., Jr., The biography of A. A. Markov, *in* A. A. Markov's «Избранные труды» ("Izbrannye trudy"; "Selected Works"), p. 1. Acad. Sci. USSR, Moscow, 1951.

124. Marx, K., «Капитал» ("Kapital"; "Das Kapital"), Vols. 1 and 2. Moscow, 1949.

125. "Marx and Engels Archives," Vol. 14 (1935).

126. Marx, K., and Engels, F., «Сочинения» ("Sochineniya"; "Collected Works"), Vol. 26. Moscow, 1935.

127. Chebyshev, P. L., On mean values, *Mat. Sb.* **II**, 1–9 (1867).

128. Kolmogorov, A. N., and Fomin, V. S., eds., «Математика, ее содержание, методы и значение» ("Matematika, ee soderžanie, metody i znachenie"; "Mathematics, Its Content, Methods and Significance"), Vol. 2. Moscow, 1956.

129. Bunyakovskiĭ, V. Ya., Some thoughts on the misconceptions of certain notions related to society, particularly lotteries and games. *Mayak (Beacon)* Pt. II (1840).

130. Medvedev, F. A., The development of the notion of the Stieltjes integral, *Istor.-Mat. Issled.* **15**, 171–224 (1963).

131. von Mises, R., «Вероятность и статистика» ("Veroyatnost' i statistika"; "Probability and Statistics"). Moscow-Leningrad, 1930 (English ed., Academic Press, New York, 1964).

132. de Moivre, A., De mensura sortis, *Philos. Trans. Roy. Soc. London Ser. A* **27** (329), 213–264 (1711).

133. de Moivre, A., "The Doctrine of Chances," 2nd ed. London, 1740. (The posthumous 3rd edition was reprinted by Chelsea, New York, 1967.)

133a. de Moivre, A., "Miscellanea Analytica." London, 1730.

134. Montmort. P., "Essai d'Analyse sur les Jeux de Hasard," 2nd ed. Quilau. Paris, 1713/14.

135. Mueller, F. W., "Batak-Sammlung." Berlin, 1893.

136. Nekrasov. P. A., «Философия и логика науки о массовых проявлениях человеческой деятельности» ("Filosofiya i logika nauki o massovyh proyavleniyah chelovecheskoï deyatel'nosti"; "The Philosophy and Logic of the Science of Mass Manifestations of Human Activity"). Moscow, 1902.

137. Nekrasov, P. A., «Московская философско-математическая школа и ее основатели» ("Moskovskya filisofsko matematicheskaya shkola i ee osnovateli"; "The Moscow Philosophical–Mathematical School and Its Founders"). Moscow, 1904.

138. Nekrasov, P. A., «Теория вероятностей» ("Teoriya Veroyatnosteï"; "Probability Theory"), rev. ed. St. Petersburg, 1912.

139. Nekrasov, P. A., «Средняя школа, математика и научная подготовка учителей» ("Spednyaya Shkola, matematika i nauchnaya podgotovka uchiteleï"; "High School Mathematics and Scientific Teacher Training"), 1916.

140. Bernoulli, D., *Novi Commentarii Academiae Scientiae Imp. Petropolitanae* **XIV** (1769).

141. Ore, O., "Cardano: The Gambling Scholar." Princeton Univ. Press. Princeton. New Jersey, 1953. (Cf. [40].)

142. Ore, O., Pascal and the invention of probability theory, *Amer. Math. Monthly* **67**, 409–419 (1960).

143. Ostrogradskiï, M. V., «Полное собрание трудов» ("Polnoe sobranie trudov"; "Complete Collection of Works"), Vol. 3. Academy of Sciences of URRSSR, Kiev, 1961.

144. Ostrogradskiï, M. V., «Педагогическое наследие» ("Pedagogicheskoe nasledie"; "Pedagogical Contributions"). Moscow, 1961.

145. Otradnych, F. B., An episode in the life of academician A. A. Markov, *Istor.-Mat. Issled.* **6**, 495–508 (1953).

146. Pannekoek, A., "A History of Astronomy." Wiley (Interscience), New York, 1961.

147. Pascal, B., "Oeuvres Complètes," Vol. 3. Paris, 1908.

148. Plackett, R. L., The principle of the arithmetic mean, *Biometrika* **45**, 130–135 (1958).

149. Planck, M., "Wege zur physikalische Erkenntnis." Stuttgart, 1944.

150. Poincaré, H., "Science and Method" (with a Preface by B. Russel). Thomas Nelson and Sons, London, 1910.

151. Poincaré, H., "Calcul des Probabilités." Paris, 1912.

152. Poincaré, H., «Последние мысли» ("Poslednie Mysli"; "Dernières Pensées"). Petrograd, 1923. [Dernières Pensées" ("Last Essays"), Flammarion, Paris, 1913; English translation (J. W. Bolduc, transl.), Dover, New York, 1963.]

153. Poincaré, H., "The Foundations of Science." The Science Press, Lancaster, Pennsylvania, 1946.

154. Poincaré, H., "Science and Hypothesis." Dover, New York, 1952.

155. Poisson, S. D., "Recherches sur la Probabilité des Jugements en matière Criminelle et en Matière Civile, Précédées des Règles Générales du Calcul des Probabilités." Bachelier, Paris, 1837.

156. Pólya, G., "Patterns of Plausible Inference" (Vol. 2 of "Mathematics and Plausible Reasoning"). Princeton Univ. Press, Princeton, New Jersey, 1954; rev. ed., with Appendix, 1968.

157. *Proceedings of the First All-Russian Conference of Mathematicians in Moscow, 1927.*

158. Gnedenko, B. V., *Trudy Inst. Istor. Estestvoznan (Proc. Inst. History Natural Sciences)* **2** (1948).

159. Prudnikov, V. E.,«Русские педагоги-математики XVIII–XIX веков» ("Russkie pedagogimatematiki XVIII-XIX vekov"; "Russian Teacher-Mathematicians in the Eighteenth and Nineteenth Centuries"). Moscow, 1956.

160. Quetelet, A.,«Социальная система и законы ею управляющие»("Sočial'naya sistema i zakony eyu upravlyayuschie"; "Social Systems and the Laws Governing Them"). St. Petersburg, 1863. (Translated from "Du Système Social et des Lois Qui le Régissent." Guillaumin, Paris, 1948.)

161. Reichsberg, N. M., «А. Кетле» ("A. Quetelet"). St. Petersburg, 1894.
162. Rigaud, S. P., "Miscellaneous Works and Correspondence of the Rev. James Bradley." Oxford Univ. Press, London and New York, 1832.
163. "Scientific Legacy," Vol. 1. Moscow-Leningrad, 1948.
164. Sheynin, O. B., «К истории оценок непосредственных измерений и закона распределения случайных ошибок»("On the History of Estimation of Direct Observations and the Distribution Law of Random Errors"). (Deposited edition.) Moscow, 1963.
165. Sheynin, O. B., On D. Bernoulli's article of 1777 and Euler's commentary, *Voprosy Istor. Estestvoznan. Tehn.* **19**, 115–117 (1965).
166. Sheynin, O. B., On the works of R. Adrian in the theory of errors and its applications, *Isotr.-Mat. Issled.* **16**, 325–336 (1965).
167. Simpson, T., "The Nature and Laws of Chance." London, 1740.
168. Simpson, T., "The Doctrine of Annuities and Reversions." London, 1742, 1791.
169. Simpson, T., A letter to the President of the Royal Society, on the advantage of taking the mean of a number of observations in practical astronomy, *Philos. Trans. Roy. Soc. London Ser. A* **49** (Pt. 1), 82–93 (1755).
170. Simpson, T., An attempt to show the advantage arising by taking the mean of a number of observations in practical astronomy, *in* "Miscellaneous Tracts on Some Curious Subjects in Mechanics, Physical Astronomy and Speculative Mathematics," pp. 64–75. London, 1757.
171. Smirnov, V. I., Daniel Bernoulli (*in* D. Bernoulli's "Hydrodynamics"), *Akad. Nauk SSSR Trudy Jakutsk Filial. Ser. Fiz.* (1959).
172. "South-Russian Chronicles," Vol. 1 (1856) (Kiev).
173. Bunyakovskiĭ, V. Ya., *Sovremennik (Contemporary)* **3** (1847) (St. Petersburg).
174. Tanneri, P., The state of the sciences in Europe (1798–1814), *in* «История XIX века» ("Istoriya XIX veka"; "History of the Nineteenth Century"), Vol. 1, Chapter 11. Moscow, 1938.
175. Tartaglia, N., "General Trattato di Numeri et Misure." 1556.
176. Todhunter, I., "History of the Mathematical Theory of Probability." Cambridge, Univ. Press, London and New York, 1865. (Reprinted by Chelsea, New York, 1949, 1961.)
177. Urbain, G., and Boll, M., "La Science, Ses Progrès, Applications," Vol. 2. Paris, 1949.
178. Voroncov-Vel'yaminov, B., «Лаплас» ("Laplas"; "Laplace"). Moscow, 1937.
179. Weiman, A. A., «Шумеро-вавилонская математика, III–I тысячелетия до н.э.» (Shumero-Vavilonskaya matematika, III–I tysyacheletiya do n. é"; "Shumero-Babylonian Mathematics, 3000–1000 в.с.е"). Moscow, 1961. (Oriental Liter. Publ.)
180. Whittaker, E. T., and Robinson, G., «Математическая обработка результатов наблюдений» ("Matematicheskaya Obrabotka rezul'tatov nablyudeniĭ"; "The Calculus of Observations"). Moscow, 1933. (Original English edition published by Blackie, Glasgow and London, 1926.)
181. Wiener, N., «Кибернетика и общество» ("Kibernetika i obschestvo"; "Cybernetics and Society"). (Translated from the 2nd edition published by Doubleday, Garden City, New York, 1954.)
182. Yushkevich, A. P., «История математики в средние века» ("Istoriya matematiki v srednie veka"; "History of Mathematics in the Middle Ages"). Moscow, 1961.
183. Zernov, N. E., «Теория вероятностей», ("Teoriya Veroyatnosteĭ"; "Probability Theory"). Moscow, 1843.

Additional Bibliography

References added by the editor of the translation

184. Adler, I., "Probability and Statistics for Everyman." John Day, New York, 1963

185. Alexandrov, P. S., *et al.*, Bernstein's obituary, *Uspehi Mat. Nauk* **24**, 211–219 (1969).
186. Bell, E. T., "Men of Mathematics." Simon and Schuster, New York, 1937.
187. Bernoulli, J., The law of large numbers, *in* "The World of Mathematics" (J. R. Newman, ed.), Vol. 3, pp. 1452–1455. Simon and Schuster, New York, 1956.
188. Borel, E., "Elements of the Theory of Probability" (J. Freund, transl.). Prentice-Hall, Englewood Cliffs, New Jersey, 1965.
189. David, F. N., "Games, Gods and Gambling." Griffin, London, 1962.
190. David, F. N., Some notes on Laplace [from Bernoulli (1723), Bayes (1763), Laplace (1813)], *in* "Anniversary Volume" (L. M. LeCam and J. Neyman, eds.), pp. 30–44. Springer-Verlag, Berlin and New York, 1965.
191. Robinson, M. F. (Madame Duclaux), "The French Ideal." Chapman and Hall, London, 1911.
192. Dunnington, C. W., "Carl Friedrich Gauss: Titan of Science." Hafner, New York, 1955.
193. Eggenberger, J., *Mitt. Naturforsch. Gesellsch. Bern* Nos. 1305–1334, 110–182 (1894).
194. Eisenhart, C., The development of the concept of the best mean of a set of measurements from antiquity to the present day. Presidential address, to appear in *Journal of the American Statistical Association* (1974).
195. Feller, W., "An Introduction to Probability Theory and Its Applications," 2nd ed. Wiley, New York, 1957.
196. de Finetti, B., Probability: interpretations, *in* "International Encyclopedia of the Social Sciences," Vol. 12, pp. 496–504. Macmillan (Free Press), New York, 1968.
197. Hacking, I., Jacques Bernoulli's "Art of Conjecturing," *Brit. J. Phil. Sci.* **22** (3), 209–249 (1971).
198. Hankins, T.L., "Jean d'Alembert—Science and the Enlightenment." Oxford Univ. Press (Clarendon), London and New York, 1970.
199. Hasofer, A. M., Random mechanisms in Talmudic literature, *Biometrika* **54**, 316–321 (1967).
200. Hoeffding, W., *Math. Reviews* **29** (No. 1656) (1965).
201. Kolmogorov, A. N., *IEEE Trans. Information Theory* **14**, 662–664 (1968).
202. Kotz, S., Statistics in the USSR, *Survey* **57**, 132–141 (1965).
203. Loève, M., "Probability Theory," 3rd ed. Van Nostrand-Reinhold, Princeton, New Jersey, 1963.
204. Martin-Löf, P., The definition of random sequences, *Information and Control* **9**, 602–619 (1966).
205. Martin-Löf, P., The literature of von Mises' kollectivs revisited, *Theoria* **35**, 12–37 (1969).
206. von Mises, R., "Probability, Statistics and Truth." MacMillan, 1939.
207. Pearson, K., *Biometrika* **17**, 201–210 (1925).
208. Quetelet, L. A. J., "A Treatise on Man and the Development of His Faculties." Edinburgh, 1842. (Reprinted in 1968 by Burt Franklin, New York.)
209. Savage, L. J., Elicitation of personal properties and expectations *J. Amer. Statist. Assoc.* **66**, 783–801 (1971).
210. Shannon, C. E., A mathematical theory of communication, *Bell System Tech. J.* **27**, 379–423, 623–656 (1948).
211. Sheynin, O. B., On the early history of the law of large numbers (Studies in the History of Probability and Statistics, 21), *Biometrika* **55**, 459–467 (1968).
212. Sheynin, O. B., Daniel Bernoulli on the normal law (Studies in the History of Probability and Statistics, 23), *Biometrika* **57**, 199–202 (1970).
212a. Sheynin, O. B., On the history of some statistical laws of distribution (Studies in the History of Probability and Statistics, 25), *Biometrika* **58**, 234–236 (1971).
213. Sheynin, O. B., On the history of the limiting theorems of de Moivre and Laplace, *Istor. i Metodologiia Estestvennykh Nauk* **9**, 199–211 (1970).
214. Sheynin, O.B., Newton and the classical theory of probability, *Arch. History Exact Sci.* **7**, 217–243 (1971).

215. Sheynin, O. B., J. H. Lambert's work on Probability Theory, *Arch. Hist. Exact. Sci.* 7(3), 244–256 (1971).
216. Smith, D. E., "A Source Book for Mathematics." McGraw-Hill, New York, 1929.
217. Yuschkevich, A. P., «История математики в России до 1917 года». ("Istoriya matematiki v Rossii do 1917 goda"; "The History of Mathematics in Russia up to 1917"). Acad. Sci. USSR, Inst. History of Natural Sci. and Engineering (Technology), Nauka, Moscow, 1968.

Index

A

Achilles, 13
Addition rule for probability, 25
Adrian, R., 149, 150
Ajax, 13
Al-Kashi, 35
Anagrams, 35
Annuities, 54, 55
Applications
  to linguistics, 180
  to medicine, 166
Arithmetic mean, 51, 83–86, 106, 107,
    108, 178
Arithmetic triangle, 45
Arithmetica Integra, 35
Ars Combinatoria, 36–38
Ars Conjectandi, 2, 51, 56–75
Astragali, 8, 9, 24

Augustus (Emperor), 1
Average man, 161
Axioms of probability, 2, 248–253

B

Babylonian mathematics, 85
Bayes, T., 87–100
  essay by, 89–100
  theorem of, 100
Ben Gerson, L., 35
Bernoulli, D., 79, 106, 110, 124, 132, 162,
    171, 177
Bernoulli, J. (James), 2, 51, 56, 57–75,
    76, 77, 85, 127, 214
  Ars Conjectandi by, 2, 51, 57–75
  numbers, 65
  theorem of, 2, 74–75, 81, 147, 163, 170,
    174–176

Bernoulli, J. (John), 76, 79, 81
Bernoulli, N., 57, 67, 76, 79, 81, 104, 125
Bernstein, S. N., 2, 140, 216, 224, 249–253
  address at 1927 conference by, 253
  axiomatization of probability by,
    250–252
  "An essay on axiomatic foundations"
    by, 249
  equivalence to Kolmogorov's
    axiomatization, 252
  Kolmogorov's appraisal of, 252
  "Probability theory" by, 250–253
  "Sur l'extension du theorème limité"
    by, 224
  work on Markov chains by, 216, 253
Bertrand, J., 234
  criticism of Quetelet's theory by, 238
  inoculation problem of, 238
  paradoxes of, 234–239
  support of d'Alembert's views by, 238
  three boxes problem of, 238
Bhâskara II, 13, 34, 35
Bienaymé, J., 201, 202
Bills of mortality, 4, 54
Binomial coefficients, 35, 45, 46
Bohr, N., 217
  model of atom by, 217
Boltzmann, L., 225–229, 234
  Lectures on Gas Theory by, 227–228
Book of Happiness, 25
Bonaparte, N., 137, 138
Borel, E., 2, 242–244, 260, 262
  "Le Hasard" by, 242
    Kostitzin's appraisal of, 244
  Probability and Certainty by, 243
  Subjective approach of, 243
Bortkiewicz, L., 160
Bradley, J., 86
Brahe, T., 30, 31, 86, 152
Brashman, N.D., 169
Brown, R., 225
Bruns, H., 216
Buffon, C. L., 118–123, 132, 177
  needle problem of, 120, 146
Bunyakovskiĭ, V. Ya., 169, 173–180, 187,
    188, 191, 242
  "An essay on mortality laws" by, 179,
    180
  "Foundations of Mathematical Theory
    of Probabilities" by, 173–177

influence of Laplace on, 174
"On the approximate summation of
    numerical tables" by, 179
Russian self-calculator of, 180
"Some thoughts on misconceptions
    . . ." by, 173

C

Cantelli, F. P., 259, 260
Cardano, G., 1, 18–25, 42, 61
  classical definition of probability theory
    by, 25
Castelli, B., 30
Catherine II, 13
Cavendish, H., 195
Celestial phenomena, 140
Central limit theorem, 2, 212
  Lyapunov's conditions, 212
Chebyshev, P. L., 2, 189, 190, 218, 224
  disciples of, 190
  discussion with A. V. Vasiliev, 189
  "Elementary proof of a general
    proposition . . ." by, 195–198
  "An essay on elementary analysis" by,
    191–195
  inequality of, 201
    one-sided, 203
  lecturing at Leningrad University, 191
  "On mean values" by, 198–201
  "On two theorems concerning
    probabilities" by, 202–205, 207,
    209
    Gnedenko's appraisal of, 206
    Khinchin's appraisal of, 208
    Kolmogorov's appraisal of, 203, 206,
      207, 224
    Kolmogorov's comments on, 203
Chiaramonti, S., 31, 32
Church, 4
Chu, Shih-Chien, 35
Circuit, 19, 20
Clausius, R., 230, 231
Condorcet, J. A., 129–135, 158
Considerazione sopra il Giuco dei Dadi,
    28
Continuous distribution function, 83, 84
Copeland, H., 257
Crimean war, 180
Crusades, 13
Cybernetics, 2

D

d'Alembert, J., 104, 123–129, 143
Dante, A., 13, 16
  Divine Comedy of, 16, 17
David, F. N., 8, 14, 43, 52, 160
Davidov, A. Yu., 165, 166, 167
de Coster, C., 13
de Finetti, B., 258
  subjective approach of, 258
de Fournival, R., 16
de Méré, C., 40, 41, 42, 169, 175
Demographic statistics, 54
Demography, 108, 109, 110, 172, 173, 177
de Moivre, A., 2, 43, 80, 81, 82, 83, 105,
    110, 130, 142, 160
  Doctrine of Chances of, 43, 80, 81, 82,
    83
  de Moivre–Laplace theorem and, 142
De Nova Stella, 31
De Ratiociniis in Ludo Aleae, 48
de Roberval, G. P., 42, 48
De Vetula, 16
De Witt, J., 55
Dialectics of Nature, 7, 39
Dialogue of the Great World Systems,
    30–33
Dice, 10, 11, 13, 15–17, 18–21, 25, 28, 29,
    38, 40, 41, 42
d'Imola, B., 17
Distribution law of errors, 32, 149
Distribution of arithmetic mean, 168
Divine Comedy, 16, 17
Division of stakes, 44–47, 49, 50, 51
Doctrine of Chances, 43, 80, 81, 82, 83
Dörge, K., 257
Doomsday, 4
Due Brevi e Facili Trattati, 27
Duration of life, 122, 123, 126

E

Edward III, 13
Einstein, A., 225
Emperors of Rome, 135
Engels, F., 7, 39
Errors
  in observations, 30, 32, 33
  random, 32, 34
  systematic, 34

D

Euclid, 248
Euler, L., 65, 66, 101–106, 107, 108, 162
Exercitationes Mathematicae, 48
Expectatio, 51

F

Fair game, 25, 51
Feller, W., 224, 238
Fermat, P., 1, 39, 42, 43, 44, 45, 47, 48,
    55, 77, 78, 189
Foundations of the Mathematical Theory
    of Probabilities, 184
Friedrich II, 13, 104

G

Galilei, G., 28–34
  Dialogue of the Great World Systems
    by, 30–33
Games
  Bassette, 78
  Dice, 55
  Ferme, 79
  Her, 79
  Hounds and Jackals, 9
  Pharaon, 78, 106, 158
  Point, 79
  Rencontre, 104
  Treize, 79, 104
Gambling, 7–14
  limitation on, 13
Gauss, C. F., 2, 85, 150–158
  controversy with Legendre, 152, 153
  correspondence with Laplace, 157, 158
  letter to H. M. W. Olbers, 152
  method of least squares of, 152,
    154–155
  normal equation, 157
  normal law of random errors of, 153,
    154
    of arithmetic mean, 154
  "Tables for determination of time
    periods" by, 152
  "Theoria combinationis . . ." by, 151,
    155, 156
  "Theoria motus corporum" by, 151, 152
  Vinograd's appraisal of, 151
  "Widows pension funds" by, 152
Gebhart, C., 53

Generating function, 145, 146
Geometric mean, 85
Gibbs, J. W., 225, 232, 233, 234
  "Basic principles of statistical
      mechanics" by, 232, 233
  influence of Boltzmann on, 232
  introduction of probability into physics
      by, 233
  Millikan's appraisal of, 234
  Wiener's appraisal of, 233
Glevenko, V. I., 252
Gnedenko, B. V., 220, 257, 261
  Course in Probability Theory by, 261,
      262, 264
Golden theorem (of Bernoulli), 73
Graunt, J., 54
Grave, D. A., 189

H

Haight, F., 158
Hardy, G., 259
Harmonic progression, 37
Hausdorff, F., 259
Henry VIII, 13
Herigone, P., 35
Hermitage, 9
Hilbert, D., 248, 249
Hobbes, T., 6, 7
Hoeffding, W., 196, 203
Holland, 41, 51
Hotimskiĭ, V., 14
Huygens, C. I., 39, 42, 48–52, 54, 55, 57,
      58, 77, 80
  problems of, 52
  De Ratiociniis in Ludo Aleae by, 48–51,
      57

I

India (ancient), Hindus, 5, 6, 34
Infant mortality, 172, 177
Inoculation, 126, 173
Insurance, 5, 109, 110, 166, 172, 173, 177,
      180, 181, 186

J

Jeffreys, H., 258
  A treatise on Probability by, 258

K

Kagan, V. F., 249
Kazan University, 167, 214
Kendall, M. G., 8, 14
Kerseboom, M., 110
Keynes, J. M., 258
Kharkov University, 162
Khayyam, O., 35
Khinchin, A. Ya., 40, 215, 259–262
  law of iterated logarithm of, 259–260
  law of large numbers of, 262
  Moscow school of probability and, 261
Kolman, E., 15, 254
Kolmogorov, A. N., 2, 75, 215, 220, 261,
      262, 263
  axiomatization by, 262–264
  Foundations of the Theory of
      Probability by, 262
  metric theory of functions of, 262
  Moscow School of Probability and, 261
  proof of the law of large numbers by,
      215
  strong law of large numbers by, 262
Koopman, B. O., 252
Korkin, A. N., 189
Kummer, E. E., 167

L

Lagrange, J. L., 85, 211
Lambert, J. H., 34, 85
Laplace, P. S., 2, 66, 67, 75, 81, 100, 131,
      135, 157, 158, 160, 162, 173, 191,
      211, 249
  Buffon's needle problem and, 146
  Celestial Mechanics by, 137
  flight to Melun by, 135
  generating functions of, 145– 146
  Gnedenko's appraisal of, 145
  inoculation problem of, 143–144
  member of the Academy, 136
  member of the National Institute, 137
  "Mémoire sur la probabilité de causes"
      by, 142
  method of least squares of, 147
  on moral expectation, 145
  mortality tables by, 142
  new proof of J. Bernoulli's theorem by,
      147

normal law of, 148
  Biermann's opinion on, 148
Philosophical Essay on Probabilities by,
  138–140, 142
on ratio of sexes, 144
relations with Bailly, 136, 137
relations with Napoleon, 137, 138
St. Petersburg problem of, 145
stake division problem of, 146, 147
theorem of, 147, 148
Théorie Analytique of, 142
Laplacian determinism, 67
Law of insufficient reason, 181
Law of iterated logarithm, 259, 260
Law of large numbers, 2, 19, 25, 159,
  213–215, 261
Law of small numbers, 159, 160
  Chebyshev's form of, 214
Legendre, A. M., 152
Leibniz, G. W., 36, 38, 59, 76, 85, 86, 87
  Ars Combinatoria of, 36–38
Liber de Ludo Aleae, 18–20
Likelihood equation, 107
Likelihood function, 106
Lilavati, 34
Limit of life, 81
Lindeberg, Y. W., 223, 224
Littlewood, J. E., 259
Lobachevskiǐ, N. I., 167, 168, 169, 248
  distribution of sum of uniform variables
    and, 168
  "New elements of Geometry" by, 168
  probability of average results and, 168
  use of Russian letters by, 169
Lomonosov, M. V., 230
Lotteries, 94, 98–100, 101–103, 171, 173,
  177, 186
  Genoise, 101–103
  Royal Oaks, 43
Louis IX, 13
Louis XIV, 41
Lyapunov, A. M., 2, 189, 211, 213,
  221–223, 241
  Bernstein's appraisal of, 223
  conditions of, 212, 222, 223
  "On a theorem in probability theory"
    by, 211, 221, 222, 223
  utilization of characteristic functions
    by, 211, 220, 221

M

Mahalanobis, P. C., 6
Malthus, T. R., 172
Markov, A. A., 2, 73, 178, 179, 190,
  208–223, 241, 261
  on Bunyakovskiǐ's "Foundations," 178,
    179
  calculus of probabilities and, 178
  "Calculus of Probabilities" by, 218–220,
    241, 242
  chains of, 215–218
  as Chebyshev's disciple, 208
  condition of, 220
    comparison with Lyapunov's,
      222–223
  "An example of statistical investigation"
    by, 217, 218
  "Extension of the law of large
    numbers" by, 214, 215
  "Extension of limit theorems" by, 215,
    216
  letters to A. V. Vasiliev of, 209
  Lyapunov's comments on, 220
  proof of Lyapunov's theorem by, 212
  "On the roots of the equation" by, 210
  teaching career of, 218
  "The theorem on the limit" by, 213–214
Martin-Löf, P., 264
Marx, K., 5, 51
Mathematical expectation, 25, 39, 49, 51,
  54, 55
Maxwell, J. C., 230–233
  distribution, 230, 231
  address to London Chemical Society by,
    230
Mean utility, 116, see also Moral
  expectation
Metaphysical determinism, 77
Method of least squares, 147, 152, 173
Mocenigo, 4
Mongols, 4
Montmort, P., 76–81, 104
Moral certainty, 132
Moral expectation, 115, 116, 117, 118,
  121, 131, 171, 172, 173
Moral probability, 158
Moral sciences, 77
Mortality tables, 109, 110, 121, 164, 177,
  179

Moscow University, 165, 168, 191
Multiplication rule for probability, 23, 25, 39
Mylon, 48

N

Nârâyana, 35
Nekrasov, P. A., 240–242
    opposition from Markov and, 241–242
    "Probability Theory" by, 241
Newton, I., 40, 76
Normal (probability) law, 33, 148, 149, 150, 153, 178
Nozzolini, 30

O

Olbers, H. M. W., 152
"On insurance," 184
Opus novum de proportionibus nemerorum, 24
Origin of planets, 118–119
Ostrogradskiĭ, M. V., 38, 66, 164, 165, 180–186, 188
    approval of Revkovskiĭ's program by, 163
    criticism of Bunyakovskiĭ's "Foundations" by, 184
    "Extract from a memoir" by, 181
    "Game of Dice" by, 186
    Gnedenko's appraisal of, 186
    influence of Laplace on, 181
    "On insurance" by, 184
    lectures at Michailov College by, 186
    "A note on the Retirement Fund" by, 180
    "On a problem concerning probabilities" by, 182–184
    "A proposal for establishing pension funds" by, 180
Ovid, 16
    Publius Ovidius Naso by, 16

P

Paccioli, I., 1, 17, 18, 24, 26, 27
Palamedes, 24
Parish registers, 4
Pascal, B., 1, 36, 38, 39, 40, 41, 42, 43, 44, 45, 46, 47, 55, 65, 77, 78, 170, 189
Pascal triangle, 34, 62, 63
Pavlovsky, A. F., 162
    first note on probability in Russian by, 162
Peano, G., 249
Pearson, K., 75
Peverone, G. F., 27
"Philosophical Essay on Probabilities," 142
Planck, M., 229, 234
Poincaré, H., 237, 239, 244
    on Bertrand's paradoxes, 248
    Calculus of Probabilities by, 244–246
    on Gauss' law, 247
    generalization of Bertrand's paradoxes by, 239
    opinion on science of, 247
Poisson, S. D., 2, 158, 159, 191, 195, 214
    distribution, 160
    "Recherches sur les probabilités" by, 158, 159
Pólya, G., 131
Pompeii, 9
Population census, 1
Practica arithmeticae generalis, 24
Price, R., 88–99
Probabilité propre, 134, 135
Probability
    Laplace definition of, 168
    subjective interpretation of, 185
Probability curves, 110
Problem of duration of play, 80–81
Proteus verses, 59
Purgatorio, 16
Pythagoreans, 34
π, value of, 120

Q

Quality control, 182
Quetelet, A., 160, 161, 167, 172, 238
    average man and, 161, 167, 238
    meeting with Laplace, 160
    A Treatise on Man by, 161
    Marx on, 161

R

Random errors, 85
Random variable, 85

Recherches sur la probabilité des
    jugements . . ., 159, 160
Relative probability, 170
Religion, 179
Renaissance, 5
Retirement funds, 179, 180, 181
Revkovskiĭ, S., 162–165
    lectures on probability by, 162, 163
    treatise on probability by, 163
Roulette, 12, 13
Russian school of probability theory, 2
Russian self-calculator, 180

S

Schopenhauer, A., 229–230
Shannon's entropy, 217
Simpson, T., 82–85
Sleshinskiĭ, I., 211
Snydezkiĭ, I. A., 162
Solar spectrum, 119
South Russian Chronicles, 4
Spinoza, B., 52–54
    Reeckening van Kanssen by, 52
Statistical regularity, 19, 25
Statistics, 5
Steklov, V. A., 189, 241
Stifel, M., 24, 35, 46, 61
St. Petersburg Academy of Sciences, 182
St. Petersburg paradox, 106, 116, 124, 125,
    127, 128, 129, 171, 177
St. Petersburg University, 167, 189, 191
Suetonius Tranquillus, G., 13
Sumatra, 12
Summa de Arithmetica, 17
Syādvāda, 5, 6
Systematic errors, 85

T

Tacquet, A., 35
Tartaglia, N. I., 1, 25–27, 46
Telescope, 30
Testimony of witnesses, 178
Theoria combinationis, 151, 155–156
Theoria motus corporum, 151
"Théorie analytique des Probabilitiés,"
    142, 145, 174, 181
Theory of errors, 80, 167, 168, 172, 173,
    233

Todhunter, I., 36, 79
Tornier, E., 257
Trattato generale di numeri e misure, 25
Triangular distribution, 83
Tribunals, 130, 139, 140, 178, 181
Trojan wars, 24
Truncated variables, 213

U

University of Odessa, 211

V

Van Hudden, J., 54
Van Schooten, F., 48, 51, 59, 61
Vasiliev, A. V., 189
Venice, 4, 17
"Venus," 8
Verona, 25
Vilnus University, 162, 163, 165
Vital statistics, 54
von Mises, R. E., 2, 254–258
    collective and, 255
    Cramér's criticism of, 256
    Khinchin's criticism of, 255, 256, 257
    Kolmogorov's appraisal of, 256, 257
    Kolmogorov's criticism of, 256, 257
von Smoluchowski, M., 225
Voronoi, G. F., 189
Voters' problem, 132, 133, 179

W

Wallis, J., 59
Wiener, N., 233
William the Norman, 4

Y

Yaglom, Ya. M., 40

Z

Zernov, N. E., 169–173
    "The theory of probability with special
    applications" by, 169–173
Zolotarev, E. I., 189